AQA Physics

Revision Guide

Jim Breithaupt

OXFORD

UNIVERSITY PRESS

OXFORD
UNIVERSITY PRESS

Great Clarendon Street, Oxford, OX2 6DP, United Kingdom

Oxford University Press is a department of the University of Oxford.
It furthers the University's objective of excellence in research,
scholarship, and education by publishing worldwide. Oxford is a
registered trade mark of Oxford University Press in the UK and in
certain other countries

British Library Cataloguing in Publication Data
Data available

978 0 19 835189 4

10 9 8 7 6

Printed by CPI Group (UK) Ltd, Croydon CR0 4YY

Paper used in the production of this book is a natural,
recyclable product made from wood grown in sustainable
forests. The manufacturing process conforms to the
environmental regulations of the country of origin.

Acknowledgements

Cover: Pedro Soares Photograph / Getty images

Artwork by Q2A Media

This book has been written to support students studying for AQA A Level Physics. The sections covered are shown in the contents list, which also shows you the page numbers for the main topics within each section.

AS exam

A level exam

Year 1 content

1 Particles and radiation
2 Waves
3 Mechanics and energy
4 Electricity
5 Skills in AS Physics

Year 2 content

6 Further mechanics and thermal physics
7 Fields
8 Nuclear physics

Plus one option from the following:

- Astrophysics
- Medical physics
- Engineering physics
- Turning points in physics
- Electronics

A Level exams will cover content from Year 1 and Year 2 and will be at a higher demand.

Section 1 Particles and radiation 2

1 Matter and radiation 2
1.1   Inside the atom   2
1.2   Stable and unstable nuclei   3
1.3   Photons   4
1.4   Particles and antiparticles   5
1.5   Particle interactions   7
    Practice questions   9

2 Quarks and leptons 10
2.1   The particle zoo   10
2.2   Particle sorting   11
2.3   Leptons at work   12
2.4   Quarks and antiquarks   13
2.5   Conservation rules   15
    Practice questions   16

3 Quark phenomena 17
3.1   The photoelectric effect   17
3.2   More about photoelectricity   18
3.3   Collisions of electrons with atoms   19
3.4   Energy levels in atoms   20
3.5   Energy levels and spectra   21
3.6   Wave-particle duality   22
    Practice questions   23

Section 2 Waves and optics 24

4 Waves 24
4.1   Waves and vibrations   24
4.2   Measuring waves   25
4.3   Wave properties 1   26
4.4   Wave properties 2   27
4.5   Stationary and progressive waves   28
4.6   More about stationary waves on strings   29
4.7   Using an oscilloscope   30
    Practice questions   31

5 Optics 33
5.1   Refraction of light   33
5.2   More about refraction   33
5.3   Total internal reflection   34
5.4   Double slit interference   35
5.5   More about interference   36
5.6   Diffraction   37
5.7   The diffraction grating   38
    Practice questions   39

Section 3 Mechanics and materials

6 Forces in equilibrium 41
6.1   Vectors and scalars   41
6.2   Balanced forces   42
6.3   The principle of moments   43
6.4   More on moments   44
6.5   Stability   44
6.6   Equilibrium rules   46
6.7   Statics calculations   46
    Practice questions   47

7 On the move 49
7.1   Speed and velocity   49
7.2   Acceleration   51
7.3   Motion along a straight line at constant acceleration   52
7.4   Free fall   53
7.5   Motion graphs   54
7.6   More calculations on motion along a straight line   54
7.7   Projectile motion 1   56
7.8   Projectile motion 2   56
    Practice questions   58

8 Newton's laws of motion 60
8.1   Force and acceleration   60
8.2   Using $F = ma$ 61
8.3   Terminal speed   62
8.4   On the road   63
8.5   Vehicle safety   64
    Practice questions   65

9 Force and momentum 67
9.1   Momentum and impulse   67
9.2   Impact forces   67
9.3   Conservation of momentum   69
9.4   Elastic and inelastic collisions   69
9.5   Explosions   69
    Practice questions   71

10 Work, energy, and power 73
10.1   Work and energy   73
10.2   Kinetic energy and potential energy   73
10.3   Power   75
10.4   Energy and efficiency   75
    Practice questions   77

11 Materials 79
11.1   Density   79
11.2   Springs   81
11.3   Deformation of solids   82
11.4   More about stress and strain   82
    Practice questions   84

Section 4 Electricity 86

12 Electric current 86
12.1   Current and charge   86
12.2   Potential difference and power   86
12.3   Resistance   88
12.4   Components and their characteristics   90
    Practice questions   92

13 Direct current circuits 94
13.1   Circuit rules   94
13.2   More about resistance   95
13.3   Electromotive force and internal resistance   96
13.4   More circuit calculations   97
13.5   The potential divider   98
    Practice questions   99

Section 6 Further mechanics and thermal physics

17 Motion in a circle — 101
17.1 Uniform circular motion — 101
17.2 Centripetal acceleration — 102
17.3 On the road — 103
17.4 At the fairground — 103
Practice questions — 105

18 Simple harmonic motion — 107
18.1 Oscillations — 107
18.2 The principles of simple harmonic motion — 107
18.3 More about sine waves — 108
18.4 Applications of simple harmonic motion — 109
18.5 Energy and simple harmonic motion — 111
18.6 Forced vibrations and resonance — 113
Practice questions — 114

19 Thermal physics — 115
19.1 Internal energy and temperature — 115
19.2 Specific heat capacity — 116
19.3 Change of state — 117
Practice questions — 119

20 Gases — 121
20.1 The experimental gas laws — 121
20.2 The ideal gas law — 122
20.3 The kinetic theory of gases — 124
Practice questions — 127

Section 7 Fields

21 Gravitational fields — 128
21.1 Gravitational field strength — 128
21.2 Gravitational potential — 129
21.3 Newton's law of gravitation — 131
21.4 Planetary fields — 132
21.5 Satellite motion — 133
Practice questions — 134

22 Electric fields — 135
22.1 Field patterns — 135
22.2 Electric field strength — 135
22.3 Electric potential — 136
22.4 Coulomb's law — 137
22.5 Point charges — 137
22.6 Comparing electric fields and gravitational fields — 138
Practice questions — 139

23 Capacitors — 141
23.1 Capacitance — 141
23.2 Energy stored in a charged capacitor — 141
23.3 Charging and discharging a capacitor through a fixed resistor — 141
23.4 Dielectrics — 144
Practice questions — 145

24 Magnetic fields — 147
24.1 Current-carrying conductors in a magnetic field — 147
24.2 Moving charges in a magnetic field — 149
24.3 Charged particles in circular orbits — 149
Practice questions — 151

25 Electromagnetic induction — 153
25.1 Generating electricity — 153
25.2 The laws of electromagnetic induction — 154
25.3 The alternating current generator — 155
25.4 Alternating current and power — 155
25.5 Transformers — 156
Practice questions — 157

Section 8 Nuclear physics

26 Radioactivity — 159
26.1 The discovery of the nucleus — 159
26.2 The properties of α, β, and γ radiation — 160
26.3 More about α, β, and γ radiation — 160
26.4 The dangers of radioactivity — 161
26.5 Radioactive decay — 162
26.6 The theory of radioactive decay — 162
26.7 Radioactive isotopes in use — 164
26.8 More about decay modes — 165
26.9 Nuclear radius — 167
Practice questions — 168

27 Nuclear energy — 170
27.1 Energy and mass — 170
27.2 Binding energy — 172
27.3 Fission and fusion — 174
27.4 The thermal nuclear reactor — 176
Practice questions — 178

Section 9 Option section summaries

Astrophysics — 180
Medical physics — 184
Engineering physics — 190
Turning points in physics — 195
Electronics — 201

Answers to practice questions — 208
Answers to summary questions — 235
For reference — 269

This book contains many different features. Each feature is designed to support and develop the skills you will need for your examinations, as well as foster and stimulate your interest in physics.

Worked example

Step-by-step worked solutions.

Maths skill

A focus on maths skills.

Question and model answer

Sample answers to exam style questions.

Summary questions

1 These are short questions at the end of each topic.

2 They test your understanding of the topic and allow you to apply the knowledge and skills you have acquired.

3 The questions are ramped in order of difficulty. Lower-demand questions have a paler background, with the higher-demand questions having a darker background. Try to attempt every question you can, to help you achieve your best in the exams.

Specification references

→ At the beginning of each topic, there are specification references to allow you to monitor your progress.

Key term

Pulls out key terms for quick reference.

Synoptic link

These highlight how the sections relate to each other. Linking different areas of physics together becomes increasingly important, as many exam questions (particularly at A Level) will require you to bring together your knowledge from different areas.

Revision tip

Prompts to help you with your understanding and revision.

Chapter 17 Practice questions

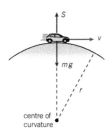

▲ Figure 1

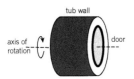

▲ Figure 2

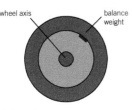

▲ Figure 3

1 The planet Jupiter orbits the Sun once every twelve years on a circular orbit that is about five times larger than the Earth's orbit.

Which one of the estimates **A–D** gives the following ratio?

the speed of Jupiter on its orbit
the speed of the Earth on its orbit

  A 0.4   **B** 0.8   **C** 2.4   **D** 5   (1 mark)

2 An object O of mass m moves in uniform circular motion at angular speed ω on a circular path of radius r.

Which one of the expressions **A–D** gives the following ratio?

the kinetic energy of the object
its centripetal acceleration

  A $\frac{m}{2\omega r}$ **B** $\frac{m}{2r}$ **C** $\frac{mr}{2}$ **D** $\frac{m\omega}{2r}$ (1 mark)

3 A vehicle of mass m travels at constant speed v over a bridge that has a radius of curvature r.

Which one of the following expressions **A–D** gives the magnitude of the support force on the vehicle when it is at the top of the bridge?

  A $mg - \frac{mv^2}{r}$ **B** $mg + \frac{mv^2}{r}$ **C** mg **D** $\frac{mv^2}{r} - mg$ (1 mark)

4 A spin drier tub of diameter 0.45 m rotates about a horizontal axis at 1400 revolutions per minute (see Figure 2).

  a Calculate:

    i the angular speed of the tub   (1 mark)

    ii the speed of a point on the wall of the tub   (1 mark)

    iii the centripetal acceleration of a point on the wall.   (1 mark)

  b Wet clothing in the tub is forced against the wall of the tub when it spins. The tub is designed with many small holes in the tube wall. Explain why water from the wet clothing leaves the tub through these holes when the tub spins.   (2 marks)

5 Vehicle wheels are fitted with balance weights to prevent the wheels wobbling at high speed. A balance weight on a wheel is positioned to make the wheel perfectly balanced. A wheel of diameter 0.60 m is fitted with a 25 g balance weight at a distance of 0.21 m from the wheel axis as shown in Figure 3.

  a Calculate the frequency of rotation of the wheel when the wheel is on a vehicle travelling at a speed of 25 m s⁻¹.   (1 mark)

  b **i** Calculate the centripetal force F on the balance weight when the wheel is travelling at this speed.   (2 marks)

    ii Draw a graph to show how the centripetal force F varies with the speed v of the vehicle.   (3 marks)

6 A cyclist travels at a speed of 8.2 m s⁻¹ on a bicycle that has wheels of diameter 50 cm.

Calculate:

  a the frequency of rotation of each wheel   (1 mark)

  b the centripetal acceleration of a point on the tyre.   (1 mark)

Practice questions at the end of each chapter, including questions that cover practical and maths skills.

1.1 Inside the atom

Specification reference: 3.2.1.1

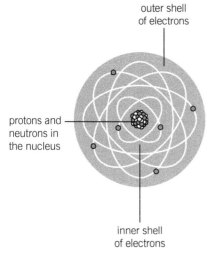

outer shell of electrons

protons and neutrons in the nucleus

inner shell of electrons

▲ **Figure 1** *Inside the atom*

▼ **Table 1** *Inside the atom*

	Charge relative to proton	Mass relative to proton
Proton	1	1
Neutron	0	1
Electron	−1	0.0005

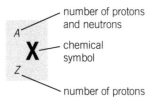

number of protons and neutrons

chemical symbol

number of protons

▲ **Figure 2** *Isotope notation*

The structure of an atom is shown in Figure 1.

- The **nucleus** contains most of the mass of the atom and its diameter is of the order of 10^{-5} times the diameter of a typical atom.
- The **atomic number Z** (or the proton number) of an atom is the number of protons in its nucleus.
- The **mass number A** (or the nucleon number) of an atom is the number of nucleons (i.e. protons and neutrons) in its nucleus.

Table 1 shows the charge and the mass of the proton, the neutron, and the electron relative to the charge and mass of the proton.

Isotopes

The atoms of an element each have the same number of protons but they can have different numbers of neutrons. Atoms of the same element with different numbers of neutrons are called **isotopes**. Figure 2 shows how we label them.

Specific charge

The **specific charge** of a charged particle is defined as its charge divided by its mass. The unit of specific charge is the coulomb per kilogram ($C\,kg^{-1}$).

Worked example

Calculate the specific charge of a nucleus of the carbon isotope $^{14}_{6}C$.

Solution

The charge of the nucleus $= 6e = 6 \times 1.60 \times 10^{-19}$ C $= 9.60 \times 10^{-19}$ C

The mass of the nucleus $= 14 \times 1.67 \times 10^{-27} = 2.34 \times 10^{-26}$ kg

Specific charge $= \dfrac{9.60 \times 10^{-19}\,\text{C}}{2.34 \times 10^{-26}\,\text{kg}} = 4.10 \times 10^{7}\,C\,kg^{-1}$

Summary questions

1 The table below gives some data about 4 different nuclei, A, B, C, and D.

	A	B	C	D
Atomic number	12	14	16	20
Mass number	24	30	32	39

   a How many neutrons and how many protons are there in D? (*1 mark*)
   b Which nucleus contains more neutrons than protons? (*1 mark*)
   c C is an isotope of sulfur (S). Another isotope E of sulfur has two fewer neutrons than C. Write down the isotope notation for E. (*1 mark*)

2 a State the number of protons and the number of neutrons in a nucleus of:
     i $^{17}_{8}O$ ii $^{9}_{4}Be$ (*2 marks*)
   b State and explain which of the two nuclei in **a** has the larger specific charge. (*2 marks*)

3 Calculate the specific charge of:
   a a uranium $^{238}_{92}U$ nucleus
   b a beryllium $^{9}_{4}Be$ ion with 2 electrons only. (*6 marks*)

1.2 Stable and unstable nuclei

Specification reference: 3.2.1.2

The strong nuclear force

The **strong nuclear force** keeps the nucleus together because it overcomes the electrostatic force of repulsion between the protons in the nucleus, and keeps the protons and neutrons together. Its key characteristics are:

- The specific size of its range, which is no more than about 3–4 fm.

- It is an attractive force from 3–4 fm to about 0.5 fm. At separations < 0.5 fm, it is a repulsive force that acts to prevent neutrons and protons being pushed into each other.

- It has the same effect between two protons as it does between two neutrons or a proton and a neutron.

Radioactive decay

Naturally occurring radioactive isotopes release three types of radiation.

1 **Alpha radiation** consists of alpha particles which each comprise two protons and two neutrons. The symbol for an alpha particle is $^4_2\alpha$ because its proton number is 2 and its mass number is 4. When an unstable nucleus of an element X emits an alpha particle, its nucleon number A decreases by 4 and its atomic number Z decreases by 2. We can represent this change by means of the equation below.

$$^A_Z X \rightarrow \,^{A-4}_{Z-2} Y + \,^4_2\alpha$$

2 **Beta radiation** from naturally occurring isotopes consists of fast-moving electrons. The symbol for an electron as a beta particle is $^0_{-1}\beta$ (or β⁻).

An unstable nucleus of an element X emits a β⁻ particle when a neutron in the nucleus changes into a proton. This type of change happens to nuclei that have too many neutrons. The beta particle is created when the change happens and is emitted instantly. In addition, an antiparticle with no charge, called an antineutrino (symbol $\bar{v}$), is emitted. We can represent this change by means of the equation below:

$$^A_Z X \rightarrow \,^A_{Z+1} Y + \,^0_{-1}\beta + \bar{v}$$

3 **Gamma radiation** (symbol γ) is electromagnetic radiation emitted by an unstable nucleus. It can pass through thick metal plates. It has no mass and no charge. It is emitted by a nucleus with too much energy, following an alpha or beta emission.

Revision tip

Remember that 1 femtometre (fm) $= 10^{-15}$ m

Revision tip

The neutrino and the antineutrino were 'hypothesised' to explain why beta particles from an isotope have a range of kinetic energies up to a maximum unlike alpha particles, which are emitted with a fixed proportion of the energy released. The existence of neutrinos and antineutrinos was confirmed about twenty years after the neutrino hypothesis was put forward.

Revision tip

There are two types of beta radiation. Nuclei that have too many protons emit beta radiation consisting of positrons. See Topic 1.4, Particles and antiparticles.

Summary questions

1 State two differences between the strong nuclear force and the electrostatic force. *(2 marks)*

2 An unstable nucleus $^A_Z X$ emits an α particle then a β⁻ particle to form a nucleus Y.
   State:
   a the mass number of Y   **b** the proton number of Y. *(2 marks)*

3 For the following radioactive decay equations, give the correct values for a, b, and c for each equation.
   a $^{64}_{29}Cu \rightarrow \,^a_b Zn + \,^c_{-1}\beta + \bar{v}$ *(2 marks)*
   b $^{228}_{90}Th \rightarrow \,^a_b Ra + \,^4_c\alpha$ *(2 marks)*

1.3 Photons

Specification reference: 3.2.1.3

Electromagnetic waves

Light is just a small part of the spectrum of **electromagnetic waves**. In a vacuum, all *electromagnetic* waves travel at the speed of light, c, which is $3.00 \times 10^8 \, \text{m s}^{-1}$. The wavelength λ of **electromagnetic radiation** of frequency f in a vacuum is given by the equation

$$\lambda = \frac{c}{f}$$

▼ **Table 1** *The main parts of the electromagnetic spectrum*

Type	Radio	Microwave	Infrared	Visible	Ultraviolet	X-rays	Gamma rays
Wavelength range	>0.1 m	0.1 m–1 mm	1 mm–700 nm	700 nm–400 nm	400 nm–10 nm	10 nm–0.001 nm	<0.1 nm

Photons

Electromagnetic waves are emitted as short bursts of waves, each burst leaving the source in a different direction. Each burst is a **photon**. The energy E of a photon depends on its frequency f in accordance with the equation

photon energy $E = hf$

where h is a constant referred to as the Planck constant. The value of h is $6.63 \times 10^{-34} \, \text{J s}$.

For a beam of photons of frequency f,

the power of the beam $= nhf$

where n is the number of photons in the beam passing a fixed point each second.

Worked example

Calculate the energy of a photon of wavelength 380 nm.

$h = 6.63 \times 10^{-34} \, \text{J s}, c = 3.00 \times 10^8 \, \text{m s}^{-1}$

Solution

To calculate the energy of a photon of this wavelength, we can use $E = \frac{hc}{\lambda}$

$E = \frac{hc}{\lambda} = \frac{6.63 \times 10^{-34} \times 3.00 \times 10^8}{380 \times 10^{-9}} = 5.23 \times 10^{-19} \, \text{J}$

Summary questions

$c = 3.00 \times 10^8 \, \text{m s}^{-1}, h = 6.63 \times 10^{-34} \, \text{J s}, e = 1.60 \times 10^{-19} \, \text{C}$

1 a State one similar property and one different property of a radio wave photon and a light photon. *(2 marks)*

   b Calculate the wavelength of an X-ray photon of energy 50 keV. *(2 marks)*

2 a Calculate the energy in J of a photon of wavelength 30 mm and state the part of the electromagnetic spectrum the photon is in. *(1 mark)*

   b Repeat part **a** for a photon of wavelength 3.0×10^{-10} m. *(2 marks)*

3 A 5.0 mW light-emitting diode emits light of wavelength approximately 600 nm. Calculate the maximum number of photons emitted per second by this LED. *(3 marks)*

1.4 Particles and antiparticles

Specification reference: 3.2.1.3

Rest energy

The equation $E = mc^2$ relates the total energy of a particle or antiparticle to its mass. When a particle is stationary, its rest mass (m_0) corresponds to **rest energy** $m_0 c^2$ locked up as mass. Rest energy must be included in the conservation of energy.

The energy of a particle or antiparticle is often expressed in millions of electron volts (MeV), where **1 MeV = 1.60×10^{-13} J**.

Antimatter

For every type of particle there is a corresponding antiparticle that:

- annihilates the particle and itself if they meet, converting their total mass into photons
- has exactly the same rest mass as the particle
- has exactly opposite charge to the particle if the particle has a charge
- can be produced together with its corresponding particle in a process known as **pair production**. This can happen when a *single* photon with sufficient energy passing near a nucleus changes into a particle–antiparticle pair. The nucleus is necessary for conservation of momentum as well as conservation of energy.

Annihilation

- A particle and a corresponding antiparticle meet and their mass, including their rest mass, is all converted into radiation energy so that energy is conserved.
- Two photons of equal energy are produced in this process (as a single photon cannot ensure a total momentum of zero after the collision).
- If a particle and corresponding antiparticle each have rest energy E_0

  Minimum energy of each photon produced, $hf_{min} = E_0$

Pair production

For a photon to produce a particle and antiparticle, each of rest energy E_0, the minimum energy of the photon must be equal to $2E_0$.

  Minimum energy of photon needed = $hf_{min} = 2E_0$

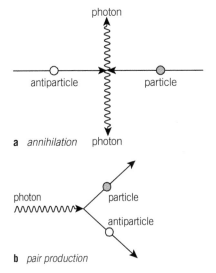

photon

antiparticle   particle

a *annihilation*   photon

photon
          particle

          antiparticle

b *pair production*

▲ **Figure 1** *Particles and antiparticles*

> ### Worked example
>
> The electron has a rest energy of 0.511 MeV. Calculate the minimum energy in J of a photon for pair production of an electron and a positron.
>
> #### Solution
>
> Minimum energy of the photon = $2E_0$ = 2 × 0.511 MeV = 1.022 MeV = 1.022 × 1.60×10^{-13} J = 1.64×10^{-13} J

Positron emission

The positron (symbol $^{0}_{+1}\beta$ or β^+) is the antiparticle of the electron, so it carries a positive charge. Positron emission takes place when a proton changes into a neutron in an unstable nucleus with too many protons. In addition, a neutrino (symbol $\nu_{(e)}$ for an electron neutrino), which is uncharged, is emitted.

$$^{A}_{Z}X \rightarrow \; ^{A}_{Z-1}Y + \; ^{0}_{+1}\beta + \nu$$

> ### Revision tip
>
> Notice in the positron emission equation that the numbers on each side
>
> along the top add up to the same total (i.e., $A = A + 0$)
>
> along the bottom add up to the same total (i.e., $Z = (Z - 1) + 1$)

Summary questions

1 MeV = 1.60×10^{-13} J

1 The rest energy of a proton is 938 MeV. Calculate the minimum energy of a photon for it to create a proton–antiproton pair. *(1 mark)*

2 The rest energy of an electron is 0.511 MeV. A positron created in an experiment has 0.250 MeV of kinetic energy. It collides with an electron at rest, creating two photons of equal energies as a result of annihilation.
   a Calculate the total energy of the positron and the electron. *(2 marks)*
   b Calculate the energy of each photon. *(1 mark)*

3 a Explain what is meant by pair production. *(1 mark)*
   b State two conditions necessary for a pair production event. *(1 mark)*

1.5 Particle interactions
Specification reference: 3.2.1.4

The electromagnetic force

This force acts between two charged objects due to the exchange of **virtual photons**. The photons are described as *virtual* because we can't detect them directly. The interaction is represented by the diagram shown in Figure 1. This is a simplified version of what is known as a **Feynman diagram**.

The weak nuclear force

The weak nuclear force is due to the exchange of particles referred to as **W bosons**. Unlike photons, these exchange particles:

- have a non-zero rest mass
- have a very short range of no more than about 0.001fm
- are positively charged (the W^+ boson) or negatively charged (the W^- boson).

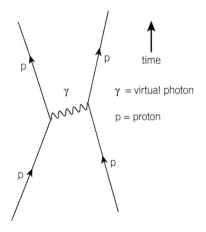

▲ **Figure 1** *Diagram for the electromagnetic force between two protons*

β decay

Figure 2 shows the diagram for each β decay process. Notice that:

- charge is conserved in both processes
- in β^- decay, a neutron changes into a proton and emits a W^- boson that decays into a β^- particle and an antineutrino
- in β^+ decay, a proton changes into a neutron and emits a W^+ boson that decays into a β^+ particle and a neutrino.

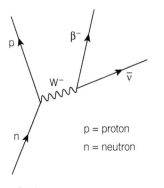

a β^- decay

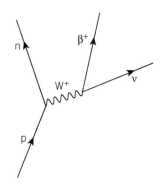

b β^+ decay

▲ **Figure 2** *W bosons in beta decay*

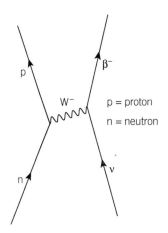

a *A neutron–neutrino interaction*

Neutrino interactions

Neutrinos and antineutrinos can interact with other particles. These rare interactions are due to W boson exchange. See Figure 3. Notice that:

- charge is conserved in both examples
- in **a**, a W^- boson is exchanged from a neutron to a neutrino so the neutron changes into a proton and the neutrino changes into a β^- particle (i.e., an electron)
- in **b**, a W^+ boson is exchanged from the proton to the antineutrino.so the proton changes into a neutron and the antineutrino changes into a β^+ particle (i.e., a positron).

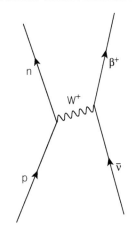

b *A proton–antineutrino interaction*

▲ **Figure 3** *The weak interaction*

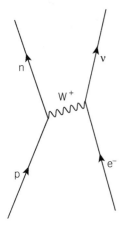

▲ **Figure 4** *Electron capture*

Revision tip

There are four fundamental forces in nature – the force of gravity, the electromagnetic force, the strong nuclear force, and the weak nuclear force. Remember that the photon and the W boson are the exchange particles of the electromagnetic force and the weak nuclear force respectively. The exchange particle of the strong nuclear force is the π meson. Scientists think the carrier of the force of gravity is the graviton – but it has yet to be observed!

Electron capture

Sometimes a proton in a proton-rich nucleus turns into a neutron as a result of interacting through the weak interaction with an inner-shell electron from outside the nucleus (**electron capture**). The W⁺ boson emitted by the proton changes the electron into a neutrino.

The same overall change can happen when a proton and an electron collide at very high speed. In addition, for an electron with sufficient energy, the overall change could also occur as a W⁻ exchange from the electron to the proton.

Summary questions

$c = 3.00 \times 10^8 \, \text{m s}^{-1}$

1 State the exchange particle involved when:
   a two electrons interact (*1 mark*)
   b a neutron and a neutrino interact (*1 mark*)
   c β⁺ decay occurs. (*1 mark*)

2 Sketch a diagram to represent:
   a β⁻ decay (*2 marks*)
   b an interaction between a proton and an antineutrino in which the proton changes to a neutron and emits a W⁺ boson which then interacts with the antineutrino. (*1 mark*)

3 a Sketch a diagram to represent electron capture. (*2 marks*)
   b Describe the process of electron capture. (*3 marks*)

Chapter 1 Practice questions

1   $^{23}_{11}\text{Na}$ is a neutral atom of sodium.

   a How many of the following does this atom contain?

      i   protons

      ii neutrons

      iii electrons. *(1 mark)*

   b Calculate the specific charge of an $^{23}_{11}\text{Na}^+$ ion. *(3 marks)*

2   A neutral chlorine (Cl) atom contains 38 nucleons.

   The specific charge of its nucleus is $4.30 \times 10^7\,\text{C\,kg}^{-1}$.

   a Determine the number of protons in its nucleus. *(3 marks)*

   b The nucleus is unstable and decays by emitting a β^- particle to become an argon (Ar) nucleus. Write down the equation for this decay. *(3 marks)*

   c Calculate the specific charge of the nucleus after it emits the β^- particle. *(2 marks)*

3   The equation below represents the decay of a bismuth (Bi) isotope by the emission of an α particle to form an isotope of thallium (Th).

$$^{210}_{83}\text{Bi} \rightarrow {}^{a}_{b}\text{Th} + {}^{c}_{d}\alpha$$

   a Determine the values of *a*, *b*, *c*, and *d* shown in the equation. *(2 marks)*

   b State which type of force X is responsible for holding a stable nucleus together and what particles the force acts between.

   c State which type of force Y has to be overcome by X to hold the nucleus together and what particles Y acts between. *(2 marks)*

4   **a** A laser emits a narrow beam of light of wavelength 630 nm. Calculate the energy of a photon of wavelength 630 nm in electron volts. *(2 marks)*

   b The power of the beam of light is 4.0 mW. Calculate the number of photons per second emitted by the laser source. *(2 marks)*

5   **a** State two characteristics of the electromagnetic force. *(2 marks)*

   b The electromagnetic force acts between two objects due to the exchange of virtual photons. With the aid of a diagram, explain what is meant by this statement. *(2 marks)*

6   In a radioactive decay of a certain nucleus, a positron is emitted.

   a Explain what a positron is. *(1 mark)*

   b **i** Describe how the number of protons and the number of neutrons in a nucleus change when a positron is emitted. *(1 mark)*

      ii Name the fundamental force responsible for positron emission and name the exchange particle of this force. *(1 mark)*

      iii Figure 1 represents positron emission. Identify the particles or antiparticles represented by the letters *a*, *b*, *c*, *d*, and *e* on the diagram. *(3 marks)*

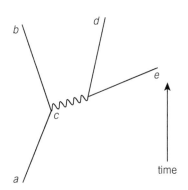

▲ **Figure 1**

7   The equation below represents electron capture.

$$\text{p} + \text{e}^- \rightarrow \text{n} + \nu_\text{e}$$

   a Describe the process of electron capture. *(2 marks)*

   b Copy and complete Figure 2 to represent this process. *(2 marks)*

8   **a** Describe what happens in pair production and give *one* example of this process. *(3 marks)*

   b A particle of rest energy 0.51 MeV meets its corresponding antiparticle and they annihilate each other producing two photons as a result. Calculate the least possible energy of each photon produced in this process. *(2 marks)*

   c Discuss whether or not annihilation may be considered as pair production in reverse. *(3 marks)*

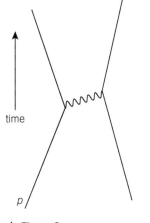

▲ **Figure 2**

2.1 The particle zoo

▲ **Figure 1** *Creation and decay of a pion. The short spiral track is a π^+ meson created when an antiproton from the bottom edge annihilates a proton. The π^+ meson decays into an antimuon that spirals and decays into a positron.*

Synoptic link

For more about electron and muon neutrinos and strange particles, see Topics 2.3, Leptons at work and 2.4, Quarks and antiquarks.

Revision tip

Particles such as K mesons that are produced through the strong interaction and decay through the weak interaction producing leptons are called **strange particles**.

Synoptic link

Decays always obey the conservation rules for energy, momentum, and charge. See Topic 2.5, Conservation rules.

Muons and mesons

Cosmic rays are high-energy particles from stars including the Sun. When cosmic rays enter the Earth's atmosphere, they collide with atoms creating photons and new short-lived particles and antiparticles, including:

- the **muon** or heavy electron (symbol μ), a negatively charged particle with a rest mass over 200 times the rest mass of the electron
- the **pion** or π **meson**, a particle which can be positively charged (π^+), negatively charged (π^-), or neutral (π^0), and has a rest mass greater than a muon but less than a proton
- the **kaon** or **K meson**, which also can be positively charged (K^+), negatively charged (K^-), or neutral (K^0), and has a rest mass greater than a pion but still less than a proton.

Decay modes

The particles listed above can also be created using accelerators in which protons collide head-on with other protons at high speed. The kinetic energy of the protons is converted into mass in the creation of these new particles.

K mesons and π mesons are produced through the strong interaction. However, the decay of K mesons takes longer than expected and produces π mesons which means they must decay via the weak interaction.

The rest masses, charge (if they were charged), and lifetimes of the new particles have been measured. Their antiparticles, including the **antimuon**, have been detected. Their decay modes are:

- A K meson can decay into π mesons, or a muon and an antineutrino, or an antimuon and a neutrino.
- A charged π meson can decay into a muon and an antineutrino, or an antimuon and a neutrino. A π^0 meson decays into high-energy photons.
- A muon decays into a muon neutrino, an electron, and an electron antineutrino. An antimuon decays into a muon antineutrino, a positron, and an electron neutrino.

Summary questions

1 a Which of the particles below are negatively charged? *(1 mark)*

 b Which one of the particles below is uncharged and unstable?

  electron K^0 meson   antimuon   neutron
  π^+ meson   antiproton *(1 mark)*

2 State whether the strong or the weak interaction acts when:

 a a muon decays *(1 mark)*

 b a K^0 meson is produced *(1 mark)*

 c a π^+ meson decays. *(1 mark)*

3 a A muon can travel much further than a π meson before they decay. What does this tell you about a muon compared with a π meson? *(1 mark)*

 b In terms of their properties, state one similarity and one difference between a muon and a π^- meson. *(2 marks)*

2.2 Particle sorting

Specification reference: 3.2.1.5

Classifying particles and antiparticles

All the particles we have discussed so far except exchange particles are classified as **hadrons** and **leptons**, according to whether or not they interact through the strong interaction.

Hadrons are further classified into two groups:

1 **Baryons** are protons and all other hadrons (including neutrons) that decay into protons, either directly or indirectly.
2 **Mesons** are hadrons that do *not* include protons in their decay products. In other words, kaons and pions are not baryons.

Energy matters

When a particle or antiparticle collides with another particle or antiparticle:

$$\text{the rest energy of the products} = \text{total energy before} - \text{the kinetic energy of the products}$$

Synoptic link

Conservation of momentum

Momentum is always conserved in a collision. *If the particles collide head-on with equal and opposite momentum*, their total momentum is zero. So all the total energy is available to create new particles and antiparticles. See Topic 9.3, Conservation of momentum, for more information.

Summary questions

1 a i   What property distinguishes a hadron from a lepton?   (*1 mark*)
     ii What property distinguishes a baryon from a meson?   (*1 mark*)
   b State whether each of the following particles or antiparticles is a baryon, a meson, or a lepton:
      i a π meson   ii an antiproton   iii a neutrino.   (*2 marks*)

2 A K$^+$ meson can decay into two π mesons.
   a Complete the following equation for this decay: K$^+ \longrightarrow$ (*1 mark*)
   b Use the rest energy values below to calculate the maximum kinetic energy of the π mesons, assuming the K$^+$ meson is at rest before it decays.   (*2 marks*)
      Rest energies : K$^+$ = 494 MeV, π$^+$ and π$^-$ = 140 MeV, π^0 = 135 MeV

3 Two protons, X and Y, moving in opposite directions each with 1.00 GeV of kinetic energy collide and produce a further proton and an antiproton.
   a Calculate the kinetic energy of the 3 protons and the antiproton after the collision.   (*3 marks*)
   b Explain why a further proton and antiproton could not be produced if X or Y was at rest and the total initial kinetic energy was still 2.00 GeV.   (*2 marks*)
      Rest energies: proton and antiproton = 0.94 GeV

Key term

Leptons are particles and antiparticles that do not interact through the strong interaction (e.g., electrons, muons, neutrinos, and their antiparticles). They can interact through the weak interaction, the gravitational interaction, and through the electromagnetic interaction (if charged).

Key term

Hadrons are particles and antiparticles that can interact through the strong interaction (e.g., protons, neutrons, π mesons, K mesons, and their antiparticles). They can also interact through the force of gravity, the weak interaction, and the electromagnetic interaction (if charged). Apart from the proton, which is stable, hadrons tend to decay through the weak interaction.

Synoptic link

Baryons and mesons are composed of smaller particles called **quarks** and **antiquarks**. See Topic 2.4, Quarks and antiquarks.

Revision tip

Rest energies are usually expressed in MeV (= 10^6 eV) or GeV (= 1000 MeV). For example, the rest energy of the proton is 938 MeV or 0.938 GeV.

2.3 Leptons at work

Specification reference: 3.2.1.5

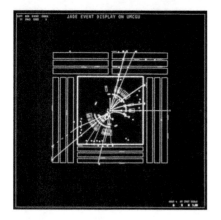

▲ Figure 1 *A two-jet event after an electron–positron collision*

All the experiments on leptons indicate they don't break down into non-leptons. A lepton can change into other leptons by emitting or absorbing a W boson. They can also interact in high-energy collisions to produce hadrons.

Neutrino types

Neutrinos and antineutrinos produced in beta decays are different from those produced by muon decays. We use the symbol v_μ for the **muon neutrino** and v_e for the **electron neutrino** (and similarly for the two types of antineutrinos).

Lepton rules

Some lepton changes are possible and some changes are never observed even though charge is conserved. To explain this, we need to assign +1 to a lepton, −1 to an antilepton, and 0 for any non-lepton and require lepton numbers to be conserved separately for muon lepton (and for electron leptons).

1   When an electron neutrino interacts with a neutron, it can produce a proton and an electron: $v_e + n \rightarrow p + e^-$

Electron lepton numbers are conserved: $1 + 0 = 0 + 1$ ✓
BUT the following change is not possible even though charge is conserved.
$$v_e + n \nrightarrow \bar{p} + e^+$$
Electron lepton numbers are not conserved: $1 + 0 \neq 0 + -1$ ✗

2   In muon decay, the muon changes into a muon neutrino. In addition, an electron is created to conserve charge and a corresponding antineutrino is created to conserve lepton number. For example: $\mu^- \rightarrow e^- + \bar{v}_e + v_\mu$

Electron lepton numbers are conserved: $0 = +1 - 1 + 0$ ✓
Muon lepton numbers are conserved: $+1 = 0 + 0 + 1$ ✓

Note: a muon *cannot* decay into a muon antineutrino, an electron, and an electron antineutrino even though charge is conserved. This is because the muon lepton number is not conserved. $\mu^- \nrightarrow e^- + \bar{v}_e + v_\mu$

Electron lepton numbers are conserved: $0 = 1 - 1 + 0$ ✓
Muon lepton numbers are *not* conserved: $+1 \neq 0 + 0 - 1$ ✗

Common misconception

Always consider charge conservation first because the individual charges are easy to remember. Also, when considering muons and antimuons, remember the muon is a particle not an antiparticle.

▼ Table 1

	Charge Q	Lepton number L
Electron	−1	+1
Muon		
Positron		
Muon antineutrino		
Antiproton		

Summary questions

1   Complete Table 1 to show the relative charge and lepton number of each particle or antiparticle in the table.   *(3 marks)*

2   A muon decays into an electron, a neutrino, and an antineutrino.
   a   Complete the equation below representing this decay: *(2 marks)*
   $$\mu^- \rightarrow \quad + \quad v_\mu + $$
   b   Show that electron lepton numbers and muon lepton numbers are conserved in the above equation.   *(2 marks)*

3   State whether or not each of the following reactions is permitted, giving a reason if it is not permitted:
   a $\bar{v}_e + p \rightarrow n + e^-$ *(2 marks)*
   b $v_e + p \rightarrow n + e^+$ *(2 marks)*
   c $e + p \rightarrow n + \bar{v}_e$ *(2 marks)*
   d $\mu^+ \rightarrow \bar{v}_e + e^+ + \bar{v}_\mu$ *(2 marks)*

2.4 Quarks and antiquarks
Specification reference: 3.2.1.6; 3.2.1.7

Strangeness

Strange particles such as K mesons are produced in pairs by making π mesons collide with protons and neutrons. One of the pair (the K meson) always decays into π mesons only whereas the other one always decays into π mesons and protons.

To explain the interactions and decays of strange particles,

- a **strangeness number** S was introduced for each particle and antiparticle (starting with +1 for the K$^+$ meson) so that strangeness is always conserved in strong interactions.
- non-strange particles (i.e., the proton, the neutron, pions, leptons) were assigned zero strangeness.

Strangeness is always conserved in a strong interaction, whereas strangeness can change by 0, +1, or −1 in weak interactions.

Look at the reactions in Table 1.

Reaction 1 is observed: the initial strangeness is zero so the Σ^0 strangeness must be −1 as the K$^+$ strangeness by definition is +1.

Reaction 2 is not observed: the initial strangeness is zero and the Σ^0 strangeness is −1. So the K$^-$ strangeness cannot be +1.

▼ **Table 1** *Some reactions that were predicted*

1	observed	$\pi^+ + N \longrightarrow K^+ + \Sigma^0$
2	*not* observed	$\pi^- + N \longrightarrow K^- + \Sigma^0$

The quark model

The properties of the hadrons, such as charge, strangeness, and rest mass can be explained by assuming they are composed of smaller particles known as quarks and antiquarks. The three types of quarks and antiquarks you need to study in this course are: the up, down, and strange quarks (u, d, s) and their antiquarks ($\overline{u}, \overline{d}, \overline{s}$). The charge Q, strangeness S, and baryon number B of the u, s, and d quarks are shown in Table 2. Their corresponding values for their antiquarks have the opposite sign (Q for $\overline{u} = -\frac{2}{3}$).

> **Revision tip**
> You may be asked to work out the charge or strangeness of any baryon or antibaryon from its quark composition.

▼ **Table 2** *Quark properties*

	u	d	s
Q	$+\frac{2}{3}$	$-\frac{1}{3}$	$-\frac{1}{3}$
S	0	0	−1
B	$+\frac{1}{3}$	$+\frac{1}{3}$	$+\frac{1}{3}$

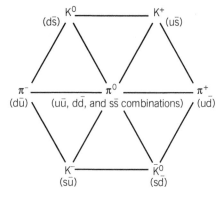

▲ Figure 1 *Quark combinations for the mesons*

Quark combinations

The rules for combining quarks to form baryons and mesons are astonishingly simple.

A **meson** consists of a quark and an antiquark. Figure 1 shows all nine different quark–antiquark combinations and the meson in each case. Notice that:

- each pair of charged mesons (e.g., π^+ and π^- or K^+ and K^-) is a particle–antiparticle pair
- each type of K meson contains one strange quark or antiquark.

Baryons and **antibaryons** are hadrons that consist of three quarks for a baryon or three antiquarks for an antibaryon.

- A proton is the uud combination, an antiproton is the $\overline{u}\,\overline{u}\,\overline{d}$ combination.
- A neutron is the udd combination.
- The Σ particle is a baryon containing a strange quark.

The proton is the only stable baryon. A free neutron decays into a proton, releasing an electron and an electron antineutrino, as in β^- decay.

Quarks and beta decay

In β^- decay, a down quark in a neutron changes to an up quark. See Figure 2a.

In β^+ decay, an up quark in a proton changes to a down quark. See Figure 2b.

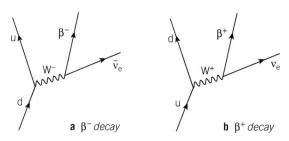

▲ Figure 2 *Quark changes in beta decay*

Summary questions

1 Determine the quark composition and strangeness of each of these hadrons:
   a i a π^+ meson *(1 mark)*
      ii an antineutron *(1 mark)*
      iii a K^+ meson. *(1 mark)*
   b Show that the antiparticle of a K^- meson is a K^+ meson. *(2 marks)*

2 a In terms of quarks, draw a diagram to represent β^- decay. *(2 marks)*
   b Describe the changes that are represented in your diagram. *(3 marks)*

3 a A Σ^- particle is a baryon which has a strangeness of -1. Determine its quark composition. *(3 marks)*
   b Show that strangeness is conserved in the reaction below. *(3 marks)*
   $$\pi^- + p \rightarrow K^+ + \Sigma^-$$

2.5 Conservation rules

Specification reference: 3.2.1.6; 3.2.1.7

Conservation of energy, conservation of momentum, and conservation of charge apply to all changes in science. Particles and antiparticles obey further conservation rules when they interact. These further rules are based on what reactions are observed and what reactions are not observed.

Conservation of leptons numbers: The total lepton number for each lepton branch before the change is equal to the total lepton number for that branch after the change.

Conservation of strangeness: In any strong interaction, strangeness is always conserved.

In addition, we know from observation that meson numbers are not conserved and baryon numbers are conserved if we assign a baryon number of:

- +1 to any baryon and −1 to any antibaryon
- 0 to any meson or lepton.

Quarks, antiquarks, and their baryon numbers

We can apply the baryon conservation rule to quarks, antiquarks, and leptons by assigning $+\frac{1}{3}$ to any quark, $-\frac{1}{3}$ to any antiquark, and 0 to any lepton.

The first reaction in Table 1 is shown in Figure 1 in terms of quarks and by the equation below. The baryon numbers under the equation show that the total baryon number is the same on each side.

$$u\,u\,d \;+\; \bar{u}\,\bar{u}\,\bar{d} \;\rightarrow\; u\,\bar{d} \;+\; \bar{u}\,d$$

$$\left(+\frac{1}{3}+\frac{1}{3}+\frac{1}{3}\right)+\left(-\frac{1}{3}-\frac{1}{3}-\frac{1}{3}\right)=\left(+\frac{1}{3}-\frac{1}{3}\right)+\left(+\frac{1}{3}-\frac{1}{3}\right)$$

The s quark decays through the weak interaction by changing into a u quark and emitting a W^- boson which can then decay into an electron and an antineutrino.

▼ **Table 1** *Conservation of baryon numbers*

Example 1	observed
Reaction	$p + \bar{p} \rightarrow \pi^+ + \pi^-$
Baryon numbers	$1 - 1 = 0 + 0$

Example 2	not observed
Reaction	$p + \bar{p} \rightarrow p + \pi^-$
Baryon numbers	$1 - 1 \neq 1 + 0$

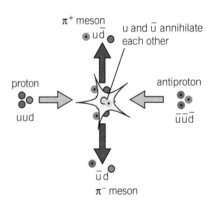

▲ **Figure 1** *Using the quark model*

Chapter 2 Practice questions

1 $\bar{p}$ e^+ K^- v_e μ^+

From the list of particles, identify the:

   a   hadrons

   b   leptons

   c   antibaryons

   d   mesons.   (2 marks)

2 a   State the distinction between hadrons and leptons.   (1 mark)

   b   A positive muon may decay in the following way:

$$\mu^+ \rightarrow e^+ + v_e + \bar{v}_\mu$$

      i   State the type of interaction that acts in the above decay process.

      ii   Show that conservation of leptons applies to the above equation.

      iii   Give *one* difference and *one* similarity between a negative
            muon and an electron.   (3 marks)

3 a   The three-quark model originated from a theory that explained
       the properties of baryons and mesons. State the quark composition
       of a neutron and explain why the neutron is uncharged.   (2 marks)

   b   Show that in the three-quark model, the antiparticle of a non-strange
       charged meson is another non-strange charged meson.   (2 marks)

4 a   A particle is made up of an up quark and a down antiquark.

      i   Name the particle that has this type of structure.

      ii   Explain its charge in terms of quarks.   (2 marks)

   b   A negatively charged baryon X has a charge of −1, a baryon
       number of 1, and a strangeness of −2.

      i   Deduce its quark composition and explain your reasoning.

      ii   Which baryon will this particle eventually decay into?   (5 marks)

5 a   With the aid of a diagram, describe the changes that take place in
       terms of quarks when a β^+ particle is emitted from a nucleus.   (3 marks)

   b   A suggested decay for the positive muon (μ^+) is

$$\mu^+ \rightarrow e^+ + \bar{v}_e + v_\mu$$

       Showing your reasoning clearly, deduce whether this decay
       satisfies the conservation rules that relate to baryon number,
       lepton number, and charge.   (3 marks)

6 The following equation represents the collision of a K^- meson with
   a proton, resulting in the production of a baryon X and a π^- meson:

$$K^- + p \rightarrow \pi^- + X$$

   a   The K^- has a strangeness of −1. Deduce the strangeness S,
       charge Q, and baryon number B of X.   (3 marks)

   b   Discuss whether or not X and a π meson could be produced in a
       collision between a π^- meson and a proton.   (2 marks)

7 State one similarity and one difference between:

   a   an antimuon and a proton

   b   a π^0 meson and a K^0 meson

   c   a π^+ meson and a neutron.   (3 marks)

8 Determine the charge Q and the strangeness S of:

   a   a uss baryon

   b   a uds baryon

   c   a $\overline{dds}$ baryon

   d   a $u\bar{s}$ meson.   (4 marks)

3.1 The photoelectric effect

Specification reference: 3.2.2.1

The **photoelectric effect** is the emission of electrons from a metal surface when electromagnetic radiation above a certain frequency is directed at the surface. The electrons emitted in this process are called **photoelectrons**.

The main observations about the photoelectric effect are listed below:

1 Photoelectric emission of electrons from a metal surface does not take place if the frequency of the incident electromagnetic radiation is less than a minimum value known as the **threshold frequency** f_{min}. This minimum frequency depends on the type of metal.

2 The number of electrons emitted per second is proportional to the intensity of the incident radiation, provided its frequency $f > fmin$.

3 Photoelectric emission occurs instantly when incident radiation is directed at the surface and regardless of intensity.

Einstein's explanation of the photoelectric effect

The wave theory of light cannot explain either the existence of a threshold frequency or why photoelectric emission occurs without delay. Einstein explained the photoelectric effect by assuming that:

- light is composed of wavepackets or **photons**, each of energy equal to hf, where f is the frequency of the light and h is the Planck constant

- an electron at or near the metal surface absorbs a *single* photon from the incident light and so gains energy equal to hf, (the photon's energy)

- an electron can leave the metal surface if the energy gained from a single photon is equal to, or greater than, the **work function** ϕ of the metal.

The *maximum* kinetic energy of an emitted electron is therefore $E_{Kmax} = hf - \phi$ so $E_{Kmax} > 0$.
Rearranging this equation gives $hf = E_{Kmax} + \phi$

Emission can take place from a metal surface provided $hf > \phi$. Therefore, threshold frequency $f_{min} = \dfrac{\phi}{h}$.

Stopping potential

The minimum potential needed to stop photoelectric emission from a metal is called the **stopping potential** V_s. At this potential, the maximum kinetic energy E_{Kmax} of an emitted electron is reduced by an amount equal to eV_s to zero. Hence $E_{Kmax} = eV_s$.

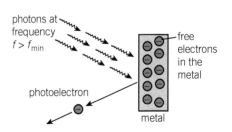

photons at frequency $f > f_{min}$ — free electrons in the metal

photoelectron

metal

▲ **Figure 1** *Explaining the photoelectric effect – one electron absorbs one photon*

Summary questions

$h = 6.63 \times 10^{-34}$ J s, $c = 3.00 \times 10^8$ m s^{-1}, $e = 1.60 \times 10^{-19}$ C

1 What is meant by the work function of a metal? *(1 mark)*

2 Explain why photoelectric emission from a metal surface only takes place if the frequency of the incident radiation is greater than or equal to a certain value. *(3 marks)*

3 A metal surface at zero potential emits electrons from its surface if light of wavelength less than or equal to 390 nm is directed at it.
   a Determine the work function of the metal. *(2 marks)*
   b Explain why light of wavelength greater than 390 nm will not cause photoelectric emission from this metal. *(2 marks)*

3.2 More about photoelectricity

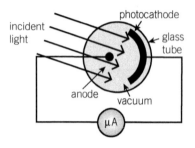

▲ **Figure 1** *Using a vacuum photocell*

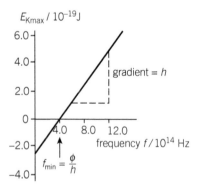

▲ **Figure 2** *A graph of E_{Kmax} against frequency*

$$\sqrt{v_x^2 + v_y^2} = \sqrt{16^2 + 30.4^2}$$

More about conduction electrons

The conduction electrons in a metal move about at random. The work function ϕ of a metal is much greater than the average kinetic energy of a conduction electron in a metal at 300 K.

- When a conduction electron absorbs a photon of energy hf, its kinetic energy increases by hf.
- If the energy of the photon $hf \geq \phi$, the conduction electron can leave the metal unless it collides with other electrons and positive ions and loses its extra kinetic energy.

The vacuum photocell

The photocell in Figure 1 contains a metal plate (the photocathode) and a smaller metal electrode, the anode. When photoelectric emission takes place, electrons are emitted from the cathode and are attracted to the anode. The microammeter in the circuit is used to measure the photoelectric current, which is proportional to the number of electrons per second that transfer from the cathode to the anode.

The photoelectric current is proportional to the intensity of the light incident on the cathode because:

- the light intensity is proportional to the number of photons per second incident on the cathode
- each photoelectron must have absorbed one photon to escape from the metal surface.

Summary questions

$h = 6.63 \times 10^{-34}$ J s, $c = 3.00 \times 10^8$ m s^{-1}, $e = 1.60 \times 10^{-19}$ C

1 Explain what is meant by the stopping potential of a metal. *(1 mark)*

2 A vacuum photocell is connected to a microammeter. When light is directed at the cathode of the photocell, the microammeter reads 0.85 µA.
   a i Explain why the microammeter gives a non-zero reading when light is directed at its cathode. *(2 marks)*
   ii Calculate the number of photoelectrons emitted per second by the photocathode of the photocell.
      Hint: Remember $Q = It$. *(2 marks)*
   b The intensity of the incident light is gradually reduced to zero. Describe and explain how the microammeter reading changes as a result. *(3 marks)*

3 The cathode of a photocell has a work function of 1.30 eV. A beam of light of wavelength 430 nm is directed at the photocathode of a vacuum photocell. Calculate:
   a the energy of a single light photon of this wavelength in eV *(1 mark)*
   b the maximum kinetic energy of a photoelectron in eV. *(2 marks)*

3.3 Collisions of electrons with atoms

Specification reference: 3.2.2.2

Ionisation

An **ion** is a charged atom. Adding electrons to an uncharged atom makes the atom into a negative ion. Removing electrons from an uncharged atom makes the atom into a positive ion (see Figure 1).

The electron volt and potential difference

The **electron volt** is a unit of energy which is defined as equal to the work done when an electron is moved through a potential difference (pd) of 1 V. For a charge q moved through a pd V, the work done = qV. Therefore, the work done when an electron moves through a potential difference V is equal to $e \times V$.

Excitation by collision

Gas atoms can absorb energy from colliding electrons without being ionised. This process, known as **excitation**, happens at certain values of energy, which are characteristic of the atoms of the gas. If a colliding electron:

- loses all its kinetic energy when it causes excitation, the current due to the flow of electrons through the gas is reduced

- does not have enough kinetic energy to cause excitation, it is deflected by the atom, with no overall loss of kinetic energy.

When excitation occurs, the colliding electron makes an electron inside the atom move from an inner shell to an outer shell (see Figure 2). Energy is needed for this process, because the atomic electron moves away from the nucleus of the atom. The excitation energy is always less than the ionisation energy of the atom, because the atomic electron is not removed completely from the atom when excitation occurs.

> **Key term**
>
> Any process of creating ions is called **ionisation**.

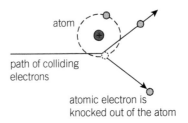

▲ **Figure 1** *Ionisation by collision*

> **Synoptic link**
>
> You met the electron volt briefly in Topic 2.1, The particle zoo.

> **Revision tip**
>
> Don't confuse atomic electrons with electrons that hit the atom.

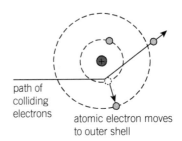

▲ **Figure 2** *A simple model of excitation by collision*

Summary questions

$e = 1.60 \times 10^{-19}$ C

1. Explain the difference between ionisation and excitation. *(2 marks)*

2. **a** The hydrogen atom has an ionisation energy of 13.6 eV. Calculate this ionisation energy in joules. *(1 mark)*
   b An electron with 16.2 eV of kinetic energy collides with a hydrogen atom and ionises it. Calculate the kinetic energy, in eV, of the electron after the collision if the collision causes an electron in the atom to be emitted from the atom with 1.0 eV of kinetic energy. *(2 marks)*

3. **a** Describe what happens to a gas atom when an electron from outside the atom collides with it and causes it to absorb energy from the electron without being ionised. *(2 marks)*
   b Explain why a gas atom cannot absorb energy from a slow-moving electron that collides with it. *(1 mark)*

3.4 Energy levels in atoms

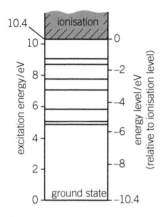

▲ **Figure 1** *The energy levels of the mercury atom*

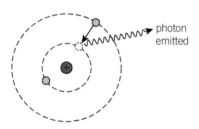

▲ **Figure 2** *De-excitation by photon emission*

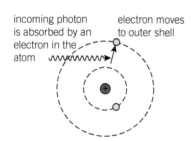

▲ **Figure 3** *Excitation by photon absorption*

Electrons in atoms

Figure 1 is an energy level diagram for an atom showing the allowed energy values of the atom. Each allowed energy corresponds to a certain electron configuration in the electron shells of the atom. The ground state of the atom is its lowest energy state. When an atom in the ground state absorbs energy, one of its electrons moves to a higher energy level so the atom becomes excited. The ionisation energy is the energy needed to ionise the atom from its ground state.

De-excitation

An excited atom is unstable because an electron that moves to an outer shell leaves a vacancy in the shell it moves from. The atom de-excites when the vacancy is filled by an electron from an outer shell. When this happens, the electron moves to a lower energy level and emits a photon (see Figure 2). When an electron moves from energy level E_1 to a lower energy level E_2, **the energy of the emitted photon, $hf = E_1 - E_2$**

Excitation using photons

An electron in an atom can absorb a photon and move to an outer shell where a vacancy exists – but only if the energy of the photon is *exactly* equal to the difference between the final and initial energy levels of the atom (see Figure 3).

Fluorescence

Certain substances **fluoresce** or glow with visible light when they absorb ultraviolet radiation. This is what happens in a **fluorescent tube**. The **tube** is a glass tube with a fluorescent coating on its inner surface. The tube contains mercury vapour at low pressure. When the tube is on, it emits visible light

- because the mercury atoms collide with each other and with electrons in the tube and become excited
- the mercury atoms emit ultraviolet photons (and photons with less energy) when they de-excite
- the ultraviolet photons are absorbed by the atoms of the fluorescent coating, causing excitation of the atoms
- the coating atoms de-excite in steps and emit visible photons.

Summary questions

$e = 1.60 \times 10^{-19}$ C

1 Figure 1 shows some of the energy levels of the mercury atom.
   a Estimate the energy needed to excite the atom from the ground state to the 4th excited level above the ground state. (*1 mark*)
   b Mercury atoms in this excited state can de-excite directly or indirectly to the ground state.
      i State and explain how many different photon energies are possible from this excited state. (*3 marks*)
      ii Which de-excitation from this state releases the highest photon energy? (*1 mark*)

3.5 Energy levels and spectra

Specification reference: 3.2.2.3

A continuous spectrum has a continuous range of wavelengths. For example, the spectrum of light from a filament lamp covers the visible spectrum from deep violet at less than 400 nm to deep red at about 650 nm.

A line emission spectrum has discrete lines of different wavelengths. The wavelengths of the lines of a line spectrum of an element are characteristic of the atoms of that element. This is because the energy levels of each type of atom are unique to that atom.

- Each line in a line spectrum is due to light of a certain colour and therefore a certain wavelength.
- The photons that produce a line all have the same energy, which is different from the energy of the photons that produce any other line.
- Each photon is emitted when an atom de-excites due to one of its electrons moving to an inner shell.
- If the electron moves from energy level E_1 to a lower energy level E_2.

the energy of the emitted photon $hf = E_1 - E_2$

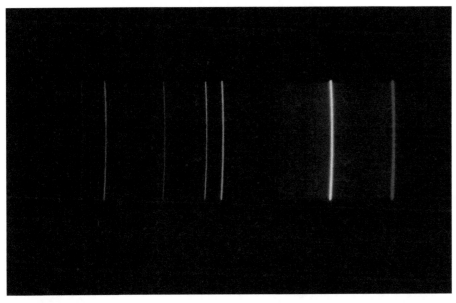

▲ **Figure 1** *A line spectrum*

Summary questions

$h = 6.63 \times 10^{-34}$ J s, $c = 3.00 \times 10^8$ m s^{-1}, $e = 1.60 \times 10^{-19}$ C

1 Describe what a line emission spectrum is and explain how it is produced. *(3 marks)*

2 The line spectrum of hydrogen includes two lines at 656 nm and 486 nm.
   a Calculate the energy of a photon of wavelength 656 nm. *(1 mark)*
   b The lines are due to electron transitions from two different energy levels X and Y to the same lower level. Level X is higher than level Y. Calculate the wavelength of a photon emitted by an electron that moves from X to Y.
   Hint: The first step is to calculate the energy of the 486 nm photon.
   (3 marks)

3.6 Wave–particle duality

Specification reference: 3.2.2.4

Synoptic link

See Topic 5.6, Diffraction, for more about diffraction and Topic 3.1, The photoelectric effect, for more about photoelectricity.

Key term

The **momentum** of a particle is defined as its mass multiplied by its velocity.

Common misconception

Don't mix up matter waves and electromagnetic waves and don't confuse their equations.

The dual nature of light

1 The *wave-like nature of light* is observed when **diffraction** of light takes place. This happens, for example, when light passes through a narrow slit.

2 The *particle-like nature of light* is observed in the photoelectric effect.

The dual nature of matter

1 The *particle-like nature of matter* is observed, for example, when electrons in a beam are deflected by a magnetic field.

2 The hypothesis that matter particles also have a *wave-like nature* was put forward by Louis de Broglie in 1923. His theory was that matter particles have a dual wave–particle nature and that they have a wavelength that depends on their momentum, p, according to the equation

$$\lambda = \frac{h}{p}$$

Evidence for de Broglie's hypothesis

A narrow beam of electrons can be diffracted by a metal foil as shown in Figure 1.

- The rows of atoms in each tiny crystal in the metal cause the electrons in the beam to be diffracted. The electrons are diffracted in certain directions only to form rings on a fluorescent screen at the end of the tube.

- Increasing the speed of these electrons makes their de Broglie wavelength smaller. So less diffraction occurs and the rings become smaller.

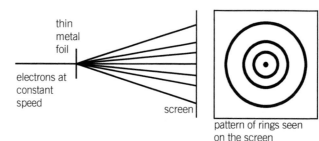

thin metal foil

electrons at constant speed

screen

pattern of rings seen on the screen

▲ **Figure 1** *Diffraction of electrons*

Summary questions

$h = 6.6 \times 10^{-34}$ J s, the rest mass of an electron $= 9.11 \times 10^{-31}$ kg, the rest mass of a proton $= 1.67 \times 10^{-27}$ kg

1 Explain what is meant by the dual wave–particle nature of matter particles. *(2 marks)*

2 a Calculate the de Broglie wavelength of an electron moving at a speed of 2.0×10^7 m s^{-1}. *(2 marks)*

b Calculate the speed at which a proton would have the same de Broglie wavelength as the electron in **a**. *(2 marks)*

Chapter 3 Practice questions

$h = 6.63 \times 10^{-34}\,\text{J s}$, $e = 1.60 \times 10^{-19}\,\text{C}$, $c = 3.00 \times 10^{8}\,\text{m s}^{-1}$

rest mass of an electron = $9.11 \times 10^{-31}\,\text{kg}$

rest mass of a proton = $1.67 \times 10^{-27}\,\text{kg}$

1 **a** Describe the photoelectric effect. *(1 mark)*

   b State and explain why the photoelectric effect cannot be observed with a particular metal if the frequency of the incident radiation is too small. *(4 marks)*

2 The threshold frequency of light incident on a certain metal surface is $2.9 \times 10^{14}\,\text{Hz}$.

   a Explain what is meant by threshold frequency. *(1 mark)*

   b Calculate the maximum kinetic energy of the photoelectrons emitted from this metal surface when light of wavelength 560 nm is directed at the metal surface. Give your answer to an appropriate number of significant figures. *(3 marks)*

   c Sketch a graph on the axes shown in Figure 1 to show how the maximum kinetic energy E_{Kmax} of the photoelectrons emitted from a metal surface varies with the frequency f of the incident light. Indicate the threshold frequency f_{min} on your graph. *(2 marks)*

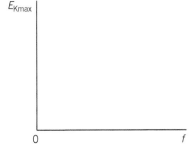

▲ **Figure 1**

3 The work function of a certain metal plate is 0.85 eV.

   a Calculate the threshold frequency of the incident radiation for this metal. *(2 marks)*

   b Calculate the maximum kinetic energy of photoelectrons emitted from this plate when light of wavelength 420 nm is directed at the metal surface. *(3 marks)*

4 Electrons are emitted from a certain metal surface when blue light is directed at its surface but not when red light is used instead. Explain why photoelectrons are emitted using blue light but not using red light. *(3 marks)*

5 A certain type of atom has excitation energies of 0.5 eV, 2.1 eV, and 4.6 eV.

   a Sketch an energy level diagram for the atom using these energy values. *(2 marks)*

   b Calculate the possible photon energies from the atom when it de-excites from the 4.6 eV level. Use a downward arrow to indicate on your diagram the energy change responsible for each photon energy. *(3 marks)*

6 Explain why the line spectrum of an element is unique to that element and can be used to identify it. *(4 marks)*

7 **a** State whether or not each of these experiments demonstrates the wave nature or the particle nature of matter or of light.

   i The photoelectric effect.

   ii Electron diffraction. *(1 mark)*

   b Calculate the momentum and velocity of:

   i an electron that has a de Broglie wavelength of 500 nm

   ii a proton that has the same de Broglie wavelength. *(4 marks)*

4.1 Waves and vibrations

Specification reference: 3.3.1.1; 3.3.1.2

Synoptic link

You have met the full spectrum of electromagnetic waves in more detail in Topic 1.3, Photons.

▲ **Figure 1** *Longitudinal waves on a slinky*

Revision tip

Vibrations of the particles in a longitudinal wave are in the same direction as that along which the wave travels.

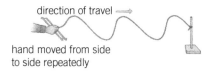

direction of travel →

hand moved from side to side repeatedly

▲ **Figure 2** *Making rope waves*

Key term

Plane-polarised waves are transverse waves that vibrate in one plane only.

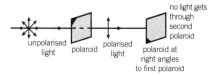

unpolarised light — polaroid — polarised light — polaroid at right angles to first polaroid — no light gets through second polaroid

▲ **Figure 3** *Explaining polarisation*

Types of waves

Mechanical waves are vibrations of the particles of a substance which pass through the substance. Sound waves, seismic waves, and waves on strings are examples of mechanical waves that pass through a substance.

Electromagnetic waves are electric and magnetic waves that progress through space without the need for a substance. Electromagnetic waves include radio waves, microwaves, infrared radiation, light, ultraviolet radiation, X-rays, and gamma radiation.

Longitudinal and transverse waves

Longitudinal waves are waves in which the direction of vibration of the particles is *parallel* to (along) the direction in which the wave travels. Sound waves, primary seismic waves, and compression waves on a slinky toy are all longitudinal waves. Figure 1 shows how to send longitudinal waves along a slinky.

Transverse waves are waves in which the direction of vibration is *perpendicular* to the direction in which the wave travels. Electromagnetic waves, secondary seismic waves, and waves on a string or a wire are all transverse waves. Figure 2 shows how to send transverse waves along a rope.

Polarisation

Transverse waves are **plane-polarised** if the vibrations stay in one plane only. If the plane of the vibrations changes, the waves are **unpolarised**. Longitudinal waves cannot be polarised.

The plane of polarisation of an electromagnetic wave is defined as the plane in which the electric field oscillates. Light from a filament lamp is unpolarised and can be polarised by passing it through a polaroid filter because the filter only allows through light that vibrates in a certain direction, according to the alignment of its molecules. If the polarised light is passed through another polaroid filter, and the molecules in the two filters are aligned in perpendicular directions, the transmitted intensity is zero.

Summary questions

1. Classify the following types of waves as either longitudinal or transverse and mechanical or electromagnetic:
   a. light
   b. infrared radiation
   c. primary seismic waves
   d. sound waves. *(3 marks)*

2. a. Explain what is meant by a 'longitudinal' wave. *(1 mark)*
   b. Explain why a series of compressions and rarefactions travel along a slinky coil when one end of the coil is moved forwards and backwards repeatedly. *(3 marks)*

3. a. State two differences between a transverse wave and a longitudinal wave. *(2 marks)*
   b. A transverse wave travels along a horizontal rope from left to right.
      i. Sketch a snapshot of the wave, indicating the direction in which the wave is travelling.
      ii. Mark a point X on the crest of a wave and indicate the direction in which the rope is moving at X. *(2 marks)*

4.2 Measuring waves

Specification reference: 3.3.1.1

The following terms are used to describe waves. See Figure 1 also.

- The **displacement** of a vibrating particle is its distance and direction from its equilibrium position.
- The **amplitude** of a wave is the maximum displacement of a vibrating particle.
- The **wavelength** λ of a wave is the least distance between two vibrating particles with the same displacement and velocity at the same time.
- One complete **cycle** of a wave is from maximum displacement to the next maximum displacement.
- The **frequency** f of a wave is the number of complete waves passing a point per second. The unit of frequency is the hertz (Hz).
- **wave speed** $c = f\lambda$ for waves of frequency f and wavelength λ.

The **period** of a wave is the time for one complete wave to pass a fixed point. For waves of frequency f, the period of the wave $= \frac{1}{f}$

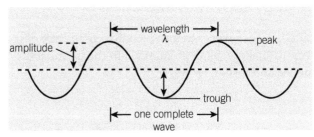

▲ Figure 1 *Parts of a wave*

Phase difference

For two points at distance d apart along a wave of wavelength λ

$$\text{the phase difference in radians} = \frac{2\pi d}{\lambda}$$

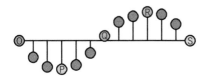

▲ Figure 2 *Progressive waves*

Summary questions

1 a Electromagnetic waves travel through the air at a speed of $3.00 \times 10^8 \, \text{m s}^{-1}$. Calculate the frequency of light waves of wavelength 320 nm. *(1 mark)*

b Sound waves travel through water at a speed of approximately $1500 \, \text{m s}^{-1}$. Estimate the wavelength of sound waves of frequency 20 kHz in water. *(1 mark)*

2 a State what is meant by the frequency of a wave. *(1 mark)*

b A motor boat travelling at a constant speed creates waves that travel at a speed of $2.9 \, \text{m s}^{-1}$ and have a frequency of 1.6 Hz. Calculate the period and the wavelength of the waves. *(2 marks)*

3 Figure 2 represents a progressive wave travelling from left to right.

a Determine the phase difference between adjacent particles in:
i degrees **ii** radians. *(1 mark)*

b Determine the phase difference between Q and S in radians. *(1 mark)*

c Compare P and R in Figure 2 and half a cycle later in terms of their amplitude, phase difference, and displacement. *(3 marks)*

4.3 Wave properties 1

Specification reference: 3.3.1.1; 3.3.1.2

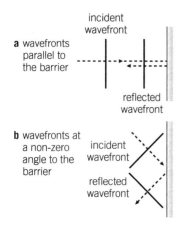

▲ **Figure 1** *Reflection of plane waves*

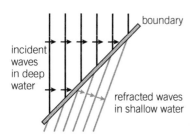

▲ **Figure 2** *Refraction*

Revision tip

In Figure 2, the refracted wavefront doesn't keep up with where it would have got to if its speed had not decreased at the boundary.

▲ **Figure 3** *The effect of the gap width*

Synoptic link

You will meet reflection, refraction, and diffraction of light in Topics 5.1, Refraction of light, 5.2, More about refraction, and 5.3, Total internal reflection.

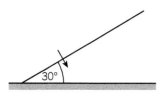

▲ **Figure 4**

Wave properties such as reflection, refraction, and diffraction occur with many different types of waves. A **ripple tank** may be used to study these wave properties. The waves observed in a ripple tank are referred to as **wavefronts**, which are lines of constant phase (e.g., crests). The direction in which a wave travels is at right angles to the wavefront.

Reflection

Straight waves directed at a certain angle to a hard flat surface (the reflector) reflect off at the same angle, as shown in Figure 2. The direction of the reflected wave is at the same angle to the reflector as the direction of the incident wave.

Refraction

When waves pass across a boundary at which the wave speed changes, the wavelength also changes. If the wavefronts approach at an angle to the boundary, they change direction as well as changing speed. This effect is known as **refraction**.

Figure 2 shows the refraction of water waves in a ripple tank when they pass across a boundary from deep to shallow water at an angle to the boundary. Note the wavelength is smaller in the shallow water.

Diffraction

Diffraction occurs when waves spread out after passing through a gap or round an obstacle. The effect can be seen in a ripple tank when straight waves are directed at a gap, as shown in Figure 3. The narrower the gap, the more the waves spread out. The longer the wavelength, the more the waves spread out.

Dish design

The bigger a satellite dish is, the stronger the signal it can receive, because more radio waves are reflected by the dish onto the aerial. But a bigger dish diffracts the waves less so it needs to be aligned more carefully than a smaller dish, otherwise it will not focus the radio waves onto the aerial.

Summary questions

1 A straight wave in a ripple tank is directed at an angle of 30° to a flat reflector, as shown in Figure 4.
   Copy the diagram accurately and show the position and direction of the wavefront after it reflects from the reflector. *(1 mark)*

2 A straight wave is directed at an angle of 30° to a straight boundary where the wave speed increases.
   a Draw the straight wave and the boundary before the wave crosses the boundary. *(1 mark)*
   b Draw the straight wave when half of it has travelled across the boundary. Indicate on your diagram the direction of the part of the wavefront that has crossed the boundary. *(1 mark)*

4.4 Wave properties 2

Specification reference: 3.3.1.1; 3.3.1.2

The principle of superposition

When waves meet, they pass through each other. Where they meet, they combine for an instant before moving apart. This effect is called **superposition**.

> **The principle of superposition states that when two waves meet, the total displacement at a point is equal to the sum of the individual displacements at that point.**

Stationary waves on a rope

Stationary waves are formed on a rope if two people send waves continuously along a rope from either end, as shown in Figure 1. The two sets of waves are referred to as **progressive waves** to distinguish them from stationary waves. They combine at fixed points along the rope to form points of no displacement or **nodes** along the rope. At each node, the two sets of waves are always 180° out of phase, so they cancel each other out.

Water waves in a ripple tank

Figure 2 shows two sets of circular waves produced in a ripple tank.

* Points of cancellation are created where a crest from one dipper meets a trough from the other dipper. At each point, the waves from one dipper are 180° out of phase with the waves from the other dipper.

* Points of reinforcement are created where a crest meets a crest, or where a trough meets a trough. At each point, the waves from one dipper are in phase with the waves from the other dipper.

Provided the two sets of waves have the same frequency and have a constant phase difference, cancellation and reinforcement occurs at fixed positions. This effect is known as **interference**. The dippers act as **coherent sources** because they vibrate at the same frequency with a **constant phase difference**.

▲ **Figure 1** *Making stationary waves*

▲ **Figure 2** *Interference of water waves*

Common misconception

A common mistake is to say coherent sources must have a phase difference of zero. This is wrong – the phase difference must be constant!

Key term

Superposition is the combining of waves when they meet for an instant before they move apart.

Summary questions

1 **a** State the principle of superposition of waves. *(1 mark)*
    b Describe the difference between a progressive wave and a stationary wave on a rope. *(1 mark)*

2 State and explain how you would expect the interference pattern in Figure 2 to change if the frequency of the waves produced by the dippers is increased? *(2 marks)*

3 Microwaves from a transmitter are directed at two parallel slits in a metal plate (see Figure 3). The two slits act as coherent sources of microwaves.
    a State what is meant by coherent sources. *(1 mark)*
    b A detector is placed on the other side of the metal plate on a line XY parallel to the plate. When the detector is moved steadily along the line, the detector signal decreases then increases again several times. Explain why the signal changes in this way. *(3 marks)*

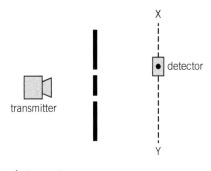

▲ **Figure 3**

4.5 Stationary and progressive waves

Specification reference: 3.3.1.3

Synoptic link

Stationary waves on a string are discussed in more detail in Topic 4.6 , More about stationary waves on strings.

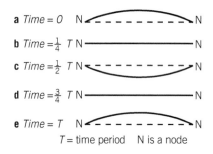

a Time = 0   N⟵- - - - - - - -⟶N

b Time = $\frac{1}{4}$ T N————————N

c Time = $\frac{1}{2}$ T N⟵- - - - - - - -⟶N

d Time = $\frac{3}{4}$ T N————————N

e Time = T   N⟵- - - - - - - -⟶N

T = time period   N is a node

▲ **Figure 1** First harmonic vibrations

Formation of stationary waves

A stationary wave is formed when two progressive waves of the same frequency pass through each other. The stationary wave formed has equally-spaced **nodes** (points of no displacement) and **antinodes** (points of maximum displacement) at fixed positions.

1   Stationary waves can be formed on a string in tension by fixing both ends and making the string vibrate. As a result, progressive waves travel towards each end, reflect at the ends, and then pass through each other to form a stationary wave pattern on the string. Different stationary wave patterns occur if the string is made to vibrate at different frequencies.

2   Sound resonates at certain frequencies in an air-filled tube or pipe. The sound waves in the pipe reflect at the ends and the reflected waves pass through each other. In a pipe closed at one end, resonance occurs if there is an antinode at the open end and a node at the closed end.

3   Microwaves form a stationary wave pattern when they are reflected by a metal plate back towards the transmitter. When a detector is moved between the transmitter and the metal plate, the detector signal is least at the nodes.

Explanation of stationary waves

Figure 2 shows two progressive waves passing through each other.

- When the two waves reinforce each other, they produce a large wave.

- A quarter of a cycle later, the two waves have each moved one-quarter of a wavelength in opposite directions. They are now in antiphase so the two waves cancel each other.

- After a further quarter cycle, the two waves reinforce each other again.

Note that the position of the points of no displacement, the nodes, does not change.

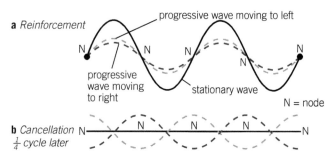

a Reinforcement

progressive wave moving to left

progressive wave moving to right

stationary wave

N = node

b Cancellation $\frac{1}{4}$ cycle later

▲ **Figure 2** Explaining stationary waves

Revision tip

In general, in any stationary wave pattern:

1   The amplitude of a vibrating particle varies with position from zero at a node to maximum amplitude at an antinode.

2   The phase difference between two vibrating particles is:

- zero if the two particles are between adjacent nodes or separated by an even number of nodes

- 180° (= π radians) if the two particles are separated by an odd number of nodes.

3   The distance between adjacent nodes = $\frac{1}{2}\lambda$ where λ is the wavelength of the progressive waves that form the nodes.

Summary questions

1   **a**   State the conditions necessary for the formation of a stationary wave. *(1 mark)*

  b   The stationary wave pattern in Figure 1 was set up on a string of length 0.60 m when it was made to vibrate at a frequency of 150 Hz.
  Calculate **i** the wavelength **ii** the speed of the progressive waves on the string. *(1 mark each)*

2   Describe the differences between a stationary wave and a progressive wave at different positions along each wave in terms of:
  a   the amplitude of the wave *(2 marks)*
  b   the phase difference. *(2 marks)*

3   A small loudspeaker connected to a signal generator was used to send sound waves into an air-filled glass tube which was closed at its other end. Explain why the tube resonated with sound at certain frequencies. *(3 marks)*

4.6 More about stationary waves on strings

Specification reference: 3.3.1.3

Stationary waves on a vibrating string

Figure 1 shows how stationary waves can be set up on a string in tension. As the frequency of the generator is increased from a very low value, different stationary wave patterns are seen on the string. In every case, the string has a node at either end.

The first harmonic pattern of vibration is the lowest possible frequency that gives the pattern shown in Figure 1a. The distance between adjacent nodes is $\frac{1}{2}\lambda_1$, so for a string of length L:

- the first harmonic wavelength $\lambda_1 = 2L$
- the first harmonic frequency $f_1 = \dfrac{c}{\lambda_1} = \dfrac{c}{2L}$, where c is the speed of the progressive waves on the wire.

The mth harmonic (where m is a whole number) has m equal loops along its length which means:

- the mth harmonic wavelength $\lambda_m = \dfrac{2L}{m}$ because each loop has a length of half a wavelength
- the mth harmonic frequency $f_m = \dfrac{c}{\lambda_m} = \dfrac{mc}{2L} = mf_1$.

In general, stationary wave patterns occur at frequencies f_1, $2f_1$, $3f_1$, $4f_1$, and so on, where f_1 is the first harmonic frequency of the first harmonic vibrations.

The frequency of the first harmonic of a vibrating string or wire is given by

$$f_1 = \frac{1}{2L}\sqrt{\frac{T}{\mu}}$$

where T is the tension in the wire and μ is its mass per unit length

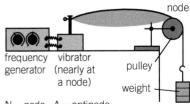

N = node A = antinode
(dotted line shows string half a cycle earlier)

a *First harmonic*

b *Second harmonic*

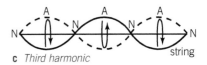

c *Third harmonic*

▲ **Figure 1** *Stationary waves on a string*

Maths skill

For a uniform wire of length L and area of cross-section A, its volume $V = LA$.

So its mass m = density $\rho \times$ volume $V = \rho LA$.

Therefore, its mass per unit length $\mu = \dfrac{m}{L} = \rho A$.

Synoptic link

See Topic 11.1, Density, for more about density

Revision tip

The data is given to 2 significant figures so the final answer is given to 2 significant figures. To avoid a 'rounding' error in the final answer, data from each step of the calculation should be carried forward with one more significant figure and only rounded off in the final answer.

Summary questions

1. A stretched wire of length 0.760 m at a tension of 64 N vibrates at its first harmonic with a frequency of 280 Hz. Calculate:
   a. the wavelength of the waves on the wire *(1 mark)*
   b. the mass per unit length of the wire. *(2 marks)*

2. For the wire in **Q1** at the same tension, calculate the length of the wire to produce a frequency of 384 Hz. *(3 marks)*

3. Two stretched wires X and Y that have the same length are under the same tension. The diameter of X is half that of Y and the density of X is 8 times that of Y. The frequency of the 1st harmonic of X is 100 Hz. Calculate the frequency of the 1st harmonic of Y. *(3 marks)*

Using an oscilloscope

Although the use of an oscilloscope is in the full A Level specification and is not part of the AS specification, it is included in this Revision Guide and in the Student Book to give you information about how to measure sound waves and determine their frequency and amplitude (i.e., the peak value).

Displaying a waveform

- The oscilloscope's **time base circuit** is switched on to make the spot move at constant speed left to right across the screen, then back again much faster. Because the spot moves at constant speed across the screen, the x-scale is calibrated, usually in milliseconds or microseconds per centimetre.

- The pd to be displayed is connected to the Y-plates via the Y-input so the spot moves up and down as it moves left to right across the screen. As it does so, it traces out the waveform on the screen. Because the vertical displacement of the spot is proportional to the pd applied to the Y-plates, the Y-input is calibrated in volts per centimetre (or per division if the grid on the oscilloscope screen is not a centimetre grid). The calibration value is usually referred to as the Y-sensitivity or Y-gain of the oscilloscope.

Figure 1 shows the trace produced when an alternating pd is applied to the Y-input. The screen is marked with a centimetre grid.

To measure the peak pd, observe that the waveform height from the bottom to the top of the wave is 3.2 cm. The amplitude (i.e., peak height) of the wave is therefore 1.6 cm. As the Y-gain is set at 5.0 V cm^{-1}, the peak pd is therefore 8.0 V (= 5.0 V cm^{-1} × 1.6 cm).

We can see from the waveform that one full cycle corresponds to a distance of 3.8 cm across the screen horizontally. As the time base control is set at 2 ms cm^{-1}, the time period T is therefore 7.6 ms (= 2 ms cm^{-1} × 3.8 cm). Therefore, the frequency of the alternating pd is 132 Hz.

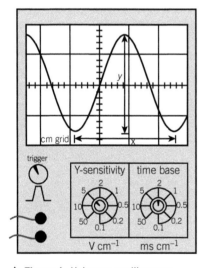

▲ **Figure 1** *Using an oscilloscope*

Summary questions

1 An alternating pd is applied to the Y-input of an oscilloscope. The height of the waveform from the bottom to the top is 5.2 cm when the Y-gain is 0.2 V cm^{-1}. Calculate the peak value of the alternating pd. *(1 mark)*

2 The time base control of an oscilloscope is set at 5.0 ms cm^{-1} and an alternating pd is applied to the Y-input. The horizontal distance across four complete cycles is observed to be 6.4 cm. Calculate the frequency of the alternating pd. *(2 marks)*

Chapter 4 Practice questions

1 **a** Explain why a transverse wave can be polarised whereas a longitudinal wave cannot be polarised. *(2 marks)*

   b A filament lamp is observed through two pieces of polaroid which are initially aligned parallel to each other. Describe and explain what you would expect to observe when one of the polaroids is rotated through 360°. *(3 marks)*

2 A microwave transmitter directs waves towards a metal plate, as shown in Figure 1. When a microwave detector is moved along a line normal to the transmitter and the plate, the detector signal increases and decreases as it moves through a sequence of equally spaced maxima and minima.

▲ **Figure 1**

   a Explain why the detector signal varies in this way as the detector is moved along the line. *(4 marks)*

   b The detector is placed at a position where the intensity is a minimum. When it is moved through a distance of 82 mm, it passes through four other minima and reaches a further minimum.

   Calculate the wavelength of the microwaves. *(3 marks)*

3 A stretched string of length L fixed at both ends is made to vibrate so that it forms a stationary wave consisting of four equal loops.

   a i Sketch the pattern of vibration of the string and mark the position of the nodes and antinodes on the string.

   ii Compare the amplitude and state the phase difference of the vibrations of points X and Y on the string at distances $\frac{L}{3}$ from each ends. *(4 marks)*

   b The length L of the string is 800 mm and its mass per unit length is 7.4×10^{-4} kg m^{-1}.

   i Calculate the wavelength of the progressive waves on the string.

   ii The string vibrates at a frequency of 200 Hz. Calculate the tension in the string. *(6 marks)*

4 An ultrasound probe in an ultrasound scanner emits ultrasound waves of frequency 1.6 MHz

   a The speed of ultrasound in body tissue is approximately 1500 m s^{-1}. Estimate the wavelength of the ultrasound waves in the body. *(1 mark)*

   b The probe sends ultrasound pulses of duration 10 μs into the body.

   i Calculate the number of ultrasound waves in each pulse.

   ii The pulses from the probe are reflected back to the probe by tissue boundaries in the body. Give *two* reasons why the amplitude of each pulse that returns to the probe is smaller than its amplitude when it entered the body. *(4 marks)*

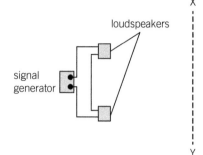

loudspeakers

X

signal generator

Y

▲ Figure 2

string at maximum displacement

node

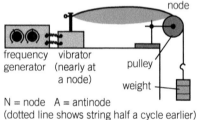

frequency generator

vibrator (nearly at a node)

pulley

weight

N = node   A = antinode
(dotted line shows string half a cycle earlier)

▲ Figure 3

5 Two small loudspeakers connected to a signal generator act as coherent sources of sound waves, as shown in Figure 2.

a Explain what is meant by coherent sources. *(1 mark)*

b A student walks along the straight line XY in front of the loudspeakers and notices the sound intensity due to the loudspeakers is a minimum at regular intervals along the line. Explain why the sound intensity is a minimum at these positions. *(2 marks)*

6 A small loudspeaker connected to a signal generator is placed near the open end of a pipe of length 380 mm, which is closed at its other end. When the output frequency of the signal generator is increased from zero, the pipe resonates with sound at different frequencies.

a i The minimum frequency at which the pipe resonates with sound is 225 Hz. At this frequency, a stationary wave is set up in the pipe with one node only which is at the closed end and an antinode very close to the open end. Calculate the wavelength of the sound in the pipe and determine the speed of sound in the pipe.

ii Describe how the amplitude of vibration of the air particles in the pipe changes along the pipe. *(3 marks)*

b Explain why the pipe resonates with sound at a frequency of about 675 Hz. *(2 marks)*

7 Figure 3 shows an arrangement used to set up stationary waves on a stretched string.

a Describe how you would use this arrangement to investigate how the frequency of vibration of the first harmonic, f_1, of a stretched string varies with the tension T in the string. *(4 marks)*

b The following results were obtained in the investigation in part **a** using a string of length 0.64 m.

T / N	1.0	1.5	2.0	2.5	3.0	3.5	4.0
f_1 / Hz	42	52	60	67	73	79	84

i The first harmonic frequency is given by the equation

$$f_1 = \frac{1}{2L}\sqrt{\frac{T}{\mu}}$$

Use the equation to show that a graph of f_1^2 against T should be a straight line through the origin. *(2 marks)*

ii Plot a graph of f_1^2 against T to determine μ, the mass per unit length of the string. *(6 marks)*

5.1 Refraction of light
5.2 More about refraction
Specification reference: 3.3.2.3

Refraction is the change of direction that occurs when light passes non-normally across a boundary between two transparent substances. Figure 1 shows the change of direction of a light ray when it enters and when it leaves a rectangular glass block in air. For a light ray travelling from air into a transparent substance

$$\text{the refractive index of the substance, } n = \frac{\sin i}{\sin r}$$

Notice that **partial reflection** can also occur when a light ray is refracted.

A general rule for refraction

Figure 2 shows a light ray that passes across a straight boundary from a transparent substance of refractive index n_1 into another transparent substance of refractive index n_2. If θ_1 and θ_2 are the angles of incidence and refraction respectively, it can be shown that $n_1 \sin \theta_1 = n_2 \sin \theta_2$

Explaining refraction

Refraction occurs because the speed of the light waves is different in each substance. For light waves entering a transparent substance from air at a plane boundary, it can be shown that

$$\frac{\sin i}{\sin r} = \frac{c}{c_s}$$

where c is the speed of the waves in a vacuum and c_s is the speed in the substance.

Therefore, the refractive index of the substance

$$n_s = \frac{c}{c_s}$$

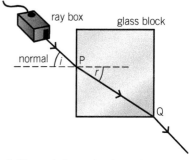

▲ **Figure 1** *Investigating refraction*

> **Revision tip**
> Remember that the normal is an imaginary line perpendicular to a boundary between two materials or a surface.

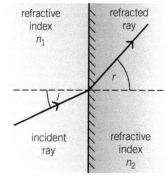

▲ **Figure 2** *The n sin θ rule*

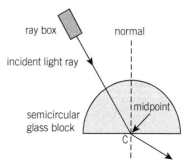

▲ **Figure 3** *Explaining refraction*

> **Revision tip**
> During refraction, the speed and wavelength both change, but frequency stays constant.

> **Revision tip**
> The refractive index of air is 1.0003. For most purposes, the refractive index of air may be assumed to be 1.

Summary questions

1 A glass block has a refractive index of 1.52. A light ray enters the block from air at point X at an angle of incidence of 30.0°. Calculate the angle of refraction of the light ray. (*2 marks*)

2 The light ray in **Q1** is refracted again at point Y where it leaves the glass block and enters the air. The angle of refraction of the light ray at Y is 70.0°. Calculate the angle of incidence of the light ray at Y. (*2 marks*)

3 A light ray in air was directed into a semicircular glass block at the midpoint C of the flat side at an angle, as shown in Figure 3. The refractive index of the glass block was 1.54.
   a Explain why the light ray was not refracted when it entered the block. (*1 mark*)
   b The angle of incidence of the light ray at C was 35°. Calculate the angle through which the light ray was deflected at C. (*3 marks*)

4 Calculate the angle of refraction for a light ray entering glass of refractive index 1.50 at an angle of incidence of 40.0° from water of refractive index 1.33. (*2 marks*)

5.3 Total internal reflection

Specification reference: 3.3.2.3

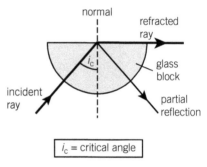

i_c = critical angle

▲ **Figure 1** *Total internal reflection*

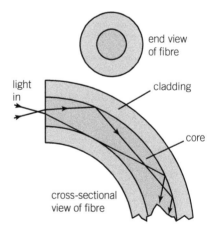

▲ **Figure 2** *Fibre optics*

When a light ray travels across a boundary between two transparent substances or between air and a transparent substance, it undergoes **total internal reflection** at the boundary if the incident substance has a larger refractive index than the other substance and the angle of incidence exceeds the critical angle.

At the critical angle i_c, the angle of refraction is 90°. Therefore, $n_1 \sin i_c = n_2 \sin 90$ where n_1 is the refractive index of the incident substance and n_2 is the refractive index of the other substance. Since $\sin 90 = 1$, then

$$\sin i_c = \frac{n_2}{n_1}$$

Optical fibres

Figure 2 shows the path of a light ray along an optical fibre. The light ray is totally internally reflected each time it reaches the fibre boundary.

1 A **communications optical fibre** transmits pulses of light from a transmitter at one end to a receiver at the other end. The material needs to be highly transparent to minimise absorption which would reduce the amplitude of light pulses travelling along the fibre. Each fibre consists of a core surrounded by cladding of lower refractive index. Total internal reflection takes place at the core–cladding boundary.

   • Without cladding, light would cross between fibres where they touch. This would reduce the amplitude of the pulses and the signals would not be secure.

   • Monochromatic light and fibres with a very narrow core are used to prevent pulse lengthening which would cause pulses to merge. **Modal** (i.e., multipath) **dispersion** is reduced by the very narrow core. In a wide core, non-axial rays undergo more total internal reflections than axial rays and therefore take longer, making the pulses longer. **Material** (or spectral) **dispersion** occurs if white light is used. Violet light is slower than red light in glass. The violet component of a white light pulse falls behind the red component so the pulse lengthens.

2 The **medical endoscope** contains two bundles of fibres, one of which is a **coherent bundle**, which means the fibre ends are in the same relative positions at each end. Light is sent through the non-coherent bundle into the body cavity to be observed. A lens over the end of the coherent bundle forms an image of the body cavity on the end of the fibre bundle. The image is observed at the other end of the coherent fibre bundle.

5.4 Double slit interference

Specification reference: 3.3.2.1

Interference of light can be observed by directing light from a suitable light source at two closely spaced parallel slits (double slits).

- The two slits act as **coherent** sources of waves, which means that they emit light waves with a constant phase difference and the same frequency.
- Alternate bright and dark fringes, referred to as **Young's fringes**, can be seen on a white screen placed where the diffracted light from the double slits overlaps. The fringes are evenly spaced and parallel to the double slits.

Where a *bright* fringe is formed, the light from one slit reinforces the light from the other slit. In other words:

- the light waves from each slit arrive *in phase* with each other
- the path difference = $m\lambda$ where m = 0, 1, 2, and so on.

Where a *dark* fringe is formed, the light from one slit cancels the light from the other slit. In other words:

- the light waves from the two slits arrive *180° out of phase*
- the path difference = $\left(m + \dfrac{1}{2}\right)\lambda$.

Warning: Never look along a laser beam, even after reflection. If a laser is used as the light source, the fringes must be displayed on a screen because a beam of laser light will damage the retina if it enters the eye.

The double slit equation

$$\text{fringe separation } w = \frac{\lambda D}{s}$$

where λ is the wavelength of light, s is the slit spacing, and D is the distance from the slits to the screen.

Summary questions

1 In a double slit experiment using red light, a fringe pattern is observed on a screen at a fixed distance from the double slits. How would the fringe pattern change if:
   a the screen is moved further from the slits   *(1 mark)*
   b blue light is used instead?   *(1 mark)*

2 The following measurements were made in a double slit experiment:
   centre-to-centre distance across 5 dark fringes = 4.5 mm,
   slit spacing s = 0.40 mm, slit–screen distance D = 0.85 m.
   Calculate the wavelength of light used.   *(3 marks)*
   Hint: Remember that the number of fringe spacings ≠ the number of fringes.

3 In Q2, state and explain two differences in the appearance of the fringes if light of a longer wavelength was used.   *(2 marks)*

Key terms

Young's fringes are parallel, bright, and dark fringes observed when light from a narrow slit passes through two closely spaced slits.

The **path difference** is the difference in the distance from each slit to a given point on the screen.

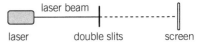

laser beam

laser   double slits   screen

▲ **Figure 1** *Using a laser to demonstrate interference*

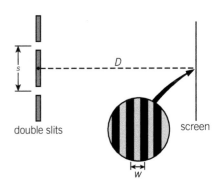

double slits   screen

▲ **Figure 2** *Diagram to show w, D, and s for a Young's double slit experiment*

Maths skills

Rearranging this equation to make λ the subject gives $\lambda = \dfrac{ws}{D}$.

Practical skill

To measure w, measure across several fringes from the centre of a dark fringe to the centre of another dark fringe and divide by the number of fringes you measured across.

35

Coherence

Interference fringes are produced using a double slit (i.e., two narrow slits close together), provided the slits emit wavefronts with a constant phase difference. The slits must act as **coherent sources** otherwise the points of cancellation and reinforcement change at random.

- A laser is a source of coherent light. So the two slits act as coherent sources when a laser beam is directed at them.

- Light from a single narrow slit illuminated by a filament lamp can be used to produce interference fringes instead of laser light. Each wave crest or wave trough from the single slit always passes through one of the double slits a fixed time after it passes through the other slit. So the waves from the double slits always have a constant phase difference.

- Light from two nearby lamp bulbs could not form an interference pattern because a filament lamp emits light waves at random. The points of cancellation and reinforcement would change at random, so no interference pattern is possible.

Wavelength and colour

In the double slit experiment, the fringe separation depends on the colour of light used. Each colour of light has its own narrow band of wavelengths. The fringe separation is greater for red light than for blue light because red light has a longer wavelength than blue light.

White light fringes

For white light, each component colour of white light produces its own fringe pattern, and each pattern is centred on the screen at the same position. As a result:

- the central bright fringe is white because every colour contributes at the centre of the pattern

- the inner bright fringes are tinged with blue on the inner side and red on the outer side. This is because the red fringes are more spaced out than the blue fringes so the two fringe patterns do not overlap exactly

- the outer bright fringes become fainter with increasing distance from the centre. Also, the fringes merge because the colours overlap.

> **Key term**
>
> **Coherent sources** emit light waves of the same frequency with a constant phase difference.

Summary questions

1 a Explain what is meant by 'coherent light sources'. (*1 mark*)
   b Explain why an interference pattern cannot be seen using two filament lamps side by side. (*2 marks*)

2 Double slit interference fringes are observed using light of wavelength 635 nm and a double slit of slit spacing 0.45 mm. The fringes are observed on a screen at a distance of 1.90 m from the double slits. Calculate the fringe separation of these fringes. (*2 marks*)

3 A double slit is used to observe the interference fringes from a 12 V filament lamp connected to a variable 0–12 V voltage supply. When the voltage across the filament lamp is increased, the colour of the filament changes from red to white. Describe how the fringe pattern alters in this change. (*3 marks*)

5.6 Diffraction

Specification reference: 3.3.2.2

Observing diffraction

Diffraction is the spreading of waves when they pass through a gap or by an edge. When waves pass through a gap:

1 The diffracted waves spread out more if the gap is made narrower or the wavelength is made larger.

2 Each diffracted wavefront has breaks either side of the centre. These breaks are due to waves diffracted by adjacent sections on the gap being out of phase and cancelling each other out in certain directions.

Diffraction is a general property of all waves and is very important in the design of optical instruments such as cameras, microscopes, and telescopes. This is because less diffraction occurs when waves pass through a wide gap than through a narrow gap. Therefore, more detail can be seen in optical images formed by wide lenses.

Diffraction of light by a single slit

When a parallel beam of light is directed at a single slit, the diffracted light forms a pattern that can be observed on a white screen, as shown in Figure 1. The pattern shows a central fringe with further fringes either side of the central fringe.

- The central fringe is twice as wide as each of the outer fringes (measured from minimum to minimum intensity).

- Each of the outer fringes is the same width.

- The fringes decreases in intensity with distance from the centre.

- The central fringe has a much higher peak intensity than the other fringes.

- The width of each fringe is proportional to $\frac{\lambda}{a}$, where a is the width of the slit.

Single slit diffraction and Young's fringes

In general, for monochromatic light of wavelength λ incident on two slits of aperture width a at slit separation s (from centre to centre), the intensity peaks of the interference fringes are affected by single slit diffraction as shown in Figure 2. If the aperture width a is made narrower, more diffraction occurs so more interference fringes are seen between adjacent single slit minima as the single slit minima will be further apart.

> ### Synoptic link
> You have met the study of water waves using a ripple tank in more detail in Topic 4.3, Wave properties 1.

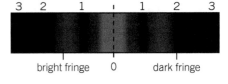

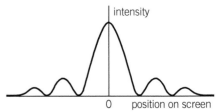

▲ **Figure 1** *Single slit diffraction*

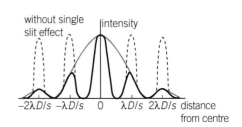

▲ **Figure 2** *Intensity distribution for Young's fringes*

Summary questions

1 A narrow beam of monochromatic light is directed normally at a single slit. The diffracted light forms a fringe pattern on a white screen.
  a Describe the appearance of the fringe pattern on the screen. (*3 marks*)
  b Sketch a graph to show how the intensity of the fringes varies with distance across the screen. (*2 marks*)

2 Single slit diffraction fringes are observed using red light. State and explain how the pattern changes if:
  a the slit is made narrower (*2 marks*)
  b blue light is used instead of red light. (*2 marks*)

5.7 The diffraction grating

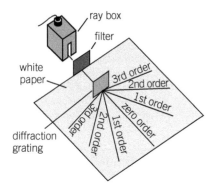

▲ **Figure 1** *The diffraction grating*

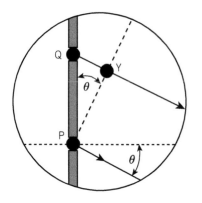

▲ **Figure 2** *The nth order wavefront*

Worked example

When monochromatic light is directed normally at a diffraction grating which has 600 lines per millimetre, a second order beam is observed at an angle of diffraction of 45.0°. Calculate the wavelength of the light.

Solution

The grating has 600 000 lines per metre. Hence $d = \dfrac{1}{600\,000}$ m

$$= 1.67 \times 10^{-6}\,\text{m}$$

Therefore $2\lambda = d\sin\theta = 1.67 \times 10^{-6} \times \sin 45.0 = 1.18 \times 10^{-6}$ which gives $\lambda = 5.90 \times 10^{-7}\,\text{m}$

Synoptic link

You have met line spectra in more detail in Topic 3.5, Energy levels and spectra.

Revision tip

Reminder: 1 degree = 60 minutes.

Testing a diffraction grating

A **diffraction grating** consists of a plate with many closely spaced parallel slits ruled on it. When a parallel beam of monochromatic light is directed normally at a diffraction grating, light is transmitted by the grating in certain directions only. This is because: the light passing through each slit is diffracted and the diffracted light waves from adjacent slits reinforce each other in certain directions only.

The diffraction grating equation

For light of wavelength λ incident normally on a grating which has N slits per metre, the angle of diffraction θ of the nth order beam is given by the equation

$$d\sin\theta = n\lambda$$

where the grating spacing $d = \dfrac{1}{N}$

Note: The maximum order number is given by the value of $\dfrac{d}{\lambda}$ rounded down to the nearest whole number.

Proof of the diffraction grating equation

Look at Figure 2. The wavefront emerging from slit P reinforces a wavefront emitted n cycles earlier by the adjacent slit Q. The distance QY is therefore equal to $n\lambda$.

The angle of diffraction of the beam θ is equal to the angle between the wavefront and the plane of the slits so QY = QP $\sin\theta$. Since QY = $n\lambda$ and QP = d (the grating spacing), it follows that $n\lambda = d\sin\theta$.

Types of spectra

1 Continuous spectra

The spectrum of light from a filament lamp is a continuous spectrum of colour from deep violet at about 350 nm to deep red at about 650 nm.

2 Line emission spectra

A glowing gas emits light that has a spectrum consisting of narrow vertical lines of different colours. The wavelengths of the lines are characteristic of the chemical element that produced the light.

3 Line absorption spectra

A line absorption spectrum is a continuous spectrum with narrow dark lines at certain wavelengths. When white light passes through a glowing gas, the gas atoms absorb light of the same wavelengths at which they can emit so the transmitted light is missing these wavelengths.

Summary questions

1 A laser beam of wavelength 630 nm is directed normally at a diffraction grating with 600 lines per millimetre. Calculate:
   a the angle of diffraction of the first order beam (*3 marks*)
   b the maximum order number. (*2 marks*)

2 Light of wavelength 430 nm is directed normally at a diffraction grating. The angle of diffraction of the 1st order beam was at 28°. Calculate the angle of diffraction of the most diffracted order. (*3 marks*)

1 A light ray enters an equilateral glass prism of refractive index 1.55 at the midpoint of one side of the prism at an angle of incidence of 35.0°.

   a **i** Sketch this arrangement and show the path of the light ray in and out of the prism.

        ii Calculate the angle of refraction of the light ray in the glass. *(3 marks)*

   b **i** Show that the angle of incidence where the light ray leaves the glass prism is about 38°.

        ii Calculate the angle of refraction of the light ray where it leaves the prism. *(3 marks)*

2 **a** Explain why the core of an optical fibre used in communications needs to be very narrow. *(2 marks)*

   b **i** The core of an optical fibre has a refractive index of 1.55. The core is surrounded by cladding of refractive index 1.45. Calculate the critical angle at the core–cladding boundary.

        ii State one advantage of cladding an optical fibre. *(4 marks)*

3 A white light ray is directed through the curved side of a semicircular glass block at the midpoint of the flat side, as shown in Figure 1. The angle of incidence of the light ray at the flat side is 40°. The refractive index of the glass for red light is 1.52 and for blue light is 1.55.

   a Calculate the angle of refraction at the midpoint of:

      i the red component of the light ray

      ii the blue component of the light ray. *(4 marks)*

   b Show the angle between the red and blue components of the refracted light ray is about 7°. *(1 mark)*

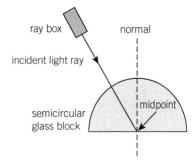

▲ **Figure 1**

4 A laser beam of wavelength 630 nm is directed normally at a double slit to produce an interference pattern of alternate bright and dark fringes on a screen.

   a Explain why alternate bright and dark fringes were seen on the screen. *(4 marks)*

   b The distance across 6 bright fringes was measured to be 65 mm. The screen was at a distance of 2.50 m from the double slit. Calculate the slit spacing. *(3 marks)*

5 Red light from a laser is directed normally at a slit that can be adjusted in width. The diffracted light from the slit forms a pattern of diffraction fringes on a white screen.

   a Sketch a graph to show how the intensity of the fringes varies across the pattern. *(3 marks)*

   b Describe how the appearance of the fringes changes if the slit is made wider. *(2 marks)*

6 Light directed normally at a diffraction grating has a wavelength range from 430 to 445 nm. The grating has 600 lines per mm.

   a How many diffracted orders are observed in the transmitted light? *(4 marks)*

   b For the highest order, calculate the angular width of the diffracted beam. *(3 marks)*

7 Young's fringes are produced on the screen using the arrangement shown in Figure 2.

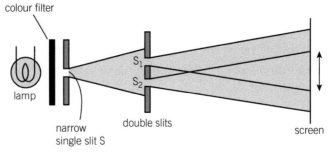

▲ **Figure 2**

    a i Explain why slit S should be narrow.

      ii Explain why slits S_1 and S_2 are coherent sources? *(6 marks)*

    b The fringes are equally spaced but some bright fringes are fainter than bright fringes further from the central fringe. Explain why some bright fringes are fainter than bright fringes further from the centre. *(3 marks)*

8 A narrow beam of monochromatic light is directed normally at a diffraction grating which has 600 lines per millimetre.

    a The angle of diffraction of the 2nd order beam is 40.0°. Calculate the wavelength of the light. *(3 marks)*

    b A semicircular glass block of refractive index 1.50 is placed with its flat side against the grating on the opposite side to the incident light. The block is positioned so that the direction of the zero order beam is unchanged as shown in Figure 3. Show that the angle of diffraction of the second order beam is now about 25°. *(4 marks)*

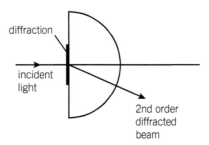

▲ **Figure 3**

Chapter 4 Practice questions

1 **a** Explain why a transverse wave can be polarised whereas a longitudinal wave cannot be polarised. (*2 marks*)

b A filament lamp is observed through two pieces of polaroid which are initially aligned parallel to each other. Describe and explain what you would expect to observe when one of the polaroids is rotated through 360°. (*3 marks*)

2 A microwave transmitter directs waves towards a metal plate, as shown in Figure 1. When a microwave detector is moved along a line normal to the transmitter and the plate, the detector signal increases and decreases as it moves through a sequence of equally spaced maxima and minima.

metal plate

transmitter   detector

▲ **Figure 1**

a Explain why the detector signal varies in this way as the detector is moved along the line. (*4 marks*)

b The detector is placed at a position where the intensity is a minimum. When it is moved through a distance of 82 mm, it passes through four other minima and reaches a further minimum.

Calculate the wavelength of the microwaves. (*3 marks*)

3 A stretched string of length L fixed at both ends is made to vibrate so that it forms a stationary wave consisting of four equal loops.

a i Sketch the pattern of vibration of the string and mark the position of the nodes and antinodes on the string.

ii Compare the amplitude and state the phase difference of the vibrations of points X and Y on the string at distances $\frac{L}{3}$ from each ends. (*4 marks*)

b The length L of the string is 800 mm and its mass per unit length is $7.4 \times 10^{-4}\,\mathrm{kg\,m^{-1}}$.

i Calculate the wavelength of the progressive waves on the string.

ii The string vibrates at a frequency of 200 Hz. Calculate the tension in the string. (*6 marks*)

4 An ultrasound probe in an ultrasound scanner emits ultrasound waves of frequency 1.6 MHz

a The speed of ultrasound in body tissue is approximately $1500\,\mathrm{m\,s^{-1}}$. Estimate the wavelength of the ultrasound waves in the body. (*1 mark*)

b The probe sends ultrasound pulses of duration 10 μs into the body.

i Calculate the number of ultrasound waves in each pulse.

ii The pulses from the probe are reflected back to the probe by tissue boundaries in the body. Give *two* reasons why the amplitude of each pulse that returns to the probe is smaller than its amplitude when it entered the body. (*4 marks*)

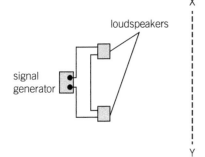

loudspeakers

signal generator

X

Y

▲ **Figure 2**

string at maximum displacement

node

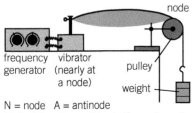

frequency generator   vibrator (nearly at a node)

pulley

weight

N = node   A = antinode
(dotted line shows string half a cycle earlier)

▲ **Figure 3**

5 Two small loudspeakers connected to a signal generator act as coherent sources of sound waves, as shown in Figure 2.

  a Explain what is meant by coherent sources. *(1 mark)*

  b A student walks along the straight line XY in front of the loudspeakers and notices the sound intensity due to the loudspeakers is a minimum at regular intervals along the line. Explain why the sound intensity is a minimum at these positions. *(2 marks)*

6 A small loudspeaker connected to a signal generator is placed near the open end of a pipe of length 380 mm, which is closed at its other end. When the output frequency of the signal generator is increased from zero, the pipe resonates with sound at different frequencies.

  a i The minimum frequency at which the pipe resonates with sound is 225 Hz. At this frequency, a stationary wave is set up in the pipe with one node only which is at the closed end and an antinode very close to the open end. Calculate the wavelength of the sound in the pipe and determine the speed of sound in the pipe.

    ii Describe how the amplitude of vibration of the air particles in the pipe changes along the pipe. *(3 marks)*

  b Explain why the pipe resonates with sound at a frequency of about 675 Hz. *(2 marks)*

7 Figure 3 shows an arrangement used to set up stationary waves on a stretched string.

  a Describe how you would use this arrangement to investigate how the frequency of vibration of the first harmonic, f_1, of a stretched string varies with the tension T in the string. *(4 marks)*

  b The following results were obtained in the investigation in part **a** using a string of length 0.64 m.

T / N	1.0	1.5	2.0	2.5	3.0	3.5	4.0
f_1 / Hz	42	52	60	67	73	79	84

    i The first harmonic frequency is given by the equation

$$f_1 = \frac{1}{2L}\sqrt{\frac{T}{\mu}}$$

       Use the equation to show that a graph of f_1^2 against T should be a straight line through the origin. *(2 marks)*

    ii Plot a graph of f_1^2 against T to determine μ, the mass per unit length of the string. *(6 marks)*

6.1 Vectors and scalars

Specification reference: 3.4.1.1

Representing a vector

A vector is any physical quantity that has a direction as well as a magnitude. Examples of vectors include displacement, velocity, acceleration, force, and momentum.

A scalar is any physical quantity that is not directional. Examples of scalars include distance, mass, density, volume, and energy.

Any vector can be represented as an arrow of length proportional to the magnitude of the vector quantity in the direction of the vector.

Addition of vectors

Any vectors of the same type can be added together by drawing a scale diagram or using a calculator.

1 Drawing a scale diagram

Figure 1 shows two displacement vectors OA and AB added together to give the resultant displacement vector OB.

$$OB = OA + AB$$

2 Using a calculator

To add two perpendicular vectors F_1 and F_2:

- the magnitude of the resultant force $F = \sqrt{F_1^2 + F_2^2}$
- the angle θ between the resultant and F_1 is given by $\tan \theta = \dfrac{F_2}{F_1}$.

Resolving a vector

Any vector can be resolved into two perpendicular components. For example. Figure 2 shows a force F resolved into components along perpendicular axes, x and y, where the direction of the force is at an angle to the x-axis.

- $F_x = F \cos \theta$ **parallel to the x-axis.**
- $F_y = F \sin \theta$ **perpendicular to x-axis.**

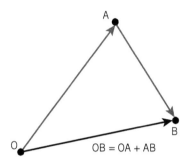

▲ **Figure 1** *Drawing a scale diagram*

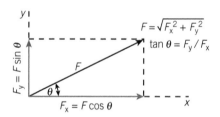

▲ **Figure 2** *Resolving a force*

Summary questions

1 An airplane takes off from airport A and lands at airport B, which is 180 km 30° due east of north from A.
   a Sketch the displacement vector AB on two perpendicular axes representing due north and due east with A at the intersection of the two axes. *(1 mark)*
   b Calculate the component of vector AB **i** due east **ii** due north. *(1 mark)*

2 Calculate the magnitude and direction of the resultant force on an object which is acted on by a force of 5.0 N and a force of 11.0 N that are:
   a in opposite directions *(1 mark)*
   b at right angles to each other. *(2 marks)*

3 A descending parachutist of total weight 900 N is acted on by a horizontal force of 150 N due to the wind. Calculate the magnitude and direction of the resultant of these two forces. *(3 marks)*

6.2 Balanced forces

Specification reference: 3.4.1.1

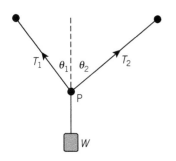

▲ **Figure 1** *A suspended weight*

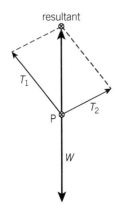

▲ **Figure 2** *The parallelogram of forces rule*

Synoptic link

The three force vectors T_1, T_2, and W in Figure 2 form a triangle. See Topic 6.6, Equilibrium rules, for more about the **triangle of forces**.

Any object at rest or moving at constant velocity is in **equilibrium**.

If *two forces only* are acting on the object, the two forces must be equal and opposite to each other for the object to be in equilibrium. The resultant of the two forces is zero and the two forces are said to be *balanced*.

If *two or more forces* are acting on the object, their combined effect (the resultant) must be zero for it to be in equilibrium. To check the resultant is zero, either:

- resolve each force along the same parallel and perpendicular lines and balance the components along each line (see the Worked example below), or
- use the parallelogram of forces rule if there are only three forces.

Worked example

An object of weight W is tied to a vertical string, which is supported by two strings at angles θ_1 and θ_2 to the vertical, as shown in Figure 1. Suppose the tension in the string at angle θ_1 to the vertical is T_1 and the tension in the other string is T_2. Derive two equations relating T_1, T_2, and W.

Solution

At the point P where the strings meet, the forces T_1, T_2, and W are in equilibrium.

Resolving T_1 and T_2 vertically and horizontally gives:

1 horizontally: $T_1 \sin \theta_1 = T_2 \sin \theta_2$ **2** vertically: $T_1 \cos \theta_1 + T_2 \cos \theta_2 = W$.

The parallelogram of forces

In Figure 1, the resultant of the three forces is zero. Therefore, the resultant of any *two* of the forces is equal and opposite to the *third* force. We can show this by using the parallelogram of forces rule to draw a scale diagram as follows:

1 Measure the angles θ_1 and θ_2 between each of the upper strings and the lower string which is vertical.

2 Draw a scale diagram of a parallelogram as shown in Figure 2, using the two force vectors T_1 and T_2 as adjacent sides.

3 The resultant of T_1 and T_2 is equal and opposite to W. The resultant is the diagonal of the parallelogram *between* the two force vectors.

Summary questions

1 A point object of weight 3.2 N is acted on by a horizontal force of 5.8 N.
   a Calculate the resultant of these two forces. *(3 marks)*
   b Determine the magnitude and direction of a third force acting on the object for it to be in equilibrium. *(1 mark)*

2 A small object of weight 8.6 N is at rest on a rough slope, which is at an angle of 40° to the horizontal, as shown in Figure 3. Calculate:
   a the frictional force F on the object
   b the support force S from the slope on the object. *(3 marks)*

3 A string fixed at each end at different heights a fixed distance apart supports a stationary object of weight 2.0 N. Calculate the tension in each section of the string when the object is stationary and the angles of each section of the string to the horizontal are 35° and 30°. *(4 marks)*

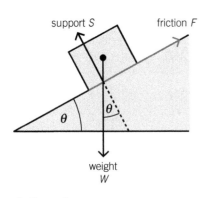

▲ **Figure 3**

6.3 The principle of moments

Specification reference: 3.4.1.2

The **moment** of a force about any point is defined as the force × the perpendicular distance from the line of action of the force to the point.

The unit of the moment of a force is the newton metre (N m).

For a force F acting along a line of action at perpendicular distance d from a certain point, **the moment of the force =** $\boldsymbol{F \times d}$.

Revision tip

Distance d is the **perpendicular** distance from the line of action of the force to the point.

The principle of moments

If a body acted on by forces is in equilibrium,

the sum of the clockwise moments about any fixed point = the sum of the anticlockwise moments about that point

Consider a uniform metre rule balanced at its centre, as in Figure 2.

- Weight W_1 provides an anticlockwise moment about the pivot = $W_1 d_1$.
- Weight W_2 provides a clockwise moment about the pivot = $W_2 d_2$.

For equilibrium, applying the principle of moments,

$$\boldsymbol{W_1 d_1 = W_2 d_2}$$

Note: The support force $S = W_1 + W_2$

▲ **Figure 1** *A turning force*

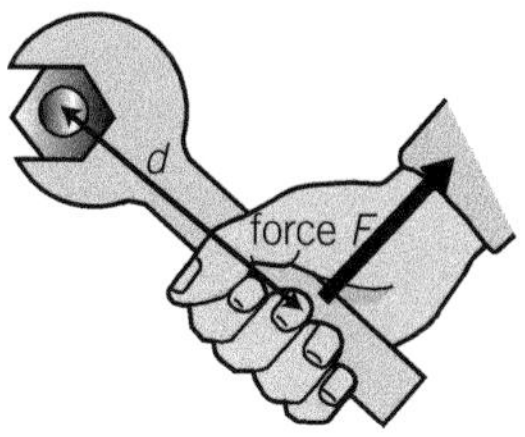

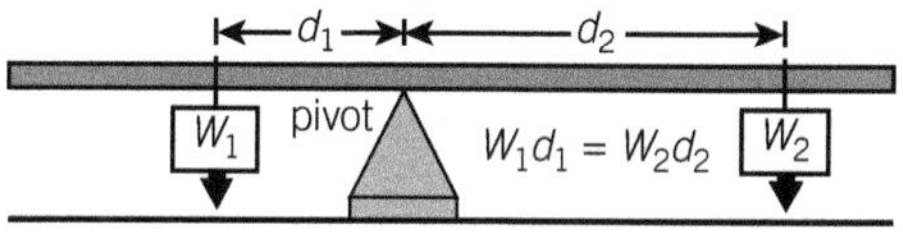

▲ **Figure 2** *Using the principle of moments*

Centre of mass

The **centre of mass** of a body is the point through which a single force on the body has no turning effect.

Figure 3 shows a metre rule that has been balanced off-centre on a knife-edge by adjusting the position of a known weight W_1.

- The known weight W_1 provides an anticlockwise moment about the pivot = $W_1 d_1$.
- The weight of the rule W_0 provides a clockwise moment = $W_0 d_0$.

Applying the principle of moments therefore gives $W_0 d_0 = W_1 d_1$.

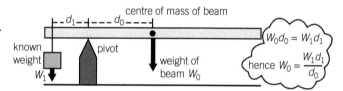

▲ **Figure 3** *Finding the weight of a beam*

Summary questions

1. A metre rule, pivoted at its centre of mass, supports a 2.0 N weight at its 10.0 cm mark, a 3.0 N weight at its 25 cm mark, and a weight W at its 90 cm mark.
   - **a** Sketch a diagram to represent this situation. *(1 mark)*
   - **b** Calculate the weight W. *(3 marks)*
2. A child of weight 200 N sits at one end of a seesaw at a distance of 1.2 m from the pivot at the centre. The seesaw is balanced by a second child of weight 120 N sitting on it at a distance of 0.8 m from the centre and an adult holding the seesaw at one end 0.4 m from the second child. Calculate the force exerted by the adult on the seesaw. *(3 marks)*
3. A uniform metre rule of weight 1.2 N is balanced horizontally on a horizontal knife-edge at its 250 mm mark. The ruler supports a 2.5 N weight at its 150 mm mark and is supported by a vertical string attached to a stand at one end and to the ruler at the 950 mm mark at the other end. Sketch the arrangement and calculate the tension in the string. *(4 marks)*

Revision tip

When there are several unknown forces, take moments about a point through which one of the unknown forces acts. This will simplify your calculation.

6.4 More on moments
6.5 Stability
Specification reference: 3.4.1.1; 3.4.1.2

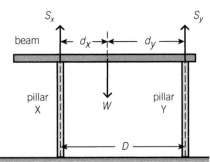

▲ **Figure 1** *A two-support problem*

Revision tip
In calculations, you can always eliminate the turning effect of a force by taking moments about a point through which it acts.

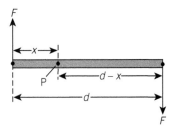

▲ **Figure 2** *A couple*

Support forces

When an object in equilibrium is supported at one point only, the support force on the object is equal and opposite to the total downward force acting on the object.

Consider a uniform beam supported on two pillars X and Y, which are at distance D apart. The weight of the beam is shared between the two pillars according to how far the beam's centre of mass is from each pillar. For example:

- Suppose the centre of mass of the beam is at distance d_x from pillar X and distance d_y from pillar Y, as shown in Figure 1, then taking moments about where X is in contact with the beam gives,

  $S_y D = W d_x$, where S_y is the support force from pillar Y.

  Therefore, $S_y = \dfrac{W d_x}{D_1}$.

Note: To determine S_x, use $S_x + S_y = W$.

Couples

A **couple** is a pair of equal and opposite forces acting on a body, but not along the same line. Figure 2 shows a couple acting on a beam. The couple turns or tries to turn the beam.

The moment of a couple = force × perpendicular distance between the lines of action of the forces.

The moment of a couple about any point is always the same. In Figure 2 the total moment about an arbitrary point P along the beam $= Fx + F(d - x)$
$= Fx + Fd - Fx = Fd$.

Stable and unstable equilibrium

- If a body in **stable equilibrium** is displaced then released, it returns to its equilibrium position. This happens because when it is displaced, its centre of mass is not directly below the point of support. So the weight of the object returns it to equilibrium.

- If an object in **unstable equilibrium** is displaced slightly from equilibrium then released, it will move away from equilibrium. The reason is that when the object is displaced slightly, the centre of mass is no longer above the point of support. The object's weight therefore acts to turn it further from the equilibrium position.

Tilting and toppling

Tilting is where an object at rest on a surface is acted on by a force that makes it turn about a point or a line where it is in contact with the surface. A tilted object will topple over if it is tilted too far.

Summary questions

1 A uniform beam of length 2.0 m and weight 120 N rests horizontally on two supports X and Y, 0.20 m and 0.25 m respectively from each end. Sketch the arrangement and calculate the support force on the beam at each end. *(5 marks)*
Hint: Mark the relevant distances as well as the forces on your diagram.

2 A bridge crane consists of a horizontal span of weight 6400 N and length 11.0 m fixed at each end to vertical pillars P and Q as shown in Figure 3. The span supports a load of 500 N at a distance of 4.0 m from pillar P.
Calculate the support force on the span from each pillar. *(4 marks)*

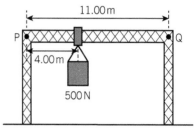
▲ Figure 3

3 A horizontal uniform diving board AB of weight 200 N and of length 2.50 m overhangs the edge of the swimming pool by 1.50 m. The board is bolted to the ground at the edge of the pool P and at end A on the ground. A diver of weight 640 N stands on the board over the pool at end B.
 a Sketch a diagram of the arrangement and show the direction of the forces on the board. *(1 mark)*
 b i Show that the force of the bolt at A on the board is 1010 N. *(3 marks)*
  ii Calculate the force of the bolt at P on the board. *(1 mark)*

4 Explain why side winds on a high-sided vehicle can cause it to topple over. *(2 marks)*

5 A filing cabinet is usually designed so only one drawer can be pulled out at any time. Explain why the cabinet would be unstable if more than one drawer was pulled out at the same time. *(3 marks)*

6 Figure 4 shows a vehicle being driven slowly across a slope. The distance between the wheels on each side of the vehicle is 1.7 m and its centre of mass is 0.9 m from the ground. Calculate the maximum angle of a slope on which the vehicle would not topple over. *(3 marks)*

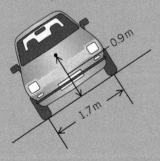
▲ Figure 4

6.6 Equilibrium rules
6.7 Statics calculations
Specification reference: 3.4.1.1; 3.4.1.2

Free body force diagrams

When two objects interact, they always exert equal and opposite forces on one another. A **free body force diagram** shows only the forces acting on one object.

The triangle of forces

For an object acted on by three forces to be in equilibrium, the three forces must give an overall resultant of zero. The three forces as vectors should form a triangle. In other words, for three forces F_1, F_2, and F_3 to give zero resultant:

their vector sum $F_1 + F_2 + F_3 = 0$

- Their lines of action must intersect at the *same* point, otherwise the object cannot be in equilibrium, as the forces will have a net turning effect.

- A scale diagram of the triangle of forces can be drawn to find an unknown force or angle, given the other forces and angles in the triangle.

The conditions for equilibrium of a body

For a body in equilibrium:

1 The resultant force must be zero. If there are only three forces, they must form a closed triangle.

2 The principle of moments must apply (i.e., the moments of the forces about the same point must balance out).

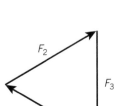
▲ **Figure 1** *The triangle of forces for a point object*

Synoptic link

See Topics 6.3, The principle of moments, and 6.4, More on moments, again if necessary.

Summary questions

1 A uniform plank of length 4.0 m and of weight 180 N rests horizontally on two bricks X and Y at 0.10 m from either end. A child C of weight 150 N stands at a distance of 0.90 m from one end.
   a Draw a free body force diagram of the forces acting on the plank. *(1 mark)*
   b Calculate the support forces acting on the plank from the supporting bricks. *(4 marks)*

2 A uniform ladder of weight 220 N rests against a smooth vertical wall at an angle of 20° and with its lower end on flat ground. The force on the ladder due to the wall is perpendicular to the wall.
   a Draw a free body force diagram of the forces acting on the ladder. *(1 mark)*
   b Determine the magnitude and direction of the force of the ground on the ladder. *(4 marks)*

3 A rectangular picture of weight 21 N is supported on a wall by a cord attached to the top corners of the picture which hangs over a wall hook directly above the midpoint of the picture. Each section of the cord is at an angle of 25° to the top edge of the picture, which is horizontal. The centre of mass of the picture is at its geometrical centre.
   a Draw a free body force diagram of the forces acting on the picture. *(1 mark)*
   b Calculate the tension in each section of the cord. *(3 marks)*

Chapter 6 Practice questions

1 A point object in equilbrium is acted on by a 3 N force, a 5 N force, and a 6 N force, which is horizontal.

   a Draw a triangle of forces diagram to determine the angle of each of the lines of action of the other forces to the 6 N force. *(3 marks)*

   b What is the resultant force on the object if the 5 N force is removed? *(1 mark)*

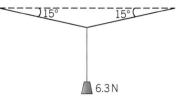

▲ **Figure 1**

2 An object of weight 6.3 N hangs on the end of a cord, which is attached to the midpoint of a wire stretched between two points on the same horizontal level, as shown in Figure 1. Each half of the wire is at 15° to the horizontal. Calculate the tension in each half of the wire. *(4 marks)*

3 A ship is towed at constant velocity by two tugboats which pull the ship with forces of 8.0 kN and 7.2 kN respectively. The angle between the tugboat cables is 45°, as shown in Figure 2.

   a Determine the angle between each cable and the direction of motion of the ship.

   (Hint: draw a parallelogram of forces and use it to establish the direction of motion on the diagram.) *(3 marks)*

   b Calculate:

     i the magnitude of the resultant of the two tugboat forces

     ii the drag force on the ship. *(3 marks)*

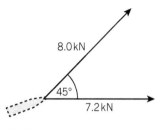

▲ **Figure 2**

4 A uniform metre rule of weight 1.20 N is pivoted on a metal rod at its 350 mm mark and is also supported in a horizontal position by a string inclined at 60° to the horizontal, as shown in Figure 3. One end of the string is tied to a stand and the other end is attached to the 950 mm mark on the ruler.

   a Sketch the arrangement (without showing the stand) and calculate the tension in the string. *(3 marks)*

   b Calculate the support force on the ruler from the pivot as in Figure 3. *(4 marks)*

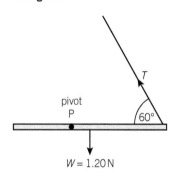

▲ **Figure 3**

5 A uniform plank of weight 350 N and of length 4.00 m is supported horizontally on two horizontal scaffolding tubes X and Y at 0.20 m from each end. A person P of weight 700 N stands on the plank at a distance of 1.50 m from end X.

   a Sketch a free body force diagram of the beam. *(1 mark)*

   b Calculate the support force on the beam from each scaffolding tube. *(4 marks)*

6 A uniform curtain pole of weight 22 N and of length 2.80 m is supported horizontally by two wall-mounted brackets A and B, which are 0.05 m from each end. The pole supports a pair of curtains of total weight 86 N.

   a Calculate the support force of each bracket on the curtain pole when the curtains are drawn along the full length of the pole between the brackets, *(3 marks)*

   b With the curtains initially as in **a**, the curtain nearest A is pulled back to A without moving the other curtain. Describe how the support force at A changes as the curtain nearest A is pulled back to A without moving the other curtain. *(2 marks)*

   c Calculate the support force on bracket A when the curtain nearest A has been pulled back to A without moving the other curtain. *(5 marks)*

   (Hint: Draw a free body diagram to show the forces on the curtain pole assuming the centre of mass of each curtain is at the centre of the curtain.)

7 A wire cable of length 6.5 m is fixed at its ends to two clamps X and Y which are about 6.3 m apart at the same height. The cable supports an object of weight 30 N at a point P as shown in Figure 4. At this position, section XP is inclined at an angle of 20° below XY and the other section YP is inclined an angle of 15° below XY.

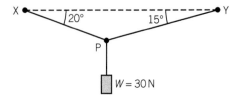

▲ Figure 4

a Show that the tension in section XP is 1.03 times the tension in section YP. (*2 marks*)

b Calculate the tension in each section of the cable. (*4 marks*)

8 A suitcase has a small pair of wheels at one of its lower corners and an extended handle on its top side. Figure 5 shows the suitcase when it is upright and stationary on a flat surface.

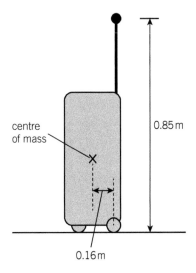

▲ Figure 5

The suitcase has a weight of 220 N and its centre of mass is a horizontal distance of 0.16 m from the axis of the wheels. The handle of the suitcase is 0.85 m directly above the axis of the wheels.

a Calculate the horizontal force that must be applied to the handle to raise one side of the suitcase off the ground. (*2 marks*)

b Explain why the position of the suitcase is unstable when it is tilted so its centre of mass is directly above the wheel axis. (*2 marks*)

c If the suitcase is tilted more than in part **b**, explain why an upward force on the handle is necessary to keep it stationary. (*2 marks*)

7.1 Speed and velocity
Specification reference: 3.4.1.3

Speed

Speed and distance are scalar quantities. Velocity and displacement are vector quantities.

The unit of speed and of velocity is the metre per second ($\mathrm{m\,s^{-1}}$).

Motion at constant speed

For an object which travels distance s in time t at constant speed:

- speed $v = \dfrac{s}{t}$
- a graph of distance against time is a straight line with a constant gradient which is equal to the speed of the object.

For an object moving round a circle of radius r at constant speed, its speed $v = \dfrac{2\pi r}{T}$ where T is the time to move round once and $2\pi r$ is the circumference of the circle.

Motion at changing speed

For an object moving at changing speed:

- over a distance s in time t, its average speed $= \dfrac{s}{t}$
- its speed v at any instant is given by $v = \dfrac{\Delta s}{\Delta t}$ where Δs is the distance it travels in a short time interval Δt
- a graph of distance against time has a gradient that changes with time. The gradient of the line at any point can be found by drawing a tangent to the line at that point and then measuring the gradient of the tangent. See Figure 1.

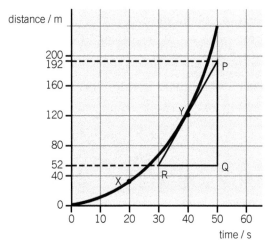

$$\text{speed at Y} = \frac{PQ}{QR} = \frac{192 - 52}{20}$$

$$= 7\,\mathrm{m\,s^{-1}}$$

▲ **Figure 1**

Velocity

An object moving at **constant velocity** moves at the same speed without changing its direction of motion. If an object changes its direction of motion or its speed or both, its velocity changes.

Figure 2 shows a displacement–time graph for an object thrown into the air. Upward directions are considered as positive and downward directions as negative. Initially, the object has zero displacement and a positive velocity.

During its ascent, its displacement and velocity (= gradient of the displacement–time graph) are both positive.

At maximum height, its displacement is at its maximum value and its velocity is zero.

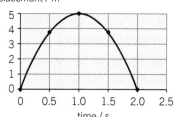

▲ **Figure 2** *A displacement–time graph*

During its descent, its velocity is negative and its displacement is positive until it returns to its initial position.

Displacement–time graphs are covered in more detail in Topic 7.5, Motion graphs.

Summary questions

1 A vehicle joins a motorway and travels a distance of 42 km in 25 minutes and a further distance of 20 km at a speed of 20 m s^{-1} before leaving the motorway.
   a Calculate the speed of the vehicle in the first part of the journey. (1 mark)
   b Calculate the average speed of the vehicle for the whole journey. (3 marks)

2 The Earth has an equatorial radius of 6 360 km and it rotates about its axis once every 24 hours.
   a Calculate the speed of an object on the Earth's surface at the equator. (2 marks)
   b A satellite orbits the Earth once every two hours in a circular orbit that passes over the Earth's poles. The satellite passes directly over a certain point P on the equator.
      i State the time taken to the next time the satellite passes directly over the equator in the same direction. (1 mark)

      ii Calculate the distance along the equator from P to the point Q on the equator directly under the satellite at the time calculated in i. (3 marks)

3 a Explain the difference between speed and velocity. (2 marks)
   b The table gives the distance moved from rest by an aircraft on a runway at different times during take-off.

Displacement / m	0	200	400	800	1200	1600	2000
Time / s	0	10	20	28	35	40	46

      i Use the data in the table to plot a displacement–time graph. (3 marks)

      ii Determine the speed of the aircraft at 30 s. (2 marks)

7.2 Acceleration

Specification reference: 3.4.1.3

Uniform acceleration

Uniform acceleration is where the velocity of an object moving along a straight line changes at a constant rate. In other words, the acceleration is constant. Consider an object that accelerates uniformly from velocity u to velocity v in time t along a straight line, as shown in Figure 1.

The acceleration, a, of the object $= \dfrac{\text{change of velocity}}{\text{time taken}} = \dfrac{v - u}{t}$

Rearranging this equation gives $v = u + at$

Non-uniform acceleration

Non-uniform acceleration is where the direction of motion of an object changes, or its speed changes, at a varying rate. Figure 2 shows how the velocity of an object increases for an object moving along a straight line with an increasing acceleration. This can be seen directly from the graph because the gradient increases with time (the graph becomes steeper and steeper) and the gradient represents the acceleration. The acceleration at any point is the gradient of the tangent to the curve at that point.

acceleration = gradient of the line on the velocity–time graph.

Summary questions

1 A car travelling at a velocity of 26 m s⁻¹ brakes sharply to a standstill in 6.7 s. Calculate its deceleration, assuming its velocity decreases uniformly. *(2 marks)*

2 An electron in a vacuum tube is accelerated in a straight line from rest to a velocity of 1.8×10^7 m s⁻¹ in a time of 51 ns. Calculate the acceleration of the electron. *(2 marks)*

3 Figure 3 shows the velocity of an object released at the surface of a liquid in a vertical tube.
   a Describe how: i the velocity of the object changed with time,
      ii the acceleration of the object changed with time. *(2 marks)*
   b Use the graph in Figure 3 to determine the acceleration of the object 0.50 s after it was released. *(2 marks)*

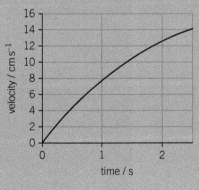

▲ Figure 3

Revision tip

- The unit of acceleration is the metre per second per second (m s⁻²).
- Acceleration is a vector quantity.
- For a moving object that does not change direction, its acceleration at any point can be worked out from its rate of change of velocity, because there is no change of direction.

Key term

Acceleration is defined as change of velocity per unit time.

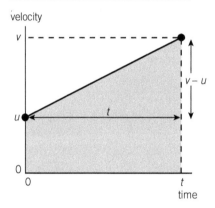

▲ Figure 1

Revision tip

Use of the equation $a = \dfrac{v - u}{t}$ for *non-uniform* acceleration gives an average value for the acceleration during time t.

Revision tip

Remember that acceleration is a vector quantity. It is dependent on changes in both speed and direction.

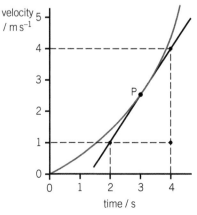

▲ Figure 2

7.3 Motion along a straight line at constant acceleration

Specification reference: 3.4.1.3

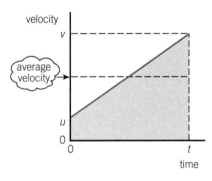

▲ **Figure 1**

Revision tip

Remember the acceleration must be constant for these equations to apply.

Worked example

A driver of a vehicle travelling at a speed of $31.0\,\mathrm{m\,s^{-1}}$ on a motorway brakes to a speed of $8.0\,\mathrm{m\,s^{-1}}$ in a distance of 105 m. Calculate the deceleration of the vehicle.

Solution

$u = 31.0\,\mathrm{m\,s^{-1}}$, $v = 8.0\,\mathrm{m\,s^{-1}}$,
$s = 105\,\mathrm{m}$, $a = ?$

To find a, use $v^2 = u^2 + 2as$

Rearranging this equation gives
$2as = v^2 - u^2$

Therefore
$a = \dfrac{v^2 - u^2}{2s} = \dfrac{8.0^2 - 31.0^2}{2 \times 105} = -4.3\,\mathrm{m\,s^{-2}}$
The deceleration is $4.3\,\mathrm{m\,s^{-2}}$.

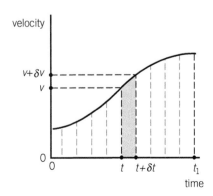

▲ **Figure 2**

The dynamics equations for constant acceleration

Consider an object that accelerates uniformly without change of direction from initial velocity u to final velocity v and displacement s in time t. Figure 1 shows how its velocity changes with time. The following equations may be used to solve problems involving uniform (i.e., constant) acceleration.

1   $v = u + at$

2   $s = \dfrac{(u + v)t}{2}$

3   $s = ut + \dfrac{1}{2}at^2$

4   $v^2 = u^2 + 2as$

These four equations are used in any situation where the acceleration is constant. Always begin a *suvat* calculation by identifying which three variables are known, then choose the equation which has these three variables in it together with the variable to be calculated.

Using a velocity–time graph to find the displacement

1   For an object moving at *constant* acceleration, a, from initial velocity u to velocity v at time t, as shown in Figure 1, its displacement, $s = \dfrac{(u + v)t}{2}$.
   This is represented on the graph by the area under the line between the start and time t. This is a trapezium which has an area (= average height × base) corresponding to $\dfrac{1}{2}(u + v)t$, which is equal to the displacement.

2   Consider an object moving with a *changing* acceleration, as shown in Figure 2. For a velocity change from v to $v + \delta v$ in a short time interval, its displacement $\delta s = v\,\delta t$. This is represented on the graph by the area of the shaded strip under the line. By considering the whole area under the line in strips of similar width, the total displacement from the start to time t is represented by the sum of the area of every strip, which is therefore the total area under the line.

Whatever the shape of the line of a velocity–time graph,

> **displacement = area under the line of a velocity–time graph.**

Summary questions

1   A vehicle accelerates uniformly from rest along a straight road and reaches a velocity of $28.0\,\mathrm{m\,s^{-1}}$ after 170 m. Calculate:
   - **a**   its acceleration   *(2 marks)*
   - **b**   the time taken.   *(2 marks)*

2   An aircraft lands on a runway at a velocity of $86\,\mathrm{m\,s^{-1}}$ and brakes to a halt 28 s later. Calculate:
   - **a**   the distance travelled in this time   *(2 marks)*
   - **b**   the deceleration of the aircraft.   *(2 marks)*

3   Figure 3 in Topic 7.2, Acceleration, shows how the velocity of an object falling in a liquid changes with time after being released at the liquid surface. Estimate the distance fallen by the object in 2.5 s.   *(2 marks)*

7.4 Free fall

Specification reference: 3.4.1.3

For a falling object on which no external forces act, apart from the force of gravity, its acceleration is constant and is referred to as the **acceleration of free fall**, g. Accurate measurements give a value of $9.81\,\mathrm{m\,s^{-2}}$ near the Earth's surface.

The 'suvat' equations from Topic 8.3, Terminal speed may be applied to any free fall situation where air resistance is negligible using the value of g as the acceleration. As a general rule, apply the direction code + for **upwards** and – for **downwards** when values are inserted into the *suvat* equations.

Straight line graphs

The value of g can be determined by measuring the distance which a suitable object falls through in different times. In time t, the distance fallen $s = ut + \frac{1}{2}at^2$, where u = the initial speed and a = acceleration.

If the object is released from rest and s is measured from the point of release, then $u = 0$ so $s = \frac{1}{2}at^2$. A graph of s against t^2 should give a straight line through the origin which has a gradient $\frac{1}{2}a$. Hence, $g = 2 \times$ the gradient of the graph.

If the distances are measured from a point below where the object is released, then $u \neq 0$ so a graph of $s = ut + \frac{1}{2}at^2$ would not be a straight line. Dividing the equation $s = ut + \frac{1}{2}at^2$ by t gives $\frac{s}{t} = u + \frac{1}{2}at$. Therefore, a graph of $\frac{s}{t}$ against t can be plotted to give a straight line with a y-intercept equal to u and a gradient $\frac{1}{2}a$.

Worked example

$g = 9.81\,\mathrm{m\,s^{-2}}$

A tennis ball thrown vertically upwards reaches a height of 14.0 m then descends to the ground below. Calculate: **a** the initial speed of the ball, **b** its displacement 3.00 s after it was thrown into the air.

Solution

a $u = ?$, $v = 0$ at maximum height, $s = 14.0\,\mathrm{m}$, $a = -9.81\,\mathrm{m\,s^{-2}}$, $t = ?$

To find u, use $v^2 = u^2 + 2as$. Note that a is negative because the direction of g is downwards.

Rearranging this equation gives $u^2 = v^2 - 2as = 0 - (2 \times -9.81 \times 14.0) = 275\,\mathrm{m^2\,s^{-2}}$

Hence, $u = \pm 16.6\,\mathrm{m\,s^{-1}}$ and since its initial velocity was upwards, $u = 16.6\,\mathrm{m\,s^{-1}}$

b To find s at 3.00 s, use $s = ut + \frac{1}{2}at^2 = (16.6 \times 3.00) + (0.5 \times -9.81 \times 3.00^2) = 5.66\,\mathrm{m}$

The position of the ball 3.00 s after being released was 5.66 m above the point at which it left the thrower's hand.

Revision tip

The general equation for a straight line graph is $y = mx + c$ where m is the gradient of the line and c is the y-intercept.

7.5 Motion graphs
7.6 More calculations on motion along a straight line

Specification reference: 3.4.1.3

Distance–time and displacement–time graphs

The gradient of the line:

- on a distance–time graph represents the speed of the object
- on a displacement–time graph represents the velocity of the object
  - a positive gradient means the velocity at that point is in the direction assigned as the positive direction
  - a negative gradient means the velocity is in the opposite direction.

The difference between a speed–time graph and a velocity–time graph

The gradient of the line:

- on a speed–time graph represents the magnitude of the acceleration
- on a velocity–time graph represents the magnitude and direction of the object's acceleration.

The area under the line:

- of a speed–time graph represents the distance travelled
- of a velocity–time graph represents the displacement of the object from its starting position. The area between the positive section of the line and the time axis represents a positive displacement. The area under the negative section of the line and the time axis represents a negative displacement.

Worked example

Figure 1 shows the velocity–time graph for a ball that is released from rest at a height h_1 above a concrete floor, hits the floor at time t_1, and rebounds to a maximum height h_2 at time t_2.

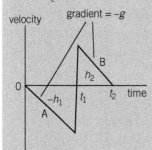

▲ **Figure 1**

a Explain why the graph line has two sections A and B with the same gradient.

b Explain why the line is almost vertical between A and B.

c Explain why the area between the time axis and each section is different.

Solution

a In both sections, the acceleration of the ball is due to gravity only and therefore the gradient of both sections is the same.

b The ball undergoes a large upward force in a very short time when it hits the floor at time t_1, causing a sudden reversal of its velocity.

c The displacement in each section = the area between the line and the time axis. The displacement is $-h_1$ for A and $+h_2$ for B. Since height h_2 is less than height h_1, the area between B and the time axis is less than the area between A and the time axis.

Using the equations for constant acceleration

The equations for motion at constant acceleration, a, are only valid when the acceleration of an object does not change.

Note:

- If a certain direction is assigned as the positive direction, then any calculated value that is negative is in the opposite direction.

- If the motion of an object is in stages where each stage has a different constant acceleration, the equations from Topic 7.3, Motion along a straight line at constant acceleration, may be used at each stage. The final velocity at the end of each stage is the initial velocity at the start of the next stage.

Synoptic link

Notice in Figure 1 that h_2 is less than h_1 because the ball loses some energy due to the impact when it rebounds. So its potential energy relative to the floor at maximum height after the rebound is less than its initial potential energy. See Topic 10.2, Kinetic energy and potential energy, for more about potential energy.

Summary questions

$g = 9.81\,\mathrm{m\,s^{-2}}$

1 A bungee jumper jumps from a platform at 0 and is in free fall until she reaches a point A when the rope attached to her starts to slow her descent and reduces her speed momentarily to zero at B. She then ascends and completes the jump.

    a Sketch a velocity–time graph for her descent from 0 to B. Label the points A and B on the graph. *(2 marks)*

    b Describe how her acceleration changed from 0 to B. *(3 marks)*

2 A small object released from rest above a level surface falls freely and hits the surface at a speed of $2.4\,\mathrm{m\,s^{-1}}$.

    a **i** Calculate its time of descent.

       ii Sketch a graph to show how its velocity changes with time during its descent. *(4 marks)*

    b Calculate the distance fallen by the object. *(2 marks)*

3 A rocket launched vertically from rest on the ground accelerated at a constant acceleration of $3.70\,\mathrm{m\,s^{-2}}$ for 30 seconds when the rocket engines were switched off. The rocket continued to ascend and then fell to the ground.

    a Calculate the velocity of the rocket and its displacement after 30 seconds. *(4 marks)*

    b **i** Calculate how long the rocket took to reach its maximum height after the rockets were switched off. *(2 marks)*

       ii Sketch a graph to show how the velocity of the rocket changed with time during its ascent to maximum height. *(3 marks)*

    c Calculate the velocity of the rocket just before it hit the ground. *(3 marks)*

7.7 Projectile motion 1
7.8 Projectile motion 2
Specification reference: 3.4.1.4

A **projectile** is acted upon only by the force of gravity. Three key principles apply to all projectiles:

- The acceleration of the object is always equal to g and is always downwards because the force of gravity acts downwards. The acceleration therefore only affects the vertical motion of the object.
- The horizontal velocity of the object is constant because the acceleration of the object does not have a horizontal component.
- The motions in the horizontal and vertical directions are independent of each other.

1 Vertical projection

An object that moves vertically has no horizontal motion. See Topic 7.4, Free fall, for notes on vertical projection.

2 Horizontal projection

An object projected horizontally falls in a parabolic arc towards the ground as shown in Figure 1. If its initial velocity is U, then at time t after projection:

- Its displacement has:
  - a *horizontal* component $x = Ut$ (because it moves horizontally at a constant speed)
  - a *vertical* component $y = \frac{1}{2}gt^2$ (because its acceleration g is vertically downwards).
- Its velocity has a horizontal component $v_x = U$, and a vertical component $v_y = -gt$.
- At time t, its speed $= (v_x^2 + v_y^2)^{\frac{1}{2}}$ and its velocity direction is at angle θ to the horizontal where $\tan\theta = \frac{v_y}{v_x}$.

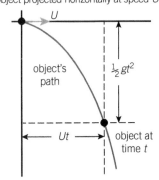

object projected horizontally at speed U

▲ **Figure 1** *Horizontal projection*

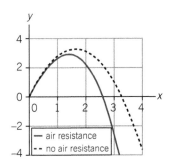

▲ **Figure 2** *Projectile paths*

3 Projection at angle θ above the horizontal

An object that is projected in a direction above the horizontal rises and falls in a parabolic arc. Because its acceleration, g, is vertically downwards, the horizontal component of its velocity is constant and the vertical component changes at a constant rate equal to $-g$. The dotted arc in Figure 2 shows the path of such a projectile when air resistance is negligible.

Worked example

$g = 9.81\,\text{m s}^{-2}$

An object is projected horizontally from the top of a tower of height 25.0 m and it hits the flat ground below 40.0 m away from the base of the tower. Calculate the speed of projection of the object.

Solution

Consider the vertical motion: $y = -25\,\text{m}, u_y = 0, a_y = -9.81\,\text{m s}^{-2}$ ($-$ for downwards)

Rearranging gives $y = \frac{1}{2}a_y t^2$ gives $t^2 = \frac{2y}{a_y} = \frac{2 \times -25.0}{-9.81} = 5.10\,\text{s}^2$ so $t = 2.26\,\text{s}$

Consider the horizontal motion: $x = 40.0\,\text{m}, a_x = 0, t = 2.26\,\text{s}$

Rearranging $x = Ut$ gives $U = \frac{x}{t} = \frac{40.0}{2.26} = 17.7\,\text{m s}^{-1}$

The effects of air resistance

A projectile moving through air experiences a **drag force** that opposes the motion of the projectile and which increases as the projectile's speed increases. The lower arc in Figure 2 shows that drag on a projectile:

- reduces the horizontal speed of the projectile and its range
- makes its descent steeper than its ascent
- reduces the maximum height if its initial direction is above the horizontal.

The shape of the projectile is important because it affects the drag force and may also cause a **lift force** in the same way as the cross-sectional shape of an aircraft wing creates a lift force. A spinning ball moving through the air also experiences a force due to the same effect. However, this force can be downwards, upwards, or sideways, depending on how the ball is made to spin.

Projectile-like motion

Any form of motion where an object experiences a constant acceleration in a different direction to its velocity will be like projectile motion. For example:

- The path of a ball rolling across an inclined board will be a projectile path. Its path curves down the board because the object is subjected to constant acceleration acting down the board, and its initial velocity is across the board.
- The path of a beam of electrons directed between two oppositely charged parallel plates is similar in shape. This is because each electron in the beam experiences a constant acceleration towards the positive plate and its initial velocity is parallel to the plates.

Summary questions

$g = 9.81 \, \text{m s}^{-2}$

1 An object is projected horizontally at a speed of $16 \, \text{m s}^{-1}$ into the sea from a cliff top and falls into the sea 3.1 s later. Calculate:
  a the height of the cliff top above the sea   (*2 marks*)
  b how far it travels horizontally   (*1 mark*)
  c its impact speed.   (*3 marks*)

2 A parcel was released from an aircraft travelling horizontally at a constant velocity of $110 \, \text{m s}^{-1}$ above level ground. The parcel hit the ground 7.5 s later.
  a Calculate the height of the aircraft above the ground. Assume the drag force on the parcel is negligible.   (*2 marks*)
  b Discuss whether or not the aircraft was directly above the parcel when the parcel hit the ground.   (*3 marks*)

3 An archer on a flat field fired an arrow at a speed of $29.0 \, \text{m s}^{-1}$ at an angle of $15.0°$ above the horizontal.
  a Calculate the horizontal and vertical components of velocity of the arrow at the instant it was released.   (*1 mark*)
  b The arrow was 2.60 m above the ground when it was fired. By considering its vertical motion, calculate:
    i the vertical component of the arrow's velocity when it hit the ground   (*2 marks*)
    ii how long the arrow took to fall to the ground.   (*2 marks*)
  c Calculate the horizontal displacement from the archer to the point where the arrow hit the ground.   (*1 mark*)

$g = 9.81\,\mathrm{m\,s^{-2}}$

1 A car decelerated uniformly to rest from a speed of 98 km h⁻¹ in a distance of 73 m.

    a Calculate the deceleration of the car in m s⁻². *(2 marks)*

    b Calculate the time taken by the car whilst decelerating. *(2 marks)*

2 Figure 1 shows how the velocity of an aircraft changes when it lands on a runway.

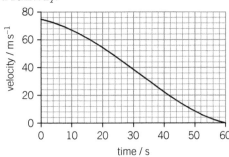

▲ **Figure 1**

    a Determine the maximum deceleration of the aircraft. *(3 marks)*

    b Estimate the distance it travelled along the runway until it stopped. *(2 marks)*

3 In a 100 m race, athlete X accelerated uniformly for 1.20 s from rest to a speed of 9.60 m s⁻¹ and completed the race at that speed.

    a i Calculate the distance travelled by the athlete in the first 1.20 s.

      ii Calculate the total time taken by the athlete to run the race. *(4 marks)*

    b A second athlete Y starting at the same time, accelerated uniformly for 1.30 s to a certain speed and maintained that speed to finish the race at the same time as X. Determine the maximum speed of Y. *(3 marks)*

4 A rail wagon moving at a speed of 1.80 m s⁻¹ on a level track reached a steady incline which slowed it down to rest in 12.0 s and caused it to reverse.

    a Calculate its rate of change of velocity on the incline. *(2 marks)*

    b Calculate:

      i its velocity after 16.0s

      ii its position on the incline 16.0s after it reached the incline. *(4 marks)*

5 A cyclist accelerated uniformly from rest at P to a speed of 8.0 m s⁻¹ in 20.0 s at Q then braked at uniform deceleration to a halt in a distance of 20.0 m at R. She then turned her cycle round in 10.0 seconds and then cycled at a speed of 4.0 m s⁻¹ for 5.0 s in the opposite direction to the original direction before she stopped at S.

    Figure 2 shows the route which was on level ground

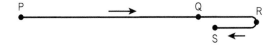

▲ **Figure 2**

    a i Calculate the distance she moved from P to Q.

      ii Calculate how long she took to cycle from Q to R. *(2 marks)*

    b Plot a velocity–time graph of her journey from P to S. Label the points on your graph corresponding to P, Q, R, and S. *(2 marks)*

c i Calculate the acceleration of the cyclist when she travelled from P to Q.

   ii Calculate her displacement from P to S. *(3 marks)*

6 A stone hit the water in a river 2.0 s after being thrown horizontally at a speed of 14 m s^{-1} from a bridge above the river.

  a Calculate the distance from the bridge to the point where the stone hit the water. *(4 marks)*

  b State and explain how the horizontal displacement would have differed if the stone had been thrown horizontally with a lesser speed. *(5 marks)*

7 An object was released from a hot air balloon when it was travelling at a speed of 2.60 m s^{-1} in a direction 30.0° above the horizontal as shown in Figure 3.

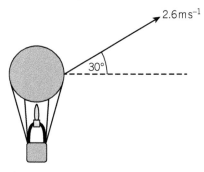

   2.6 m s^{-1}

   30°

▲ **Figure 3**

  a Calculate the horizontal and vertical components of velocity of the object at the instant it was released. *(2 marks)*

  b The object took 3.90 s to fall to the ground. Calculate:

   i the vertical distance fallen by the object from the point of release

   ii the horizontal distance travelled by the object from the point of release to where it hit the ground. *(3 marks)*

8 A javelin thrown on a level field by an athlete from a height of 1.80 m above the ground hit the ground 85.5 m away 4.24 s after it was released.

  a i Calculate the horizontal component of the javelin's velocity.

   ii Calculate the vertical component of the javelin's initial velocity. *(4 marks)*

  b i Show that the speed at which the javelin was thrown was about 29 m s^{-1}.

   ii Calculate the angle θ above the horizontal at which the javelin was thrown. *(2 marks)*

8.1 Force and acceleration

Specification reference: 3.4.1.5

Maths link

Why does a heavy object fall at the same rate as a lighter object?

All objects fall at the same rate, regardless of their weight. Newton explained their identical falling motion because, for an object of mass m in free fall,

$$\text{acceleration} = \frac{\text{force of gravity } mg}{\text{mass } m}$$

$= g$ which is independent of m.

Worked example

A vehicle of mass 800 kg accelerates uniformly from rest to a velocity of 15.0 m s^{-1} in 20 s. Calculate the force needed to produce this acceleration.

Solution

$$\text{acceleration } a = \frac{v - u}{t} = \frac{15 - 0}{20}$$

$$= 0.75 \, \text{m s}^{-2}$$

Force $F = ma = 800 \times 0.75 = 600 \, \text{N}$

Newton's first law of motion

Objects either stay at rest or move with constant velocity unless acted on by a force.

An object moving at constant velocity is either acted on by no forces, or the forces acting on it are balanced (so the resultant force is zero).

Newton's second law of motion for constant mass

When a resultant force F acts on an object of mass m, the object undergoes acceleration a such that

$$F = ma$$

Weight

The acceleration of a falling object acted on by gravity alone is equal to g. Because the force of gravity on the object is the only force acting on it, its **weight** (in newtons) $W = mg$, where m = the mass of the object (in kg).

- When an object is in equilbrium, the support force on it is equal and opposite to its weight.
- g is also referred to as the **gravitational field strength** at a given position. The gravitational field strength at the Earth's surface is 9.81 N kg^{-1}.

Inertia

The mass of an object is a measure of its **inertia**, which is its resistance to change of motion. More force is needed to give an object a certain acceleration than to give an object with less mass the same acceleration.

Summary questions

$g = 9.81 \, \text{N kg}^{-1}$

1 A heavy goods vehicle of mass 26 000 kg accelerates uniformly along a straight line from rest to a speed of 12 m s^{-1} in 50 s. Calculate:
   a the acceleration of the vehicle   (1 mark)
   b the resultant force on the vehicle   (1 mark)
   c the ratio of the accelerating force to the weight of the car.   (1 mark)

2 An aeroplane of mass 4000 kg lands on a runway at a speed of 62 m s^{-1} and stops in a distance of 1600 m. Calculate:
   a the deceleration of the aeroplane   (2 marks)
   b the braking force on the aeroplane.   (1 mark)

3 A golf ball of mass 0.027 kg was struck by a golf club with a force of 6.0 kN on level ground. The ball hit the ground 4.2 seconds later at a point 310 m away from the golfer.
   a Calculate the acceleration of the ball when it was struck.   (1 mark)
   b Estimate:
      i the average speed of the golf ball   (1 mark)
      ii the time of contact between the ball and the club.   (2 marks)
   c Discuss two assumptions made in your estimate in b ii.   (4 marks)

8.2 Using $F = ma$

Specification reference: 3.4.1.5

When an object of mass m is acted on by two unequal forces F_1 and F_2 acting in opposite directions and F_1 is greater than F_2,

resultant force $F_1 - F_2 = ma$

where a is its acceleration, which is in the same direction as F_1.

1 A car pulling a trailer

Consider the example of a car of mass M fitted with a trailer of mass m on a level road. When the car and the trailer accelerate, the car pulls the trailer forward and the trailer holds the car back. Assume air resistance is negligible.

If F is the driving force on the car pushing it forwards (from its engine thrust) and T is the tension in the tow bar, the resultant force:

- on the car $= F - T = Ma$

- on the trailer $= T = ma$ (because the tension T in the tow bar is the force pulling the trailer forwards).

Combining the two equations gives the driving force $F = Ma + ma = (M + m)a$

2 Lift forces

The resultant force on the lift $= T - mg$, where T is the tension in the lift cable and m is the total mass of the lift and occupants (Figure 2). Therefore $T - mg = ma$, where $a =$ acceleration.

Table 1 shows how the tension in the cable depends on the motion of the lift.

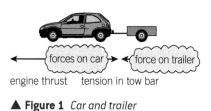

▲ **Figure 1** *Car and trailer*

▲ **Figure 2** *In a lift*

> ### Worked example
>
> $g = 9.81\,\text{m s}^{-2}$
>
> A lift of total mass 650 kg starts moving downwards with an acceleration of $1.2\,\text{m s}^{-2}$ for a brief time. Calculate the tension in the lift cable during this time.
>
> #### Solution
>
> The lift is moving down so its velocity $v < 0$. Since it accelerates, its acceleration a is in the same direction as its velocity, so $a < 0$.
>
> Therefore, inserting $a = -1.2\,\text{m s}^{-2}$ in the equation $T - mg = ma$ gives
>
> $T = mg + ma = (650 \times 9.81) + (650 \times -1.2) = 5600\,\text{N}$

▼ **Table 1**

Motion	Tension
constant velocity	$= mg$
upwards and accelerating	$> mg$
upwards and decelerating	$< mg$
downwards and decelerating	$> mg$
downwards and accelerating	$< mg$

Summary questions

$g = 9.81\,\text{m s}^{-2}$

1 A rocket of mass 1050 kg blasts vertically from the launch pad at an acceleration of $4.80\,\text{m s}^{-2}$. Calculate:

  a the weight of the rocket *(1 mark)*

  b the thrust of the rocket engines. *(2 marks)*

2 A lift and its occupants have a total mass of 1200 kg. Calculate the tension in the lift cable when the lift is:

  a descending at constant velocity *(1 mark)*

  b descending at a constant acceleration of $0.50\,\text{m s}^{-2}$. *(3 marks)*

3 A skateboarder of mass 36 kg on a slope at 20° to the horizontal accelerates 5.0 m down the slope from rest in 3.0 s. Calculate:

  a the acceleration of the skateboarder *(2 marks)*

  b **i** the component of the skateboarder's weight acting down the slope

    ii the frictional force on the skateboarder. *(3 marks)*

> **Revision tip**
>
> When you use the equations in the examples opposite, make sure the two forces are the ones that act on the object under consideration not the forces that act on other objects. It's easy to get the forces mixed up when two or more objects interact.

8.3 Terminal speed

Specification reference: 3.4.1.5

Motion of an object falling in a fluid

Any object moving through a fluid experiences a force that drags on it due to the fluid. The **drag force** depends on:

- the shape of the object
- its speed and increases with increase of speed
- the viscosity of the fluid, which is a measure of how easily the fluid flows past a surface.

For an object released from rest in a fluid, its speed increases as it falls, so the drag force on it increases. The resultant force on the object is the difference between the force of gravity on it (its weight) and the drag force. As the drag force increases, the resultant force decreases, so the acceleration becomes less. If it continues falling, it attains **terminal speed**. (Figure 1).

At any instant, the resultant force $F = mg - D$, where m is the mass of the object and D is the drag force. Therefore, the acceleration of the object = $\frac{mg - D}{m} = g - \frac{D}{m}$.

The initial acceleration = g because the speed is zero, and therefore the drag force is zero, at the instant the object is released.

Motion of a powered vehicle

For a powered vehicle of mass m moving on a level surface, if F_E represents the **motive force** (driving force) provided by the engine, the resultant force on it = $F_E - F_R$, where F_R is the resistive force opposing the motion of the vehicle. (F_R = the sum of the drag forces acting on the vehicle.)

Therefore, its acceleration $a = \dfrac{F_E - F_R}{m}$

The maximum speed (the terminal speed) of the vehicle v_{max} is reached when the resistive force becomes equal and opposite to the engine force, and $a = 0$.

The top speed of a road vehicle or an aircraft depends on its engine power and its shape. A vehicle with a streamlined shape can reach a higher top speed than a vehicle with the same engine power that is not streamlined.

Key term

Terminal speed occurs when the drag force on an object is equal and opposite to its weight – its acceleration is then zero and its speed remains constant as it falls.

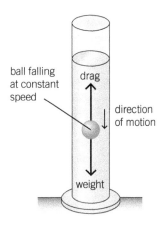

▲ **Figure 1** Terminal speed

Revision tip

The magnitude of the acceleration at any instant is the gradient of the speed–time curve.

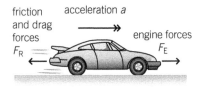

▲ **Figure 2** Vehicle power

Summary questions

$g = 9.81\,\mathrm{m\,s^{-2}}$

1 A metal ball of mass 0.045 kg, released from rest in a liquid, falls a distance of 0.32 m in 7.1 s. Assuming the ball reaches terminal speed within a fraction of a second, calculate:
  a its terminal speed   (1 mark)
  b the drag force on it when it falls at terminal speed.   (2 marks)

2 A steel ball fell vertically after it was released from rest above the surface of a liquid. The ball decelerated when it entered the liquid until its speed became constant.
  a Sketch a graph to show how the speed of the ball changed with time after it was released until it reached terminal speed. On your graph, label the point where the ball entered the water with the letter W. (2 marks)
  b Explain why the ball decelerated after it entered the water. (5 marks)

3 A vehicle of mass 22 000 kg has an engine which has a maximum engine force of 3800 N and a top speed of 36 m s⁻¹ on a level road. Calculate:
  a its maximum acceleration from rest   (1 mark)
  b the total resistive force acting on it at top speed.   (1 mark)

8.4 On the road

Stopping distances

Thinking distance is the distance travelled by a vehicle in the time it takes the driver to react. For a vehicle moving at constant speed v, the thinking distance s_1 = speed × reaction time = vt_0, where t_0 is the reaction time of the driver.

The reaction time of a driver is affected by distractions, drugs, and alcohol.

Braking distance is the distance travelled by a car in the time it takes to stop safely, from when the brakes are first applied. Assuming constant deceleration, a, to zero speed from speed u, the braking distance $s_2 = \frac{u^2}{2a}$ since $u^2 = 2as_2$.

The braking distance for a vehicle depends on the vehicle speed before the brakes are applied, on the road conditions, and on the condition of the tyres.

Stopping distance = thinking distance + braking distance = $s_1 + s_2$

> **Revision tip**
>
> Remember the values for decelerations are negative as a deceleration is in the opposite direction to the velocity.

Summary questions

$g = 9.81\,\mathrm{m\,s^{-2}}$

1 A vehicle is travelling at a speed of $20\,\mathrm{m\,s^{-1}}$ on a level road, when the driver sees a pedestrian stepping off the pavement into the road $58.0\,\mathrm{m}$ ahead. The driver reacts within $0.70\,\mathrm{s}$ and applies the brakes, causing the car to decelerate at $5.1\,\mathrm{m\,s^{-2}}$ and stop safely a short distance from the pedestrian.
   a Calculate: **i** the thinking distance **ii** the braking distance. *(3 marks)*
   b How far does the driver stop from where the pedestrian stepped into the road? *(1 mark)*

2 A vehicle travelling at a speed of $28\,\mathrm{m\,s^{-1}}$ on a dry level road brakes to a standstill in a distance of $71\,\mathrm{m}$. Calculate:
   a the deceleration of the vehicle from this speed to a standstill over this distance *(2 marks)*
   b the braking force on a vehicle of mass $15000\,\mathrm{kg}$ on this road as it stops. *(1 mark)*

3 a Explain why the stopping distance on a wet road is longer than the stopping distance at the same speed on the same road when it is dry. *(2 marks)*
   b The braking force on a vehicle travelling on a certain type of level road surface is $0.57 \times$ the vehicle's weight. For a vehicle of mass $2400\,\mathrm{kg}$, calculate the braking distance on this road for a speed of $31\,\mathrm{m\,s^{-1}}$. *(4 marks)*

Worked example

$g = 9.81\,\mathrm{m\,s^{-2}}$

A vehicle of mass $1200\,\mathrm{kg}$, travelling on a level road at a speed of $14\,\mathrm{m\,s^{-1}}$, is brought to a standstill without skidding in a distance of $25.0\,\mathrm{m}$. Calculate:

a the deceleration of the vehicle

b the braking force.

Solution

a $u = 14\,\mathrm{m\,s^{-1}}, v = 0, s = 25.0\,\mathrm{m}$

To calculate a, rearrange $v^2 = u^2 + 2as$ to give $a = \frac{v^2 - u^2}{2s} = \frac{0 - 14^2}{2 \times 25} = -3.92\,\mathrm{m\,s^{-2}}$

Therefore, the deceleration is $3.92\,\mathrm{m\,s^{-2}}$

b Braking force = mass × acceleration = $1200 \times 3.92 = 4700\,\mathrm{N}$

8.5 Vehicle safety

Specification reference: 3.4.1.5

Impact force

For a collision in which the velocity of an object of mass m changes from initial velocity u to final velocity v in a distance s, and

- the **impact time** $t = \dfrac{2s}{u + v}$ the acceleration $a = \dfrac{v - u}{t}$

- the **impact force** $F = ma$

Contact time and impact time

When objects collide and bounce off each other, they are in contact with each other for a certain time which is the same for both objects. The impact time is equal to the contact time in this case. When two vehicles collide and do *not* separate from each other after the collision, the impact time is not the same as the contact time.

Car safety features

The following vehicle safety features are designed to increase the impact time and so reduce the impact force.

- *Vehicle bumpers* give way a little in a low-speed impact and so increase the impact time.

- *Crumple zones* are designed to crumple in a front-end impact and so increase the impact time for the engine compartments.

- *Seat belts* restrain their wearer from crashing into the vehicle frame after the vehicle suddenly stops. The seat belt brings the wearer to a stop more gradually than without it.

- *Collapsible steering wheels* are designed to collapse in a front-end impact, if the driver makes contact with the steering wheel, thus increasing the impact time.

- *Airbags* reduce the force on a person, because they act as a cushion and increase the impact time on the person. Also, an airbag spreads the impact force over a wider area of the body so it has less effect on the body.

Synoptic link

The impact force can also be worked out using the equation

$$F = \frac{\text{change of kinetic energy}}{\text{Impact distance}}$$

See Topic 10.2, Kinetic energy and potential energy.

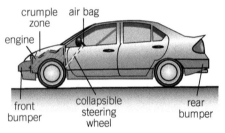

▲ **Figure 1** *Vehicle safety features*

Summary questions

$g = 9.81\,\text{m s}^{-2}$

1 A car of mass 1100 kg travelling at a speed of 12 m s^{-1} is struck from behind by another vehicle, causing its speed to increase to 18 m s^{-1} in a distance of 3.0 m. Calculate the impact force on the car. (*3 marks*)

2 The front bumper of a car of mass 900 kg is capable of withstanding an impact with a stationary object, provided the car is not moving faster than 3.0 m s^{-1}, when the impact occurs. The impact time at this speed is 0.40 s. Calculate the impact force on the car. (*2 marks*)

3 In a crash, a vehicle travelling at a speed of 23 m s^{-1} stops after 4.8 m. A passenger of mass 62 kg is wearing a seat belt, which restrains her forward movement relative to the car to a distance of 0.5 m.
   a Calculate the resultant force on the passenger. (*3 marks*)
   b Explain why the force on the passenger would have been much larger if she had not been wearing a seat belt. (*3 marks*)

1 A vehicle of mass 1500 kg towing a trailer of mass 300 kg on a level road accelerates from rest to a velocity of 8.8 m s^{-1} in 40 s, without change of direction.

    a i Calculate the force that accelerated the vehicle and the trailer.

      ii Calculate the tension in the tow bar during this time. Assume the tow bar is horizontal   (*3 marks*)

    b Calculate how long the vehicle without the trailer would take to accelerate with the same force from rest to a velocity of 8.8 m s^{-1}.   (*2 marks*)

2 A cyclist on a straight level section of a road accelerated from rest at a constant acceleration and reached a speed of 8.3 m s^{-1} in a distance of 74 m. The total mass of the cyclist and the cycle was 61 kg.

    a Calculate the resultant force on the cyclist and the cycle.   (*3 marks*)

    b The cyclist then used the same force to travel on a downhill section which was inclined at an angle of 5.0° to the horizontal. Calculate the acceleration on this section.   (*4 marks*)

3 Figure 1 shows how the velocity of a motor car increased with time as it accelerated from rest along a straight horizontal road.

    a Use the graph to determine the maximum acceleration of the car.   (*2 marks*)

    b Throughout the motion shown in Figure 1 the driving force acting on the car was reduced gradually.

      i The mass of the car and its contents was 1200 kg. Calculate the initial driving force.

      ii When the car was travelling at constant velocity, the driving force was 650 N. What was the magnitude of the resistive force acting on the car at this velocity? Give a reason for your answer.   (*7 marks*)

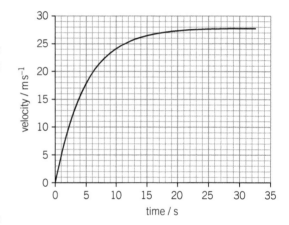

▲ **Figure 1**

4 A fairground ride ends with a car moving up a ramp of length 28 m at a slope of 32° to the horizontal then onto a flat section where it brakes and stops.

▲ **Figure 2**

    a The car enters at the bottom of the ramp at a speed of 18 m s^{-1} and leaves it at 0.5 m s^{-1}.

      i Calculate the deceleration of the car on the ramp.

      ii The mass of the car and its passengers is 750 kg. Calculate the resultant force on the car.   (*3 marks*)

    b i Calculate the component of the weight of the car and the passengers parallel to the ramp when the car is on the ramp.

      ii Calculate the average resistive force on the car as it travels up the ramp.   (*4 marks*)

5 A skydiver in free fall opens a parachute. Figure 3 shows how the speed of the skydiver changes after the parachute is opened.

a

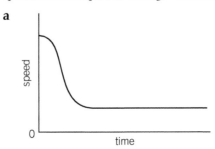

▲ **Figure 3**

Describe and explain how the acceleration changes with time. (*4 marks*)

b i Sketch a graph on a copy of Figure 3 to show how the speed would have changed if the weight of the skydiver had been less. Assume the parachute is opened at the same speed.

ii Explain why your graph differs from the one shown in Figure 3.

(*4 marks*)

6 A load of weight 250 N is lifted vertically using an overhead pulley and a cable attached to a counterweight of weight 290 N as shown in Figure 4.

a Assuming friction and air resistance are negligible:

i calculate the acceleration of the load

ii show that the tension in the cable is about 270 N. (*6 marks*)

b The total weight of the pulley and the cable is 55 N. Calculate the support force on the pulley. (*2 marks*)

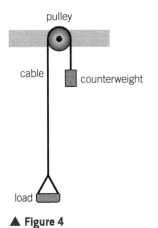

▲ **Figure 4**

7 a In a car safety test, the front end of a certain type of car of mass 1800 kg travelling at a speed of 30 m s^{-1} crumpled in a distance of 0.80 m when the car hit a wall. Calculate:

i the impact time

ii the impact force. (*3 marks*)

b A test dummy in the front passenger seat was fitted with a seat belt which allowed the dummy to move a distance of 0.40 m before it stopped the dummy. Calculate the force exerted by the seat belt during the crash on the dummy. (*3 marks*)

8 A student on a skateboard accelerated from rest down a slope through a distance of 15 m in 5.6 s to the bottom of the slope then travelled onto flat level ground and gradually stopped.

a i Calculate the acceleration of the student down the slope.

ii The mass of the student was 38 kg. Calculate the resultant force on the student over this distance. (*3 marks*)

b The slope was at an angle of 6.1° to the horizontal.

i Calculate the component of the student's weight down the slope.

ii Explain why the resultant force calculated in **a ii** was less than the component of the student's weight down the slope. (*4 marks*)

▲ **Figure 5**

9.1 Momentum and impulse
9.2 Impact forces

Specification reference: 3.4.1.6

Momentum and Newton's laws of motion

For an object of mass m moving at velocity v, its momentum $p = mv$.

Newton's first law of motion: An object remains at rest or in uniform motion unless acted on by a force.

Newton's second law of motion: The rate of change of momentum of an object is proportional to the resultant force on it.

For a change of momentum $\Delta(mv)$ in time Δt, Newton's second law can be written as

$$F = \frac{\Delta(mv)}{\Delta t}$$

where the unit of force (the **newton**) is defined as the amount of force that gives an object of mass $1\,\text{kg}$ an acceleration of $1\,\text{m s}^{-2}$.

1   If mass is constant, then $\Delta(mv) = m\Delta v$, where Δv is the change of velocity of the object.

$\therefore F = \dfrac{m\Delta v}{\Delta t} = ma$ where acceleration $a = \dfrac{\Delta v}{\Delta t} \left(= \dfrac{v-u}{t}$ for constant acceleration$\right)$

2   If mass is lost or gained at constant velocity v, then $\Delta(mv) = v\Delta m$, where Δm is the change of mass of the object.

$\therefore F = v\dfrac{\Delta m}{\Delta t}$ where $\dfrac{\Delta m}{\Delta t}$ = change of mass per second

This form of Newton's second law is used in any situation where an object gains or loses mass continuously (e.g., water from a hose pipe or hot gas from a jet engine).

Worked example

A pump forced water out of a hose pipe at a rate of 6.6 kilograms per minute with a velocity of $19\,\text{m s}^{-1}$. Calculate the force on the hose pipe due to the water.

Solution

Mass of water lost per second $= \dfrac{6.6\,\text{kg}}{60\,\text{s}} = 0.11\,\text{kg s}^{-1}$

Force $= v\dfrac{\Delta m}{\Delta t} = 19 \times 0.11 = 2.1\,\text{N}$

$$\textbf{impulse} = F\Delta t = \Delta(mv)$$

Hence, the impulse of a force acting on an object is equal to the change of momentum of the object.

Force–time graphs

A force–time graph can be used to find the change of momentum of an object. This is because the area under the line represents the impulse of the resultant force on the object and therefore gives the change of momentum. See Figure 1.

The area under the line of a force–time graph represents the change of momentum or the impulse of the force.

> **Revision tip**
> Momentum is a vector quantity. Its direction is the same as the direction of the object's velocity.

> **Revision tip**
> The unit of momentum may be either kg m s^{-1} or (more neatly) N s.

> **Key term**
> The **impulse** of a force is defined as the force × the time for which the force acts.

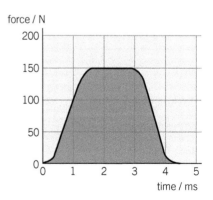

The area under the line is approximately 9 blocks. Each block represents an impulse of $0.050\,\text{N s}$ ($= 50\,\text{N} \times 1\,\text{ms}$). So the total impulse and therefore the change of momentum is approximately $0.45\,\text{N s}$

▲ **Figure 1** *Force against time for an impact*

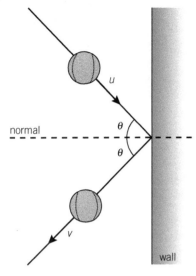

▲ **Figure 2** *An oblique impact*

Rebound impacts

Consider a ball that hits a wall at speed u in an oblique impact and rebounds at the same speed and the same angle to the wall.

The normal component of its momentum is $+mu \cos \theta$ before the impact and $-mu \cos \theta$ after the impact.

Therefore, its change of momentum $= (-mu \cos \theta) - (mu \cos \theta) = -2mu \cos \theta$

Note: If $\theta = 0$, the initial velocity is perpendicular (i.e., normal) to the wall so the change of momentum $= -2mu$.

Summary questions

1 An aircraft of total mass 25 000 kg accelerated on a runway from rest to a velocity of 160 m s⁻¹ after 58 s when it took off. Calculate the engine force during this time. *(2 marks)*

2 The velocity of a vehicle of mass 1300 kg was reduced from 20 m s⁻¹ by a constant force of 1800 N which acted for 4.0 s, then by a force that decreased uniformly to zero in a further 6.0 s.
   a Sketch the force–time graph for this situation. *(1 mark)*
   b Use the force–time graph to determine the total change of momentum and hence show that the final velocity of the vehicle was about 10 m s⁻¹. *(3 marks)*

3 A van of mass 1500 kg travelling at a speed of 2.0 m s⁻¹ was struck from behind by another vehicle. The impact lasts for 0.55 s and causes the speed of the van to increase to 6.5 m s⁻¹. Calculate the impact force. *(2 marks)*

4 A molecule of mass 5.0×10^{-26} kg moving at a speed of 420 m s⁻¹ hits a surface at 60° to the normal and rebounds without loss of speed at 60° to the normal in an impact lasting 0.22 ns. Calculate the force on the molecule. *(3 marks)*

9.3 Conservation of momentum
9.4 Elastic and inelastic collisions
9.5 Explosions

Specification reference: 3.4.1.6

Newton's third law of motion

When two objects interact, they exert equal and opposite forces on each other.

In other words, if object A exerts a force on object B, there must be an equal and opposite force acting on object A due to object B.

The principle of conservation of momentum

For any system of interacting objects, their total momentum remains constant, provided no external resultant force acts on the system.

Interactions between objects can transfer momentum between them. But their total momentum does not change. Figure 1 shows two snooker balls A and B of masses m_A and m_B that move along a straight line, before and after they collide.

The total initial momentum = $m_A u_A + m_B u_B$

The total final momentum = $m_A v_A + m_B v_B$

Because the total momentum is conserved,

$$m_A v_A + m_B v_B = m_A u_A + m_B u_B$$

If the colliding objects stick together as a result of the collision, they have the same final velocity. The above equation with V as the final velocity may therefore be written

$$(m_A + m_B)V = m_A u_A + m_B u_B$$

Elastic and inelastic collisions

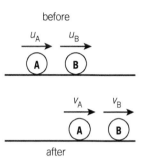
▲ **Figure 1**

Revision tip

The vector nature of momentum needs to be taken into account in straight-line collisions by defining one direction as + and the opposite direction as –.

Key terms

An **elastic collision** is one where there is no loss of kinetic energy.

An **inelastic collision** occurs where the colliding objects have less kinetic energy after the collision than before the collision.

Synoptic link

The kinetic energy of an object can be worked out using the kinetic energy equation $E_K = \frac{1}{2}mv^2$, where m is the mass of the object and v is its speed (see Topic 10.2, Kinetic energy and potential energy).

Worked example

A railway wagon X of mass 8000 kg moving at $3.0\,\mathrm{m\,s^{-1}}$ collides with wagon Y of mass 5000 kg moving towards X at a speed of $0.5\,\mathrm{m\,s^{-1}}$. After the collision, the two wagons separate and X moves at a speed of $1.0\,\mathrm{m\,s^{-1}}$ without change of direction. Calculate:

a   the speed and direction of Y after the collision

b   the loss of kinetic energy due to the collision.

Solution

a   Let the initial direction of X be the + direction so the velocity of Y before the collision = $-0.5\,\mathrm{m\,s^{-1}}$.

The total initial momentum = $(8000 \times 3.0) + (5000 \times -0.5) = 21\,500\,\mathrm{kg\,m\,s^{-1}}$.

The total final momentum = $(8000 \times 1.0) + 5000V$, where V is the speed of the 5000 kg wagon after the collision.

Using the principle of conservation of momentum

$8000 + 5000V = 21\,500$

$5000V = 21\,500 - 8000 = 13\,500$

$V = \dfrac{13\,500}{5000} = 2.7\,\mathrm{m\,s^{-1}}$

Synoptic link

Excitation by collision in a gas occurs when the gas molecules undergo collisions and are excited to higher energy states. See Topic 3.3, Collisions of electrons with atoms.

b   Kinetic energy of X E_{Kx} before the collision $= \dfrac{1}{2} \times 8000 \times 3.0^2 = 36\,000$ J

E_{Ky} before the collision $= \dfrac{1}{2} \times 5000 \times 0.5^2 = 625$ J

E_{Kx} after the collision $= \dfrac{1}{2} \times 8000 \times 1.0^2 = 4000$ J

E_{Ky} after the collision $= \dfrac{1}{2} \times 5000 \times 2.7^2 = 18\,225$ J

$\therefore$ loss of kinetic energy due to the collision $= (36\,000 + 625) - (4000 + 18\,225)$
$= 14\,400$ J

Explosions

When two objects A and B of masses m_A and m_B fly apart after being initially at rest, they recoil from each other with equal and opposite amounts of momentum.

Using the principle of the conservation of momentum, their total momentum immediately after the explosion = 0 as the total initial momentum is zero.

Therefore, $m_A v_A + m_B v_B = 0$ where v_A and v_B are their velocities when they move apart.

So, $m_B v_B = -m_A v_A$, where the minus sign means that the two masses move apart in opposite directions.

Summary questions

1   In a laboratory experiment, trolley A of mass 1.50 kg moving at a speed of 0.35 m s⁻¹ collides with trolley B moving in the opposite direction at a speed of 0.25 m s⁻¹. The two trolleys couple together on collision and move in the initial direction of A at a speed of 0.050 m s⁻¹ immediately after the collision. Calculate the mass of B. (4 marks)

2   A ball X of mass 0.60 kg moving at a speed of 3.0 m s⁻¹ along a straight line collides with a ball Y of mass 0. 40 kg which was initially stationary. As a result of the collision, X has a velocity of 0.80 m s⁻¹ in the same direction along the line. Calculate the speed and direction of Y immediately after the collision.   (4 marks)

3   A rail wagon P of mass 3000 kg moving at a velocity of 1.2 m s⁻¹ collides with a wagon Q of mass 2000 kg moving at a speed of 2.5 m s⁻¹ in the opposite direction. After the collision, Q moves at a velocity of 0.50 m s⁻¹ in the direction P was originally moving in.
  a   Calculate the speed and direction of P after the collision.   (4 marks)
  b   Show that the collision is inelastic.   (3 marks)

4   A shell of mass 2.4 kg is fired at a speed of 150 m s⁻¹ from an artillery gun of mass 900 kg.
  a   Calculate the recoil velocity of the gun.   (2 marks)
  b   Calculate the kinetic energy of the gun as a percentage of the total kinetic energy of the shell and the gun.   (2 marks)

1 **a** State the principle of conservation of momentum. *(1 mark)*

   b A vehicle of mass 1200 kg moving at a speed of 21.0 m s^{-1} collided with a vehicle of mass 1600 kg moving in the same direction at a speed of 3.0 m s^{-1}. The two vehicles locked together on impact. Calculate:

   i the velocity of the two vehicles immediately after impact

   ii the loss of kinetic energy due to the impact. *(6 marks)*

2 A football of mass 0.44 kg travelling at a speed of 24 m s^{-1} strikes a wall at an angle of 30° to the normal and rebounds at the same speed at the same angle to the normal.

   a Calculate the change of momentum of the ball *(1 mark)*

   b The ball was in contact with the wall for 95 ms. Calculate the impact force on the wall. *(2 marks)*

3 **a** State what is meant by an elastic collision. *(1 mark)*

   b A ball X of mass 0.12 kg moving at a speed of 0.58 m s^{-1} collides head-on with a second ball Y of mass 0.10 kg moving in the opposite direction at a speed of 0.50 m s^{-1}. After the impact, X continues in the same direction at a speed of 0.15 m s^{-1}.

   i Calculate the speed and direction of Y after the collision. *(3 marks)*

   ii Show that the collision was inelastic. *(3 marks)*

4 Two trolleys, A of mass 1.20 kg and B of mass 0.80 kg, are initially stationary on a level track.

   a When a trigger is pressed on one of the trolleys, a spring pushes the two trolleys apart and B moves away at a velocity of 0.15 m s^{-1}.

   i Calculate the velocity of A.

   ii Calculate the total kinetic energy of the two trolleys immediately after the explosion. *(6 marks)*

   b In part **a**, if the test had been carried out with A held firmly, calculate the speed at which B would have recoiled, assuming the energy stored in the spring before release is equal to the total kinetic energy calculated in **a ii**. *(2 marks)*

5 **a** When an α particle is emitted from a nucleus of the bismuth isotope, a nucleus of thallium (Th) is formed. Complete the equation below.

$$^{210}_{83}\text{Bi} \rightarrow \alpha + \text{Th}$$

   (2 marks)

   b The α particle in part A is emitted at a speed of 1.5×10^7 m s^{-1}.

   i Calculate the speed of recoil of the thallium nucleus immediately after the α particle has been emitted. Assume the parent nucleus is initially at rest.

   ii The mass of the α particle is 6.7×10^{-27} kg. Calculate the kinetic energy of the α particle immediately after it has been emitted as a percentage of the energy released by the bismuth nucleus. Ignore relativistic effects. *(6 marks)*

6 Figure 1 shows how the force, *F*, on a ball varied with time, *t*, when a tennis racquet was used to hit the ball horizontally when the ball was momentarily stationary.

  a State what the area under the curve represents. *(1 mark)*

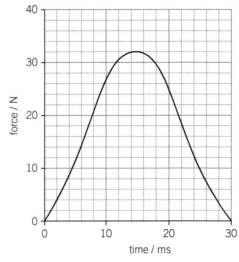

▲ **Figure 1**

  b i Use the graph to determine the change of momentum of the ball.

     ii The mass of the ball was 0.056 kg. Calculate the speed at which the ball left the racquet.

     iii Estimate the average acceleration of the ball during the impact. *(6 marks)*

7 Figure 2 shows a pile driver used to hammer a steel pile vertically into the ground. The pile driver hammer is repeatedly raised above the top end of the pile then released so it drops onto the top end of the pile and hammers the pile deep into the ground.

  a The hammer has a mass of 4000 kg and is raised to a height of 0.80 m above the top end of the pile then released.

     i Calculate the speed and the momentum of the hammer just before it makes contact with the pile.

     ii The pile has a mass of 2000 kg. Calculate the velocity of the hammer and the pile immediately after the impact. *(6 marks)*

  b Each time the hammer is dropped onto the pile, the pile is pushed 20 mm further into the ground.

     i Calculate the deceleration of the pile as it is pushed into the ground.

     ii Calculate the average force of friction on the pile as it is pushed into the ground. *(3 marks)*

8 A water jet cutter used in industry to cut materials in fine detail produces a jet of water at a speed of 400 m s⁻¹ and a flow rate of 2.7 kg per minute.

  a Calculate the momentum transferred per second by the water. *(3 marks)*

  b Calculate the force exerted by the water jet when it is directed normally onto a surface. Assuming the water does not rebound from the surface. *(2 marks)*

  c The water jet emerges from a nozzle of diameter 0.38 mm. Estimate the pressure of the water jet at normal incidence on the surface. (Hint: Pressure = force per unit area). *(2 marks)*

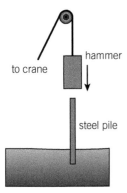

▲ **Figure 2**

10.1 Work and energy
10.2 Kinetic energy and potential energy

Specification reference: 3.4.1.7; 3.4.1.8

Work is done on an object when a force acting on it makes it move in the direction of the force. The work done is calculated using the equation

work done = force × distance moved in the direction of the force

The unit of work, the joule (J), is the work done when a force of 1 N moves its point of application by a distance of 1 m in the direction of the force.

When a force F acts on an object and moves it through a distance s along a straight line at an angle to the force direction, the work done W by the force is given by $W = Fs\cos\theta$.

Force–distance graphs

The area under the line of a force–distance graph represents the total work done.

Figure 2 shows a force–distance graph for a variable force that moves an object in the direction of the force. The work done to move the object through any distance can be determined from the area under the part of the line corresponding to that distance.

Figure 3 shows how the force needed to extend a spring varies with the extension of the spring. The graph is a straight line through the origin provided the spring obeys Hooke's Law. By considering the area under the line (which is a triangle), it can be shown that the work done and therefore

the energy stored in the spring $= \frac{1}{2}F\Delta L$

Energy

Energy is measured in joules. When a force does work, energy is transferred by the force equal to the work done.

The **kinetic energy** of an object of mass m moving at speed v is given by the equation

kinetic energy, $E_K = \frac{1}{2}mv^2$

The change of **gravitational potential energy** ΔE_p of an object of mass m raised through a vertical height Δh, is given by the equation

$$\Delta E_p = mg\,\Delta h$$

At the Earth's surface, $g = 9.81\,\text{m s}^{-2}$.

Conservation of energy

Whenever energy is transferred, the total amount of energy after the transfer is always equal to the total amount of energy before the transfer. The total amount of energy is unchanged.

Energy cannot be created or destroyed.

This statement is known as the **principle of conservation of energy**.

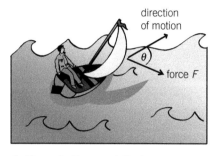

▲ **Figure 1** *Force and displacement*

> **Revision tip**
> No work is done when F and s are at right angles to each other.

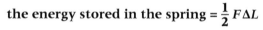

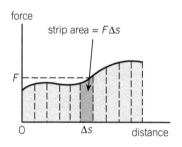

▲ **Figure 2** *Force–distance graph for a variable force*

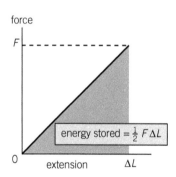

▲ **Figure 3** *Force against extension for a spring*

> **Synoptic link**
> The force needed to stretch a spring is proportional to the extension of the spring. This is known as Hooke's law. For more about springs see Topic 11.2, Springs.

Energy changes involving kinetic and potential energy

Consider an object of mass m that is released from rest. Suppose its speed is v after it descends through a vertical distance Δh. If the resistive forces on it are negligible, its gain of kinetic energy is equal to its loss of potential energy. Therefore

$$\frac{1}{2}mv^2 = mg\,\Delta h$$

If the resistive forces are *not* negligible, its gain of kinetic energy is less than its loss of potential energy because work W is done by the resistive forces. In other words,

$$\frac{1}{2}mv^2 = mg\,\Delta h - W$$

Therefore, the work done by the resistive forces $W = mg\,\Delta h - \frac{1}{2}mv^2$.

Summary questions

$g = 9.81\,\mathrm{m\,s^{-2}}$

1 Calculate the work done when:
   a a force of 14 N moves an object by a distance of 3.0 m in a direction at 60° to the direction of the force *(1 mark)*
   b a spring that obeys Hooke's Law is stretched from zero extension to an extension of 0.45 m by a force that increases to 20 N. *(1 mark)*

2 A ball of mass 0.048 kg was thrown directly downwards at a speed of $16.0\,\mathrm{m\,s^{-1}}$ from the top of a tower of height 23.0 m. Calculate:
   a its kinetic energy *(2 marks)*
   b its speed just before it hit the ground. Assume air resistance is negligible. *(2 marks)*

3 A skier of mass 80 kg (including the skis) reaches a speed of $14\,\mathrm{m\,s^{-1}}$ after skiing from rest 560 m down a slope onto level ground which is 40 m lower than the starting point. Calculate:
   a the loss of potential energy *(1 mark)*
   b the gain of kinetic energy of the skier and the skis *(1 mark)*
   c the average resistive force during the descent. *(2 marks)*

10.3 Power
10.4 Energy and efficiency
Specification reference: 3.4.1.7

Power is defined as the rate of transfer of energy.

The unit of power is the watt (W), equal to an energy transfer rate of 1 joule per second. Note that 1 kilowatt (kW) = 1000 W, and 1 megawatt (MW) = 10^6 W.

If energy ΔE is transferred steadily in time Δt,

$$\text{power } P = \frac{\Delta E}{\Delta t}$$

Engine power

Vehicle engines, marine engines, and aircraft engines are all designed to make objects move. The output power of an engine is sometimes called its **motive power**.

1 For a powered vehicle driven by a constant force F moving at speed v, the output power of the engine $P = Fv$.

The work done by the engine is transferred into the internal energy of the surroundings by the resistive forces.

2 For a powered vehicle that gains speed when moving horizontally, the work done by the engine increases the kinetic energy of the vehicle and enables the vehicle to overcome the resistive forces acting on it. Therefore,

the output power of the engine	=	energy per second wasted due to the resistive force	+	the gain of kinetic energy per second

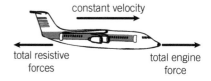

▲ **Figure 1** Engine power

Synoptic link

The electrical power supplied to a component = IV where I is the current in the component when the pd across it is V. See Topic 12.2, Potential difference and power, for more about electrical power.

Worked example

When an aircraft is in level flight at a constant velocity of 120 m s⁻¹, its engines have a total power output of 4.0 MW. Calculate the driving force of the engine at this speed.

Solution

Rearranging power = force × velocity gives the driving force,

$$F = \frac{\text{output power}}{\text{speed}} = \frac{4.0 \times 10^6}{120} = 33 \text{ kN}$$

Machines at work

If a machine exerts a force F on an object to make it move at a constant velocity v,

$$\text{the output power of the machine } P_{OUT} = Fv$$

Efficiency measures

Useful energy is energy transferred for a purpose. In any machine, some of the energy supplied to it is wasted, usually as energy transferred to the surroundings by:

- sound waves created by vibrating machinery, or
- heating due to friction or air resistance.

$$\text{The efficiency of a machine} = \frac{\text{useful energy transferred by the machine}}{\text{energy supplied to the machine}}$$
$$= \frac{\text{work done by the machine}}{\text{energy supplied to the machine}}$$

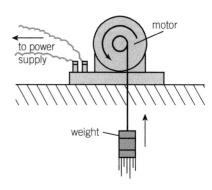

▲ **Figure 2** Efficiency

Note:

- **Efficiency** can be expressed in terms of power as

$$\frac{\text{the output power of a machine}}{\text{the input to the machine}}$$

- *Percentage* efficiency = efficiency × 100%

Renewable energy

- Most of the energy we use at present is obtained from fossil fuels. Scientists think that the use of fossil fuels is causing climate change, due to the increasing amount of carbon dioxide in the atmosphere.

- Carbon emissions can be cut by building new nuclear power stations and developing more renewable energy resources.

- The output power of many renewable energy resources can be estimated using AS/A Level Physics principles and formulae. For example, the output power of a wind turbine is proportional to v^3 where v is the wind speed. This is because:

  - the kinetic energy of air moving at speed v is proportional to v^2

  - the mass of air passing through the area swept out by the blades is proportional to v

  - so the kinetic energy per second of the air passing through a wind turbine is proportional to v^3.

Summary questions

$g = 9.81 \, \text{m s}^{-2}$

1 A heavy goods vehicle of mass 40 000 kg moving at a constant velocity of 31 m s⁻¹ has an output power of 240 kW.
     a Calculate the motive force of its engine at this speed. *(2 marks)*
     b The engine has an efficiency of 37% at this speed. Calculate the energy per second wasted at this speed. *(2 marks)*

2 The maximum power that can be obtained from a wind turbine is proportional to the cube of the wind speed. When the wind speed is 12 m s⁻¹, the power output of a certain wind turbine is 1.4 MW.
     a Calculate the power output of this wind turbine when the wind speed is 8.0 m s⁻¹. *(3 marks)*
     b The efficiency of the wind turbine at 12 m s⁻¹ is 48%. Calculate the kinetic energy per second of the wind passing through the turbine. *(2 marks)*

3 A hydroelectric power station produces electrical power at an overall efficiency of 28%. The power station is driven by water from an upland reservoir 350 m above the power station. Calculate the output power when the water passes through it at a flow rate of 52 m³ per second. The density of water is 1000 kg m⁻³. *(3 marks)*

Chapter 10 Practice questions

$g = 9.81 \text{ m s}^{-2}$

1 A skateboarder of mass 52.0 kg at the top of a slope travels a distance of 10.2 m down the slope and reaches a speed of 2.12 m s^{-1} at the bottom of the slope which is 2.50 m lower than the top.

   a Calculate the skateboarder's:

      i loss of potential energy

      ii gain of kinetic energy. *(2 marks)*

   b The average frictional force on the skateboarder during the descent.
(3 marks)

2 A child of mass 35 kg on a trampoline bounces up and down repeatedly through a vertical distance of 0.90 m between the lowest and the highest position of her centre of mass.

   a Describe the energy changes that take place when she descends from her highest position to her lowest position. *(3 marks)*

   b Each time she descends from her highest position, she accelerates due to gravity then decelerates vertically through a distance of 0.16 m to her lowest position.

      i Calculate her speed just before she starts to decelerate.

      ii Estimate the maximum energy that could be transferred to the trampoline in each bounce. *(4 marks)*

   c The trampoline is fitted with 90 springs, each with a spring constant of 3500 N m^{-1}. When the child is at her lowest position, the average extension of each spring is 42 mm. Calculate the energy stored in the springs at maximum extension. *(2 marks)*

3 A road truck of mass 12×10^3 kg travels on a straight uphill road at a constant velocity of 18 m s^{-1}.

   a Calculate the kinetic energy of the truck at this speed. *(1 mark)*

   b The gradient of the road is at an angle of 5.0° to the horizontal as shown in Figure 1.

      i Calculate the gain of potential energy of the truck each second.

      ii The output power of the truck engine at this speed is 228 kW. Calculate the output force of the engine.

      iii Show that the total resistive force on the truck is 2.4 kN. *(7 marks)*

▲ **Figure 1**

4 a A tidal power station traps seawater over a flat area of 150 km² when the tide is 5.0 m above the power station turbines. The trapped water is released gradually over a period of 6 hours. Calculate:

      i the mass of trapped water

      ii the average loss of potential energy per second of this trapped water when it is released over a period of 6 hours

      The density of sea water = 1050 kg m^{-3}. *(5 marks)*

   b A solar cell panel of area 1 m² can produces 220 W of electrical power on a sunny day. Calculate the area of panels that would generate electrical power at the same rate as the tidal power station in part **a**.
(1 mark)

5   In the fairground ride shown in Figure 2, each carriage is pulled through a vertical height of 15 m to the top of an inclined track which then descends steeply to a level section which is partly in shallow water.

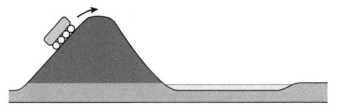

▲ Figure 2

a   A carriage of total mass 1200 kg including its riders travels at a constant speed up the incline in 20 s, pulled by a conveyor belt driven by an electric motor operating at an efficiency of 58%. Estimate the minimum power supplied to the motor when it pulls the carriage up the track.   *(3 marks)*

b   The speed of the carriage at the start of its descent is 1.2 m s⁻¹. Calculate its speed when it reaches the level section of the track.
                                                         (4 marks)

c   The carriage leaves the shallow water at a speed of 11.5 m s⁻¹.

   i   Calculate the energy transferred to the water.

   ii The length of track under water is 18.4 m. Estimate the average resistive force on the carriage when it passes through the water.
                                                         (4 marks)

6   a   A jet plane of mass 250 000 kg travels for 2400 m before it lifts off the runway at a speed of 150 m s⁻¹.

   i   Calculate the kinetic energy of the aircraft at this speed.

   ii Estimate the average engine force during the time it travels along the runway.   *(3 marks)*

   b   After taking off, the aircraft climbs to a height of 8500 m above the ground and travels at a constant velocity of 250 m s⁻¹.

   i   Calculate the gain of potential energy of the plane in reaching this height from the ground.

   ii Calculate its kinetic energy when it moves at 250 m s⁻¹.   *(2 marks)*

   c   Estimate the mass of fuel burned to take off from rest and reach the above height and speed. Assume the engines are 55% efficient and each kilogram of fuel used releases 30 MJ of energy.   *(3 marks)*

7   A hydroelectric power station generates 450 MW of electrical power at an efficiency of 80%. The power station is driven by water that has descended from an upland reservoir 390 m above the power station.

   a   Calculate the volume of water per second passing through the turbines of the power station when its generators produce 450 MW of electrical power.

      The density of water = 1000 kg m⁻³.   *(6 marks)*

   b   The turbines at the power station are designed to pump water uphill to the reservoir using electrical power from electrical power generators elsewhere when the national demand for electrical power is low.

   i   Estimate the percentage of the electrical power used for this purpose that is wasted. Assume the pumping process is also 80% efficient.   *(3 marks)*

   ii Explain why the efficiency of this pumped storage process could never be more than 80%.   *(3 marks)*

11.1 Density

Specification reference: 3.4.2.1

The **density of a substance** is defined as its mass per unit volume.

If volume V of a substance has mass m, its density ρ is given by

$$\rho = \frac{m}{V}$$

The unit of density is the kilogram per cubic metre (kg m^{-3}).

Rearranging the above equation gives $m = \rho V$ or $V = \frac{m}{\rho}$.

Density measurements

To measure the density of a substance, measure the mass and volume of a sample of the substance, and then calculate the density from $\frac{\text{mass}}{\text{volume}}$.

1 A regular solid

- Measure its mass using a top pan balance.

- Measure its dimensions using vernier callipers or a micrometer and calculate its volume using the appropriate equation. For example, for a sphere of radius r, volume $= \frac{4}{3}\pi r^3$ (see Figure 1 for other volume equations).

2 A liquid

- Measure the mass of an empty measuring cylinder. Pour some of the liquid into the measuring cylinder and measure the volume of the liquid directly. Use as much liquid as possible to reduce the percentage error in your measurement.

- Measure the mass of the cylinder and liquid to enable the mass of the liquid to be calculated.

3 An irregular solid

- Measure the mass of the object.

- Lower the object on a thread into the liquid in a measuring cylinder and measure the increase in the liquid level. This is the volume of the object.

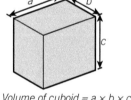

a Volume of cuboid $= a \times b \times c$

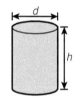

b Volume of cylinder $= \frac{\pi d^2}{4} \times h$

▲ **Figure 1** Volume equations

Density of alloys

An alloy is a solid mixture of two or more metals. For an alloy, of volume V and mass m, that consists of two metals A and B:

- if the volume of metal A $= V_A$, the mass of metal A $m_A = \rho_A V_A$, where ρ_A is the density of metal A

- if the volume of metal B $= V_B$, the mass of metal B $m_B = \rho_B V_B$, where ρ_B is the density of metal B.

Given the density of each metal and how much of each is in the alloy, these equations can be used to work out the density of the alloy.

Worked example

A brass object consists of 0.62 kg of copper and 0.29 kg of zinc. Calculate the volume and the density of this object. The density of copper = 8900 kg m^{-3}. The density of zinc = 7100 kg m^{-3}.

Solution

Volume of copper = mass of copper ÷ density of copper

$$= 0.62 \text{ kg} \div 8900 \text{ kg m}^{-3} = 7.0 \times 10^{-5} \text{ m}^3$$

Volume of zinc = mass of zinc ÷ density of zinc

$$= 0.29\,kg \div 7100\,kg\,m^{-3} = 4.1 \times 10^{-5}\,m^3$$

Total mass $m = 0.62 + 0.29 = 0.91\,kg$

Total volume $V = 1.1 \times 10^{-4}\,m^3$

Density of alloy $\rho = \dfrac{m}{v} = \dfrac{0.91\,kg}{1.1 \times 10^{-4}\,m^3} = 8270\,kg\,m^{-3}$

Summary questions

1 A concrete paving stone has dimensions $2.0\,cm \times 60\,cm \times 75\,cm$ and a mass of $22.4\,kg$. Calculate:

   a its volume *(1 mark)*

   b its density. *(1 mark)*

2 An empty tin of diameter $50\,mm$ and of height $90\,mm$ has a mass of $85\,g$. It is filled with a liquid to within $5\,mm$ of the top. Its total mass is then $248\,g$. Calculate: a the mass b the volume c the density of the liquid in the tin. *(3 marks)*

3 A metal wire has a diameter of $0.220 \pm 0.005\,mm$, a length of $2415 \pm 5\,mm$, and a mass of $0.730 \pm 0.005\,g$.

   a Calculate:

      i   the volume of the wire *(2 marks)*

      ii the density of the metal in $kg\,m^{-3}$. *(1 mark)*

   b Show that the uncertainty in the density value is about $430\,kg\,m^{-3}$. *(4 marks)*

11.2 Springs
Specification reference: 3.4.2.1

Hooke's law

Hooke's law states that the force needed to stretch a spring is directly proportional to the extension of the spring from its natural length.

Hooke's law may be written as **Force $F = k\Delta L$**

where k is the spring constant (or the stiffness constant) and ΔL is the extension from its natural length L. The unit of k is $N\,m^{-1}$.

- The graph of F against ΔL is a straight line of gradient k through the origin.
- If a spring is stretched beyond its **elastic limit**, it does not return to its original length when the applied force is removed.

Spring combinations

Springs in parallel

Figure 2 shows a weight supported by means of two springs P and Q with spring constants k_P and k_Q in parallel with each other. The extension, ΔL, of each spring is the same.

The effective spring constant $k = k_P + k_Q$.

Springs in series

Figure 3 shows a weight supported by means of two springs joined end-on. The tension in each spring is the same and is equal to the weight W.

- $\dfrac{\text{the extension in P}}{\text{the extension in Q}} = \dfrac{k_Q}{k_P}$
- The effective spring constant k is given by the equation $\dfrac{1}{k} = \dfrac{1}{k_P} + \dfrac{1}{k_Q}$.

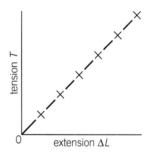

▲ **Figure 1** *Hooke's law*

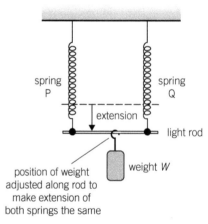

▲ **Figure 2** *Two springs in parallel*

position of weight adjusted along rod to make extension of both springs the same

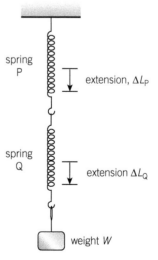

▲ **Figure 3** *Two springs in series*

Summary questions

$g = 9.81\,m\,s^{-2}$

1 A steel spring has a spring constant of $40\,N\,m^{-1}$. The spring hangs vertically and supports an $8.0\,N$ weight at rest. Calculate:
   a the extension of the spring *(1 mark)*
   b the energy stored in the spring. *(1 mark)*

2 Two identical springs of length $250\,mm$ are arranged in parallel and vertical as in Figure 2. When the springs support a $1.6\,N$ weight, they both extend to a length of $274\,mm$. Calculate:
   a for each spring: **i** its tension *(1 mark)* **ii** its spring constant. *(3 marks)*
   b the total energy stored in the springs. *(3 marks)*

3 Two unequal steel springs P and Q of length $250\,mm$ are suspended vertically in series from a fixed point as in Figure 3. A $1.6\,N$ weight is attached to the ends of the two springs. The lengths of the springs are then $280\,mm$ for P and $300\,mm$ for Q. Calculate:
   a the tension in each spring *(1 mark)*
   b the spring constant of each spring. *(4 marks)*

11.3 Deformation of solids
11.4 More about stress and strain

Specification reference: 3.4.2.2

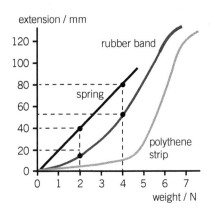

▲ **Figure 1** *Typical graphs*

Key terms

Elasticity is a property of a solid material that returns to its original shape after it has been deformed, once the forces that deformed it have been removed.

Tensile deformation occurs when an object is stretched.

Compressive deformation occurs when an object is compressed.

Tension–extension graphs

- A steel spring gives a straight line (see Figure 1), in accordance with Hooke's law (Topic 11.2, Springs).
- A rubber band at first extends easily when it is stretched then it becomes increasingly difficult to stretch further (see Figure 1).
- A polythene strip 'gives' and stretches easily after its initial stiffness is overcome then becomes more difficult to stretch (see Figure 1).

Tensile stress and tensile strain

- The **tensile stress** in the wire, $\sigma = \frac{T}{A}$, where T is the tension. The unit of stress is the **pascal** (Pa) equal to $1\,\mathrm{Nm^{-2}}$.
- The **tensile strain** in the wire, $\varepsilon = \frac{\Delta L}{L}$, where ΔL is the extension (increase in length) of the wire. Strain is a ratio and therefore has no unit.

Figure 2 shows how the tensile stress in a wire varies with tensile strain.

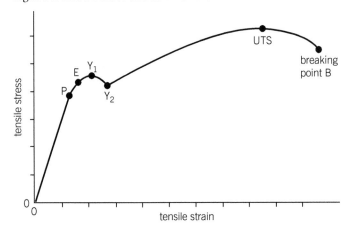

▲ **Figure 2** *Tensile stress versus tensile strain for a metal wire*

From zero to the limit of proportionality P, the tensile stress is proportional to the tensile strain. The value of $\frac{\text{stress}}{\text{strain}}$ is a constant, known as the **Young modulus** of the material.

$$\text{Young modulus } E = \frac{\text{tensile stress, } \sigma}{\text{tensile strain, } \varepsilon} = \frac{T}{A} \div \frac{\Delta L}{L} = \frac{TL}{A\Delta L}$$

- Beyond P, the **elastic limit** is the point beyond which the wire is permanently stretched and suffers **plastic deformation**.
- At the **yield point** Y_1, the wire weakens temporarily. Beyond Y_2, a small increase in the tensile stress causes a large increase in tensile strain as the material of the wire undergoes plastic flow.
- Beyond maximum tensile stress, the **ultimate tensile stress** (UTS) (or breaking stress), the wire loses its strength, extends, becomes narrower and breaks at point B.

Revision tip

A **brittle** material snaps without any noticeable yield.

A **ductile** material stretches considerably as the tension is increased before it breaks.

Loading and unloading curves

The area under a tension against extension line gives the work done. If the loading and unloading curves differ, energy transferred to the material is not all recovered as useful energy when the material unstretches.

1 For a metal wire, the loading and unloading lines are the same up to the elastic limit so all the energy stored in the wire can be recovered, provided the elastic limit is not exceeded.

2 For a rubber band, the unloading curve is below the loading curve except at zero and maximum extensions. The area between the curves gives the internal energy gained by the rubber band when it unstretches.

3 For a polythene strip, it does not return to its initial length when it is completely unloaded. The area between the loading and unloading curves represents work done to deform the material permanently, as well as internal energy retained by the polythene when it unstretches.

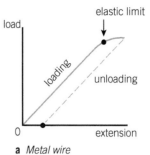

a *Metal wire*

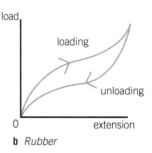

b *Rubber*

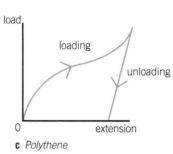

c *Polythene*

▲ **Figure 3** *Loading and unloading curves*

Worked example

A steel wire of uniform diameter 0.26 mm and of length 1540 mm is stretched to a tension of 58 N. Calculate the extension in the wire at this tension.

The Young modulus for steel $= 2.1 \times 10^{11}$ Pa.

Solution

Tension $T = 58$ N

Area of cross section of wire $= \frac{1}{4}\pi d^2 = \frac{1}{4} \times \pi \times (0.26 \times 10^{-3})^2 = 5.3 \times 10^{-8}$ m^2

To find the extension ΔL, rearrange the Young modulus equation

$E = \frac{TL}{A\Delta L}$ gives $\Delta L = \frac{TL}{AE} = \frac{58 \times 1.540}{5.3 \times 10^{-8} \times 2.1 \times 10^{11}} = 8.0 \times 10^{-3}$ m

---

Summary questions

The Young modulus of steel $= 2.1 \times 10^{11}$ Pa

1 A metal wire of diameter 0.26 mm and of unstretched length 1.524 m was suspended vertically from a fixed point. When a 48 N weight was suspended from the lower end of the wire, the wire stretched by an extension of 6.6 mm. Calculate the Young modulus of the wire material. *(3 marks)*

2 A vertical steel wire of diameter 0.22 mm and of length 1.500 m is fixed at its upper end, and has a weight of 32 N suspended from its lower end. Calculate:
   a the extension of the wire *(3 marks)*
   b the elastic energy stored in the wire. *(1 mark)*

3 A rectangular steel block of length 50 mm and cross-sectional area 3.4×10^{-4} m^2 is placed in a vice and its length is compressed by 0.13 mm when the vice is tightened. Calculate:
   a the compressive stress exerted on the block *(2 marks)*
   b the elastic energy stored in it. *(3 marks)*

1 a Define the density of a substance. (*1 mark*)

   b i A metal sphere has a diameter of 14.20 mm and a mass of 11.80 g. Calculate the density of the metal. (*2 marks*)

      ii The diameter value has an uncertainty of ±0.02 mm. The mass value has an uncertainty of ±0.04 g. Calculate the uncertainty in the density. (*3 marks*)

2 a Describe an experiment you would carry out to measure the density of the metal in a wire of diameter about 2 mm. (*3 marks*)

   b A square metal plate 50 mm by 50 mm in length and width has a mass of 14.20 g.

      i The following measurements of the plate thickness *t* were made at different places on the plate.

         2.10 mm   2.06 mm   2.14 mm   2.16 mm   2.08 mm

         Calculate the mean value of *t* and estimate the uncertainty in *t*. (*1 mark*)

      ii Calculate the volume *V* of the plate.

      iii Calculate the density ρ of the plate. (*3 marks*)

   c The uncertainty in the length of each side of the plate was +0.5 mm. The uncertainty in the mass of the plate was +0.10 g. Estimate the percentage uncertainty in ρ. (*3 marks*)

3 Figure 1 shows a steel spring suspended from a stand supporting a mass hanger of weight 0.25 N at its lower end. Different weights were added to the mass hanger and the extension of the spring was measured each time.

   The results were plotted on the graph shown in Figure 2.

   a i Use the graph to determine the spring constant of the spring. (*2 marks*)

      ii Determine the length of the spring when it was unloaded and the mass hanger was removed. (*2 marks*)

   b Calculate the energy stored in the spring when its length was 450 mm. (*3 marks*)

4 a An elastic cord of unstretched length 320 mm has a cross-sectional area of 1.23 mm². The cord is stretched to a tension of 250 N. Assume that Hooke's law is obeyed for this range and that the cross-sectional area remains constant.

      The Young modulus for the material of the cord = 3.9 GPa.

      i Calculate the extension of the cord at this tension. (*3 marks*)

      ii Calculate the work done to stretch the cord to this extension. (*1 mark*)

   b The cord is tested to investigate how much energy it absorbs when it is used to stop an object. Figure 3 shows the cord being struck at its midpoint with a hammer of mass 0.170 kg moving at an initial speed of 1.40 m s⁻¹. The hammer rebounds from the cord after hitting it.

      i Calculate the kinetic energy of the hammer before it hits the cord. (*1 mark*)

      ii The hammer rebounds from the cord at a speed of 1.15 m s⁻¹. Calculate the percentage of the kinetic energy of the hammer that it regains after the rebound. (*2 marks*)

      iii Explain why the hammer does not regain all its kinetic energy when it rebounds. (*2 marks*)

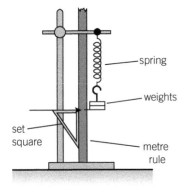

▲ Figure 1

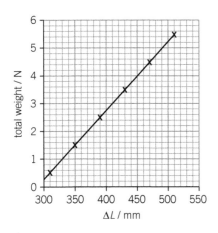

▲ Figure 2

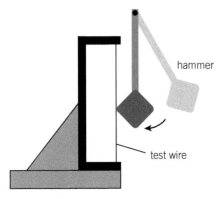

▲ Figure 3

5 A material in the form of a wire, 2.92 m long and with a diameter of 0.71 mm, is suspended from a support so that it hangs vertically. Different weights are suspended from its lower end and the extension of the wire is measured for each weight. The table shows the extension of the wire for each load as the weight is increased.

Load / N	0	9.8	19.6	29.4	39.2	49.0	58.8
Extension / mm	0	0.38	0.74	1.15	1.50	1.91	2.27

  a i Plot a graph of load (on the *y*-axis) against extension (on the *x*-axis). *(3 marks)*

    ii Use the graph to determine a value of the Young modulus for the material of the wire. *(5 marks)*

  b The material of the wire has an ultimate tensile stress of 530 MPa. Calculate the maximum load the wire could support without breaking. *(1 mark)*

6 Figure 4 shows stress–strain curves for three different materials P, Q, and R up to the fracture point of each material. Answer the following questions about Figure 4, giving a reason for your answer to each question.

  a Which of the three materials has the largest Young modulus? *(2 marks)*

  b Which of the three materials is the most: **i** brittle **ii** ductile? *(4 marks)*

  c Which material stores the most energy per unit volume up to its limit of proportionality? *(2 marks)*

7 A lift of weight 10 000 N is designed to carry a maximum passenger load of 6300 N. The lift is supported by six parallel steel cables, each of diameter 6.0 mm and of length 35.0 m.

Young modulus of steel = 210 GPa.

  a Calculate the extension of each cable when the lift is moving upwards at a steady velocity and it is fully loaded. *(5 marks)*

  b When the lift starts to ascend, it accelerates to a speed of 0.72 m s^{-1} in 8.0 s.

    i Calculate the average acceleration of the lift during this time. *(1 mark)*

    ii Estimate the additional extension of each cable during this time. *(4 marks)*

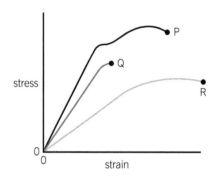

▲ **Figure 4**

12.1 Current and charge
12.2 Potential difference and power

Specification reference: 3.5.1.1

Electrical conduction

An electric current is the rate of flow of charge in the wire or component. The current is due to the passage of **charge carriers**.

- In a metallic conductor, the charge carriers are conduction electrons (i.e., electrons in the metal not attached to an atom). When a voltage is applied across the metal, these conduction electrons are attracted towards the positive terminal of the metal.
- In an insulator, each electron is attached to an atom and cannot move away from the atom. When a voltage is applied across an insulator, no current passes through the insulator, because no electrons can move through the insulator.
- In a semiconductor, the number of charge carriers increases with an increase of temperature. The resistance of a semiconductor therefore decreases as its temperature is raised.
- In a solution of salt, the charge carriers are positive and negative ions which are charged atoms or charged groups of atoms (i.e., molecules).

Charge flow

For a current I, the charge flow ΔQ in time Δt is given by $\Delta Q = I\,\Delta t$

For charge flow ΔQ in a time interval Δt, the current I is given by $I = \dfrac{\Delta Q}{\Delta t}$

The unit of current is the *ampere* (A), which is defined in terms of the magnetic force between two parallel wires when they carry the same current. The symbol for current is I.

The unit of charge is the *coulomb* (C), equal to the charge flow in one second when the current is one ampere. The symbol for charge is Q.

The magnitude of the charge of the electron $e = 1.60 \times 10^{-19}$ C. The convention for the direction of current in a circuit is from positive (+) to negative (−).

Energy and potential difference

Potential difference is defined as the work done (or energy transfer) per unit charge.

The unit of pd is the volt, which is equal to 1 joule per coulomb.

If work W is done when charge Q flows through the component, the pd across the component, V, is given by $V = \dfrac{W}{Q}$

The electromotive force (emf) of a source of electricity is defined as the electrical energy produced per unit charge passing through the source. The unit of emf is the volt, the same as the unit of pd.

For a source of emf ε in a circuit, the electrical energy produced when charge Q passes through the source $= Q\varepsilon$. The charge carriers are forced round the circuit and they transfer energy to the components in the circuit.

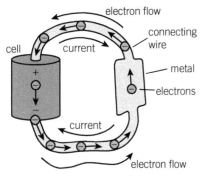
▲ **Figure 1** *Convention for current*

Revision tip
Remember 1 volt = 1 joule per coulomb.

Electrical power and current

Consider a component or device that has a potential difference V across its terminals and a current I passing through it. In time Δt:

- the work done W by the charge carriers $= IV \Delta t$
- the electrical power P supplied to the device $= IV$
- the unit of power is the watt (W). Therefore, one volt is equal to one watt per ampere.

Synoptic link

You will meet emf in more detail in Topic 13.3, Electromotive force and internal resistance.

Summary questions

$e = 1.60 \times 10^{-19}$ C

1 In an electron beam experiment, the beam current was 1.2 mA for 300 s. Calculate:
   a the charge flowing along the beam   *(1 mark)*
   b the number of electrons that passed along the beam.   *(1 mark)*

2 A 230 V microwave oven has a power rating of 800 W. Calculate:
   a the current taken by the appliance   *(1 mark)*
   b the energy transfer to the appliance in 1 minute.   *(1 mark)*

3 A battery has an emf of 12 V and negligible internal resistance. It is capable of delivering a charge of 3.6×10^6 C. Calculate:
   a how long the battery will last after it has been fully charged when the current from it is 13 A   *(1 mark)*
   b the maximum energy it could deliver without being recharged.   *(1 mark)*

12.3 Resistance

Specification reference: 3.5.1.1; 3.5.1.3

Reminder about prefixes

▼ **Table 1** *Prefixes*

Prefix	Symbol	Value
Nano	n	10^{-9}
Micro	μ	10^{-6}
Milli	m	10^{-3}
Kilo	k	10^{3}
Mega	M	10^{6}
Giga	G	10^{9}

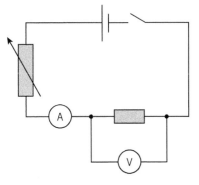

▲ **Figure 1** *Measuring resistance*

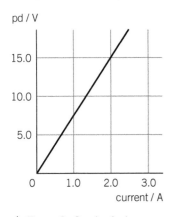

▲ **Figure 2** *Graph of pd versus current for a resistor*

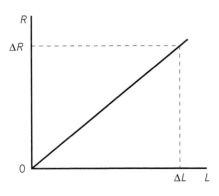

▲ **Figure 3** *Graph of resistance against length for a wire*

Definitions and laws

The resistance of a component in a circuit is caused by the repeated collisions between the charge carriers passing through the component and the positive ions in the component vibrating about fixed positions in the component.

The **resistance** of any component is defined as

$$\frac{\textbf{the pd across the component}}{\textbf{the current through it}}$$

For a component which passes current I when the pd across it is V, its resistance R is given by the equation

$$R = \frac{V}{I}$$

The unit of resistance is the ohm (Ω), which is equal to 1 volt per ampere.

Rearranging the above equation gives $V = IR$ or $I = \frac{V}{I}$.

Measurement of resistance

The resistance of a resistor can be measured using the circuit shown in Figure 1.

To investigate the variation of current with pd, the variable resistor is adjusted in steps. At each step, the current and pd are recorded from the ammeter and voltmeter, respectively. The measurements can then be plotted on a graph of pd against current, as shown in Figure 2 (or for current against pd – see Topic 12.4, Components and their characteristics).

Ohm's law states that the pd across a metallic conductor is proportional to the current through it, provided the physical conditions do not change.

The graph for a resistor is a straight line through the origin so the pd across the resistor is proportional to the current. In this case, the resistance is equal to the gradient of the graph.

Resistivity

The resistivity ρ of a sample of material of length L, uniform cross-sectional area A, and resistance R is given by

$$\textbf{resistivity } \rho = \frac{RA}{L}$$

Note:

1 The unit of resistivity is the ohm metre (Ω m).

2 Rearranging the above equation gives $R = \rho \frac{L}{A}$.

To determine the resistivity of a wire:

- Measure the diameter of the wire d using a micrometer at several different points along the wire to give a mean value for d then calculate its cross-sectional area ($A = \frac{\pi d^2}{4}$).

- Measure the resistance R of different lengths L of wire to plot a graph of R against L, see Figure 3. The resistivity of the wire is given by the graph gradient × A.

Superconductivity

A **superconductor** is a wire or a device made of material that has zero resistivity at and below a **critical temperature** that depends on the material. This property of the material is called **superconductivity**. This critical temperature is also called its **transition temperature**. A superconductor material loses its superconductivity if its temperature is raised above its critical temperature.

Superconductors are used to make high-power electromagnets that generate very strong magnetic fields in devices such as magnetic resonance scanners and particle accelerators. Superconductors with higher transition temperatures could be used in new applications such as lightweight electric motors and power cables that transfer electrical energy without energy dissipation.

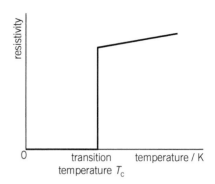

▲ **Figure 4** *Resistivity of a superconductor versus temperature near the transition temperature*

Revision tip
Resistivity is a property of a material. Don't confuse resistivity and resistance.

Summary questions

1 A rectangular strip of metal has a resistance of 0.156 Ω. When the strip is in a circuit, the current through it is 0.970 A. Calculate:
  a the pd across the strip   (*1 mark*)
  b the energy per second dissipated in the strip.   (*1 mark*)

2 The strip of metal in **Q1** has a length 0.382 m, width 35.1 mm, and thickness 0.62 mm. Calculate the resistivity of the metal. (*2 marks*)

3 **a** Calculate the resistance of a copper wire of uniform diameter 1.28 mm and length 25.0 m.
     Resistivity of copper = 1.70×10^{-8} Ω m.   (*3 marks*)
  b Calculate the energy per second dissipated in the wire when the current through it is 13.0 A.   (*1 mark*)

12.4 Components and their characteristics

Specification reference: 3.5.1.2; 3.5.1.3

Investigating the characteristics of different components

To measure the variation of current with pd for a component, use (as shown in Figure 1) either:

- a potential divider to vary the pd from zero, or
- a variable resistor to vary the current to a minimum.

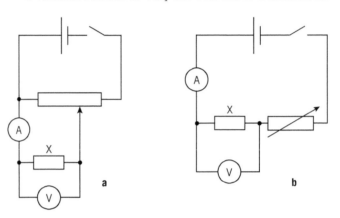

▲ **Figure 1** *Investigating component characteristics* **a** *Using a potential divider,* **b** *Using a variable resistor*

The advantage of using a potential divider is that the current through the component and the pd across it can be reduced to zero. This is not possible with a variable resistor circuit.

The measurements for each type of component are usually plotted as a graph of current (on the *y*-axis) against pd (on the *x*-axis). Typical graphs for a wire, a filament lamp, and a thermistor are shown in Figure 2. Note that the measurements are the same, regardless of which way the current passes through each of these components.

In both circuits, an ammeter sensor and a voltmeter sensor connected to a data logger could be used to capture data (i.e., to measure and record the readings) which could then be displayed directly on an oscilloscope or a computer.

- A wire gives a straight line through the origin. This means that the resistance of the wire does not change when the current changes. In this case, the gradient of the line is equal to $\frac{1}{\text{resistance } R \text{ of the wire}}$.
- A filament bulb gives a curve with a decreasing gradient because its resistance increases as it becomes hotter.
- A thermistor at constant temperature gives a straight line. The higher the temperature, the greater the gradient of the line, as the resistance falls with increase of temperature. The same result is obtained for a **light-dependent resistor** in respect of light intensity.

The diode

To investigate the characteristics of the diode, one set of measurements is made with the diode in its forward direction (i.e., forward biased) and another set with it in its reverse direction (i.e., reverse biased). The current is very small when the diode is reverse biased and can only be measured using a milliammeter.

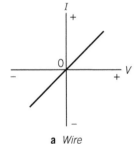

a *Wire*

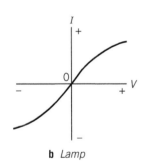

b *Lamp*

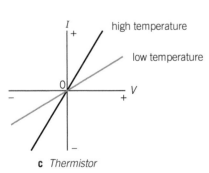

c *Thermistor*

▲ **Figure 2** *Current versus pd for different components*

Typical results for a silicon diode are shown in Figure 3. A silicon diode conducts easily in its forward direction above a pd of about 0.6 V and hardly at all below 0.6 V or in the opposite direction. This voltage is called the 'forward voltage' of the diode.

Resistance and temperature

1   The resistance of a metal increases with increase of temperature. This is because the positive ions in the conductor vibrate more when its temperature is increased so the charge carriers (conduction electrons) cannot pass through the metal as easily. A metal has a **positive temperature coefficient** because its resistance increases with increase of temperature.

2   The resistance of an intrinsic semiconductor decreases with increase of temperature. This is because the number of charge carriers (conduction electrons) increases when the temperature is increased. A thermistor made from an intrinsic semiconductor therefore has a **negative temperature coefficient**.

The resistance of the thermistor decreases non-linearly with increase of temperature, whereas the resistance of the metal wire increases much less over the same temperature range.

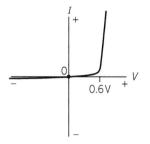

▲ **Figure 3** *Current versus pd for a diode*

Revision tip
When current passes through a forward-biased silicon diode, the pd across the diode is about 0.6 V.

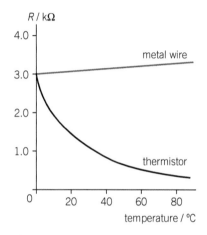

▲ **Figure 4** *Resistance variation with temperature for a thermistor and a metal wire*

Summary questions

1   **a**   Sketch a graph of current I (on the y-axis) against pd V for a filament bulb for positive and negative values of V.   *(1 mark)*
    b   Describe how and explain why the resistance of the filament bulb changes as the current is increased from zero.   *(4 marks)*

2   A light-dependant resistor has a resistance of 100 000 Ω in darkness and 400 Ω in daylight. It is connected in series with an ammeter and a 4.5 V cell. Calculate the ammeter reading when the thermistor is:
    a   **i** in darkness   **ii** in daylight.   *(2 marks)*
    b   Explain why the resistance in darkness is much greater than in daylight.   *(2 marks)*

3   A diode is connected in series with a cell, an ammeter, and a resistor.
    a   Draw a circuit diagram for this circuit, showing the diode in its forward direction.   *(1 mark)*
    b   State and explain how the ammeter reading would change if the resistor was replaced by a resistor with greater resistance.   *(2 marks)*

Chapter 12 Practice questions

1 A 6.0 V battery is connected across a wire-wound resistor. The current in the resistor is 2.6 A.

  a The wire has a diameter of 0.38 mm and a resistivity of $4.8 \times 10^{-7}\,\Omega\,m$. Calculate the length of the wire. *(4 marks)*

  b i Calculate the total charge that flows past a point in the conductor in 10 minutes.

    ii Calculate the energy transferred to the resistor by the electric current in 10 minutes. *(2 marks)*

  c Describe the energy transfers to and from the resistor due to the current. *(3 marks)*

2 Figure 1 shows a graph of V against I for a 12 V filament lamp.

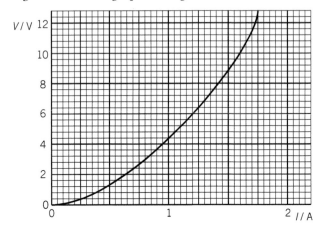

▲ Figure 1

  a Calculate the resistance of the lamp at: **i** 6.0 V **ii** 12.0 V. *(2 marks)*

  b When the pd across the lamp is 12.0 V, calculate:

    i the energy supplied to the lamp in 300 s

    ii the charge that flows through the lamp in 300 s. *(2 marks)*

  c Show that the power dissipated by the lamp at 12.0 V is about 3 times the power dissipated at 6.0 V. *(2 marks)*

3 The following measurements were made in an investigation to measure the resistivity of the material of a certain wire.

Pd across the wire / V	0.00	1.10	2.05	3.90	4.10	4.95
Length of the wire / mm	198	400	603	796	998	1202

Current = 0.41 A

Diameter of wire = 0.26 mm

  a Plot a graph of the pd against the length of the wire. *(3 marks)*

  b i Show that the pd, V, across the wire varies with the length, L, according to the equation

$$V = \frac{\rho I L}{A}$$

    where ρ is the resistivity of the wire, A is its area of cross section, and I is the current in the wire

    ii Use the graph to calculate the resistivity of the material of the wire. *(6 marks)*

  c Discuss how the graph would have differed if a thinner wire of the same material and the same current had been used. *(2 marks)*

4 **a** Describe, with the aid of a circuit diagram, how you would obtain an estimate of the ratio of the resistance of a light-dependent resistor (LDR) in darkness to when it is daylight. The apparatus available includes a battery, a variable resistor, an ammeter, and a voltmeter. *(5 marks)*

b A student uses a light-dependent resistor in series with a 4.5 V battery and an ammeter as a light meter as shown in Figure 2

   i Explain how the ammeter reading changes when the light intensity incident on the LDR increases. *(2 marks)*

   ii When the LDR is in daylight, the ammeter reads 1.2 mA. Calculate the resistance of the LDR in this situation. *(1 mark)*

   iii Explain why the resistance of an LDR is much greater in darkness than in daylight. *(3 marks)*

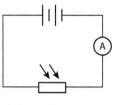

▲ **Figure 2**

5 **a** Explain what is meant by a forward voltage of a diode. *(1 mark)*

b The current in a light-emitting diode (LED) in a circuit is 0.17 A when forward voltage across it is 2.2 V.

   i Calculate the charge that flows through it in 10 ms.

   ii Calculate the power supplied to the LED described above. *(2 marks)*

c The LED in part **b** emits photons of energy 3.5×10^{-19} J.

   i Assuming all the energy supplied to the LED is emitted as light, estimate the maximum number of photons per second that the LED could emit when the current through it is 0.17 A.

   ii Give one reason why the LED emits fewer photons per second than calculated in part **ci**. *(2 marks)*

6 **a** **i** What is meant by the transition temperature of a superconductor? *(1 mark)*

   ii A certain superconducting material M has a transition temperature of 90 K. Sketch a graph to show how the resistivity of M changes when its temperature is increased from below to above 90 K. *(2 marks)*

b Figure 3 shows a sample of M in the form of a wire in series with a resistor in a circuit.

   i When the temperature of M is below its transition temperature, explain why the potential difference across M is zero even though the current through it is non-zero. *(1 mark)*

   ii Explain how the ammeter and voltmeter readings would change if the temperature of M is raised above its transition temperature. *(2 marks)*

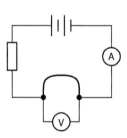

▲ **Figure 3**

7 The heating element of an electric heater consists of a metal wire W coiled round a ceramic tube as shown in Figure 4.

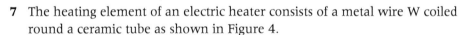

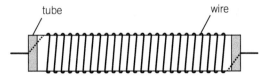

▲ **Figure 4**

a The wire has a diameter of 0.30 mm and the resistivity of the wire metal is 5.6×10^{-8} Ω m. When the pd across the element is 230 V, the current in it is 10.8 A. Calculate the length of W. *(4 marks)*

b A wire X of the same material has a diameter of 0.60 mm.

   i Calculate the length of X that would have the same resistance as W. *(3 marks)*

   ii Discuss whether X would be more suitable than W as the heating element in the electric heater. *(4 marks)*

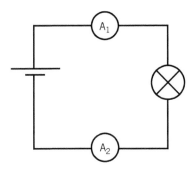

▲ **Figure 1** *Components in series*

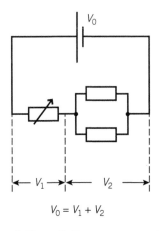

$V_0 = V_1 + V_2$

▲ **Figure 2** *Two components in parallel*

Synoptic link

Sources of emf usually possess some internal resistance. See Topic 13.3, Electromotive force and internal resistance.

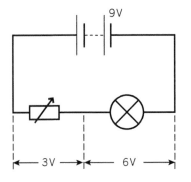

▲ **Figure 3** *The loop rule*

Current rules

1 At any junction in a circuit, the total current leaving the junction is equal to the total current entering the junction.

2 The current entering a component is the same as the current leaving it.

3 For two or more components in series, the current in them is the same.

These rules hold because the rate of charge flowing into a junction or a component is always equal to the rate flowing out.

Potential difference rules

The potential difference (abbreviated as pd), or voltage, between any two points in a circuit is defined as the energy transfer per coulomb of charge that flows from one point to the other.

1 For two or more components in series, the total pd across all the components is equal to the sum of the potential differences across each component.

2 The pd across components in parallel is the same.

3 For any complete loop of a circuit, the sum of the emfs round the loop is equal to the sum of the potential drops around the loop.

The above statements follow from the conservation of energy. For example, in Figure 3, if the pd across the light bulb is 6 V, the pd across the variable resistor is 3 V (= 9 V – 6 V). So every coulomb of charge leaves the battery with 9 J of energy and supplies 3 J to the variable resistor and 6 J to the bulb.

Summary questions

1 A battery which has an emf of 9.0 V and negligible internal resistance is connected in series with a variable resistor and a 6.0 V 24 W light bulb.
   a Sketch the circuit diagram. *(1 mark)*
   b The variable resistor is adjusted so the light bulb is at normal brightness. Calculate: **i** the current through the light bulb **ii** the pd across the variable resistor **iii** the power supplied by the battery. *(3 marks)*

2 A 4.5 V battery of negligible internal resistance is connected in parallel with a 4.5 V 0.5 W torch bulb and a 4.5 V 3.0 W torch bulb.
   a Sketch the circuit diagram for this circuit. *(1 mark)*
   b Calculate the current through each torch bulb. *(2 marks)*
   c Show that the energy supplied by the battery each second equals the energy supplied to the torch bulbs each second. *(3 marks)*

3 A 9.0 V battery of negligible internal resistance is connected in series with an ammeter, a 1.5 kΩ resistor, and an unknown resistor R.
   a Sketch the circuit diagram. *(1 mark)*
   b The ammeter reads 2.0 mA. Calculate:
      i the pd across the 1.5 kΩ resistor
      ii the pd across R **iii** the resistance of R. *(3 marks)*

13.2 More about resistance

Specification reference: 3.5.1.4

Resistors in series

Resistors in series pass the same current. The total pd is equal to the sum of the individual pds. For two or more resistors R_1, R_2, R_3, and so on, in series, the theory can easily be extended to show that the total resistance is equal to the sum of the individual resistances.

$$R = R_1 + R_2 + R_3 + \cdots$$

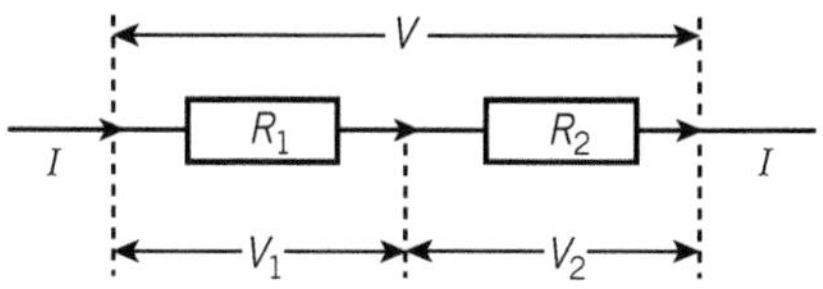

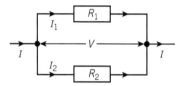

▲ **Figure 1** *Resistors in series*

Resistors in parallel

Resistors in parallel have the same pd. The current through a parallel combination of resistors is equal to the sum of the individual currents.

For two or more resistors R_1, R_2, R_3, and so on, in parallel, the theory can easily be extended to show that the total resistance R is given by

$$\frac{1}{R} = \frac{1}{R_1} + \frac{1}{R_2} + \frac{1}{R_3} + \cdots$$

▲ **Figure 2** *Resistors in parallel*

Resistance heating

The heating effect of an electric current in any component is due to the resistance of the component. The charge carriers repeatedly collide with the positive ions of the conducting material. There is a net transfer of energy from the charge carriers to the positive ions as a result of these collisions.

When a current I passes through a component of resistance R, the pd across it is V.

$$\textbf{the rate of heat transfer} = I^2 R = IV = \frac{V^2}{R}$$

Synoptic link

You learnt about resistance in Topic 12.3, Resistance, and Topic 12.4, Components and their characteristics.

Summary questions

1 A $4.0\,\Omega$ resistor and a $12.0\,\Omega$ resistor are connected in parallel with each other. The parallel combination is connected in series with a $6.0\,V$ battery and a $2.0\,\Omega$ resistor. Assume the battery itself has negligible internal resistance.
   a Sketch the circuit diagram. *(1 mark)*
   b Calculate: **i** the total resistance of the circuit
      ii the battery current. *(3 marks)*
   c Calculate the power supplied to the $4.0\,\Omega$ resistor. *(2 marks)*

2 A $2.0\,\Omega$ resistor and a $10.0\,\Omega$ resistor are connected in series with each other. The series combination is connected in parallel with a $4.0\,\Omega$ resistor and in parallel with a $9.0\,V$ battery of negligible internal resistance.
   a Sketch the circuit diagram. *(2 marks)*
   b Calculate: **i** the total resistance of the circuit
      ii the battery current. *(3 marks)*
   c Calculate the power supplied to each resistor. *(3 marks)*

3 Calculate the resistance of a heating element designed to operate at 3 kW and 230 V. *(2 marks)*

Worked example

The pd across a $1000\,\Omega$ resistor in a circuit was measured at $6.0\,V$. Calculate the electrical power supplied to the resistor.

Solution

Current $I = \dfrac{V}{R} = \dfrac{6.0\,V}{1000\,\Omega} = 6.0\,mA$

Power $P = I^2 R = [6.0 \times 10^{-3}]^2 \times 1000$
$= 0.036\,W$

13.3 Electromotive force and internal resistance

Specification reference: 3.5.1.6

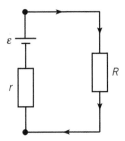

▲ **Figure 1** *Emf and internal resistance*

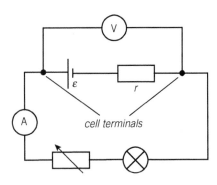

▲ **Figure 2** *Measuring internal resistance. Note that the lamp (or a fixed resistor) limits the maximum current that can pass through the cell*

Internal resistance

- The **electromotive force** (emf, symbol ε) of the source is the electrical energy per unit charge produced by the source. If electrical energy E is given to a charge Q in the source,

$$\varepsilon = \frac{E}{Q}$$

- The *pd across the terminals* of the source is the electrical energy per unit charge delivered by the source when it is in a circuit. The terminal pd is less than the emf whenever current passes through the source. The difference is due to the internal resistance of the source.

The internal resistance of a source is the loss of potential difference per unit current in the source when current passes through the source.

When a cell of emf ε and internal resistance r is connected to an external resistor of resistance R, as shown in Figure 1, the current through the cell, $I = \dfrac{\varepsilon}{R + r}$. Therefore,

$$\varepsilon = IR + Ir$$

Measurement of internal resistance

The circuit shown in Figure 2 is used to measure the terminal pd for different values of current. The current is changed by adjusting the variable resistor. The measurements of terminal pd and current for a given cell may be plotted on a graph, as shown in Figure 3.

The graph is a straight line with a negative gradient. This can be seen by rearranging the equation $\varepsilon = IR + Ir$ to become $IR = \varepsilon - Ir$. Because the terminal pd $V = IR$, then

$$V = \varepsilon - Ir$$

Comparing this with the equation for a straight line, $y = mx + c$, a graph of V on the y-axis against I on the x-axis gives a straight line with

- gradient = $-r$
- y-intercept = ε.

13.4 More circuit calculations

Specification reference: 3.5.1.4; 3.5.1.6

Circuits with a single cell and one or more resistors

1 The **cell current** = $\dfrac{\text{cell emf}}{\text{total circuit resistance}}$

   The total circuit resistance includes internal resistance if that is not negligible.

2 The **pd across each resistor in series with the cell = current × the resistance of each resistor.**

Circuits with two or more cells in series

The current through the cells is calculated by dividing the overall (net) emf by the total resistance.

- The overall emf is the difference between the sum of the emfs in each direction.

- The total internal resistance is the sum of the individual internal resistances.

Circuits with identical cells in parallel

For a circuit with n identical cells in parallel:

- the cells act as a source of emf ε and internal resistance $\dfrac{r}{n}$
- the current through each cell = $\dfrac{I}{n}$, where I is the total current supplied by the cells.

Diodes in circuits

Assume that a silicon diode has:

- a forward pd of 0.6 V whenever a current passes through it
- infinite resistance in the reverse direction or at pds less than 0.6 V in the forward direction.

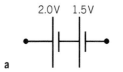

a

b

▲ **Figure 1** *Cells in series*

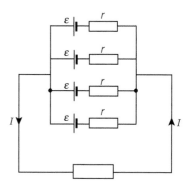

▲ **Figure 2** *Cells in parallel*

Revision tip

In problems on dc circuits that include filament lamps, if you are given data for the pd V and power P of the lamp, its current $I = \dfrac{P}{V}$.

Worked example

A silicon diode is connected in its forward direction in series with a 5.0 V cell of negligible internal resistance and a 15 kΩ resistor, as in Figure 3. Calculate the current through the diode.

Solution

The pd across the diode is 0.6 V because it is forward-biased. Therefore, the pd across the resistor is 4.4 V (= 5.0 V − 0.6 V). The current through the resistor is therefore 2.9×10^{-4} A (= 4.4 V ÷ 15 000 Ω).

Summary questions

1 A battery of emf 12.0 V with an internal resistance of 1.55 Ω is connected in series with a variable resistor and two 6.0 V 6.0 W lamps X and Y in parallel with each other. The variable resistor is adjusted so the lamps are at normal brightness.
   a Sketch the circuit diagram.   *(1 mark)*
   b Calculate:
      i the current through each lamp   *(1 mark)*
      ii the battery current   *(1 mark)*
      iii the pd across the variable resistor.   *(2 marks)*

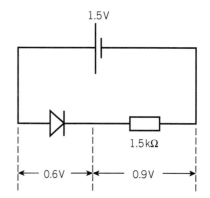

▲ **Figure 3** *Using a diode*

Synoptic link

A solar panel consists of parallel rows of identical solar cells in series. For more about solar panels, see 'Renewable energy' in Chapter 10, Work, energy, and power.

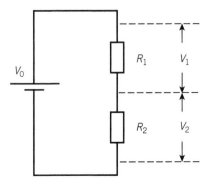

▲ **Figure 1** *A potential divider*

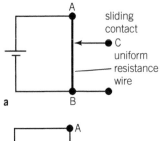

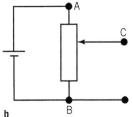

▲ **Figure 2** *Potential dividers used to supply a variable pd* **a** *A linear track using resistance wire* **b** *Circuit symbol*

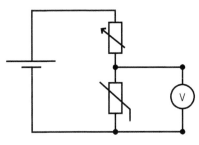

▲ **Figure 3** *A temperature sensor*

The theory of the potential divider

A **potential divider** consists of two or more resistors in series with each other and with a source of fixed potential difference.

To supply a fixed pd

Two resistors R_1 and R_2 in series are connected to a source of fixed pd V_0, as shown in Figure 1.

The ratio of the pds across each resistor is equal to the resistance ratio of the two resistors.

$$\frac{V_1}{V_2} = \frac{R_1}{R_2}$$

To supply a variable pd

The source pd is connected to a fixed length of uniform resistance wire. A sliding contact on the wire moved along the wire gives a variable pd between the contact and one end of the wire.

Sensor circuits

A *sensor circuit* produces an output pd which changes as a result of a change of a physical variable such as temperature or light intensity.

A *temperature sensor* consists of a potential divider made using a thermistor and a variable resistor, as in Figure 3. If the temperature of the thermistor is then raised, its resistance falls, so the pd across it falls.

Summary questions

1. **a** Sketch a circuit diagram to vary the brightness of a 12 V light bulb between zero and a maximum using a variable potential divider and a 12 V battery. *(1 mark)*
   b Explain why the same range of brightness could not be obtained if the light bulb was connected in series with the battery and a variable resistor. *(3 marks)*

2. **a** A potential divider consists of an $8.0\,\Omega$ resistor, R, in series with a $12.0\,\Omega$ resistor and a 6.0 V battery of negligible internal resistance. Calculate:
   i the current *(2 marks)*
   ii the pd across each resistor. *(3 marks)*
   b The $12.0\,\Omega$ resistor is replaced by a thermistor with a resistance of $16.0\,\Omega$ at 20 °C and $6.0\,\Omega$ at 100 °C. Calculate the pd across R at:
   i 20 °C **ii** 100 °C. *(4 marks)*

3. A light sensor consists of a 4.5 V battery of negligible internal resistance, an LDR, and a $2.2\,k\Omega$ resistor in series with each other. A voltmeter is connected in parallel with the resistor. When the LDR is in darkness, the voltmeter reads 1.2 V.
   a Sketch the circuit diagram for this arrangement. *(1 mark)*
   b Calculate: **i** the pd across the LDR **ii** the resistance of the LDR when the voltmeter reads 1.2 V. *(3 marks)*
   c Describe and explain how the voltmeter reading changes when the LDR is exposed to daylight. *(2 marks)*

1   A circuit consists of two resistors of resistance $3.0\,\Omega$ and $6.0\,\Omega$ in parallel with each other and connected in series with an ammeter and a $6.0\,V$ battery which has an internal resistance of $3.0\,\Omega$, see Figure 1.

   a Calculate: **i** the current through the ammeter **ii** the pd across the $3.0\,\Omega$ resistor. *(4 marks)*

   b When one of the resistors is disconnected from the circuit, describe and explain how the pd across the battery changes. *(3 marks)*

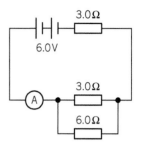

▲ **Figure 1**

2   A $9.0\,V$ battery of negligible internal resistance is connected in series with a silicon diode in its forward direction and a $6.8\,k\Omega$ resistor, see Figure 2.

   a Calculate: **i** the pd across the $6.8\,k\Omega$ resistor **ii** the battery current. *(3 marks)*

   b Another $6.8\,k\Omega$ resistor is connected in parallel with the diode to $6.8\,k\Omega$ resistor. Explain the effect this has on the battery current. *(2 marks)*

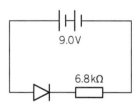

▲ **Figure 2**

3   **a** With the aid of a circuit diagram, describe an experiment to measure the internal resistance of a battery. *(5 marks)*

   b A $6.0\,V$ battery of internal resistance r is connected across a $3.0\,\Omega$ resistor in parallel with a $9.0\,\Omega$ resistor, see Figure 3. The potential difference across the battery is $5.4\,V$.

      i Calculate the current in the battery.

      ii Calculate the internal resistance of the battery.

      iii Calculate the ratio $\dfrac{\text{power dissipated in the internal resistance}}{\text{power supplied by the battery}}$. *(5 marks)*

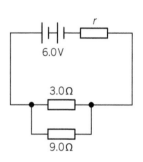

▲ **Figure 3**

4   Figure 4 shows a circuit diagram for a dc power supply connected to a mobile phone battery B in order to charge the battery. The terminals of the same polarity are connected together to achieve this.

   a The power supply has an emf of $5.0\,V$ and internal resistance $0.9\,\Omega$. When charging begins, the battery has an emf of $2.0\,V$ and internal resistance r.

      i Calculate the net emf of the circuit when charging begins.

      ii The initial charging current is $0.50\,A$. Calculate the value of r. *(3 marks)*

   b **i** As the battery charges, its emf increases and the current decreases. Explain why the current decreases.

      ii The battery takes 3.0 hours to charge and can store $1.6\,kC$ of charge. Estimate the average current in the battery when it is recharged. *(3 marks)*

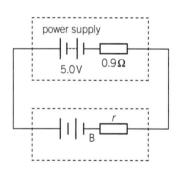

▲ **Figure 4**

5   A solar panel connected to an appliance produces a potential difference of $6.0\,V$ and a current of $3.0\,A$ when it is illuminated by light of constant intensity.

   a Calculate the energy supplied by the solar panel in 1 hour. *(1 mark)*

   b Each solar cell in the panel produces a potential difference of $1.2\,V$ and a current of $1.0\,A$.

      i Calculate how many cells are in the panel and sketch the circuit diagram to show how they are connected together. *(3 marks)*

      ii When the appliance is disconnected from the panel without changing the incident light intensity, the potential difference across its terminals increases to $6.3\,V$. Calculate the effective internal resistance of the panel at this light intensity. *(3 marks)*

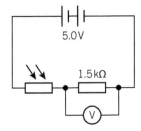

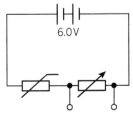

▲ Figure 5

6 Figure 5 shows a potential divider circuit that includes a light-dependent resistor (LDR) in series with a 1.5 kΩ resistor R. The battery has an emf of 5.0 V and has negligible internal resistance.

   a Explain what is meant by a potential divider. *(2 marks)*

   b The potential difference V_0 across the resistor R is 4.2 V when the LDR is exposed to light of a certain intensity. Calculate the LDR resistance at this light intensity. *(3 marks)*

   c State and explain what happens to the output voltage when the light intensity incident on the LDR is reduced. *(2 marks)*

   d Describe how you would use the above circuit with a voltmeter connected across the output terminals to investigate how the light intensity from a lamp varies as the LDR is moved around the lamp. *(6 marks)*

▲ Figure 6

7 The potential divider in Figure 6 is to be used to monitor the temperature of a cold storage room.

   a State and explain how the potential difference across the variable resistor changes when the temperature of the thermistor is increased. *(2 marks)*

   b i The thermistor T has a resistance of 2.7 kΩ at −5 °C. Calculate the resistance of the variable resistor that would give an output voltage of 2.0 V. *(3 marks)*

      ii An alarm is to be connected to the output terminals of the potential divider. With the variable resistor set at the resistance calculated in part **b i**, the alarm switches on if the temperature of the thermistor increases above −5 °C. If the variable resistor is adjusted to a lower resistance, explain how this would affect the temperature at which the alarm switches on. *(4 marks)*

17.1 Uniform circular motion

Specification reference: 3.6.1.1

An object rotating at a steady rate is said to be in **uniform circular motion**. Consider a point on the perimeter of a wheel of radius r rotating at a steady speed.

- The circumference of the wheel, $C = 2\pi r$.
- The frequency of rotation, $f = \frac{1}{T}$, where T is the time for one rotation.
- The angular speed, ω, of the object = its angular displacement per second

$$= \frac{2\pi}{T} = 2\pi f$$

- The speed, v, of a point on the perimeter $= \dfrac{\text{circumference}}{\text{time for one rotation}}$

$$= \frac{2\pi r}{T} = 2\pi r f = \omega r$$

- The angular displacement of a point, $\theta = \omega t$ where t is the time taken.

▲ **Figure 1** *In uniform circular motion*

Worked example

The wheels of a skateboard each have a diameter of 60 mm. When the skateboarder travels at a speed of 8.5 m s^{-1}, calculate:

a the frequency of rotation of each wheel

b the angular speed of each wheel

c the angle in radians that the wheel turns through in a distance of 20 m.

a Circumference of wheel $= 2\pi r = \pi \times \text{diameter} = \pi \times 0.060\,\text{m} = 0.188\,\text{m}$

Time for one wheel rotation, $T = \dfrac{\text{circumference}}{\text{speed}} = \dfrac{0.188\,\text{m}}{8.5\,\text{ms}^{-1}} = 0.0222\,\text{s}$

Frequency $f = \dfrac{1}{T} = \dfrac{1}{0.0222\,\text{s}} = 45\,\text{Hz}$ (2 s.f.)

b Angular speed $\omega = \dfrac{2\pi}{T} = 290\,\text{rad s}^{-1}$ (2 s.f.)

c Angle the wheel turns through in 20 m,

$\theta = \dfrac{\text{distance}}{\text{circumference}} = \dfrac{20\,\text{m}}{0.188\,\text{m}} = 106\,\text{rad} = 110\,\text{rad}$ (2 s.f.)

Summary questions

1 The wheels of a car each have a diameter of 0.60 m. When the car travels at a speed of 25 m s^{-1}, calculate:

   a the number of turns made by the wheel in a distance of 100 m *(1 mark)*

   b the frequency of rotation of each wheel *(1 mark)*

   c the angular speed of each wheel. *(1 mark)*

2 The International Space Station (ISS) orbits the Earth once every 93 minutes at a height of about 400 km.

   a Calculate the frequency of rotation of ISS. *(1 mark)*

   b The Earth has a radius of about 6400 km. Estimate the speed of ISS. *(1 mark)*

3 In a gear system, a gear wheel of radius 8.2 mm drives a smaller gear wheel of radius 2.5 mm.

   When the larger gear wheel turns at a frequency of 0.95 Hz, calculate:

   a the angular speed of the larger gear wheel *(1 mark)*

   b the frequency of rotation of the smaller gear wheel *(3 marks)*

   c i the angular displacement in degrees of the smaller wheel in 0.020 s *(1 mark)*

    ii the distance moved by the gear tooth of the smaller gear wheel in this time. *(1 mark)*

▲ **Figure 2** *Image of a large gear wheel and a small gear wheel in contact with each other*

17.2 Centripetal acceleration

Specification reference: 3.6.1.1

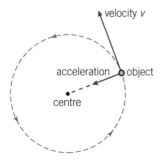

▲ **Figure 1** *Centripetal acceleration*

Revision tip

Remember the angular speed
$\omega = v/r$

Synoptic link

In Topic 24.3, Charged particles in circular orbits, you will meet the use of a magnetic field to bend a beam of charged particles (e.g., electrons) in a circular path. The magnetic force on the moving charged particles is the centripetal force.

Revision tip

The centripetal force is at right angles to the direction of the object's velocity. Therefore, no work is done by the centripetal force on an object in uniform circular motion because there is no displacement in the direction of the force. The kinetic energy of the object is therefore constant, so its speed is unchanged.

The direction of the velocity of an object in uniform circular motion changes continually as the object moves along its circular path, see Figure 1. This change in the direction of the velocity is towards the centre of the circle. So, the acceleration of the object is towards the centre of the circle and is called **centripetal acceleration**.

For an object moving at constant speed v in a circle of radius r,

$$\text{its centripetal acceleration, } a = \frac{v^2}{r} = \omega^2 r$$

where ω is the angular speed of the object.

Centripetal force

Any object that moves in uniform circular motion is acted on by a resultant force that always acts towards the centre of the circle. The resultant force causes its centripetal acceleration so is referred to as the **centripetal force** on the object.

For an object of mass m moving at constant speed v along a circular path of radius r, applying Newton's second law for constant mass in the form $F = ma$, gives

$$\text{centripetal force } F = \frac{mv^2}{r} = m\omega^2 r$$

Summary questions

1 A child of mass 28 kg stands on a playground roundabout at a distance of 1.1 m from its centre when it is turning at a steady rate once every 2.3 s. Calculate:

   a the angular speed of the roundabout *(1 mark)*

   b the child's speed *(1 mark)*

   c the centripetal force on the child. *(1 mark)*

2 Figure 2 shows a luggage conveyor belt at an airport. The radius of curvature of the outer edge of the belt at each end is 2.8 m. A suitcase of mass 22 kg is placed near the outer edge of the conveyor belt when the belt is moving at a speed of 0.78 m s^{-1}. Estimate the centripetal force on the suitcase when it moves round the end of the conveyor belt. *(1 mark)*

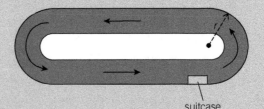

suitcase

▲ **Figure 2** *Top view*

3 A rotating doorway has a circular floor that turns at a steady rate of once every 12 seconds.

   a Calculate the speed of a person standing at a distance of 2.5 m from the centre of the rotating floor. *(1 mark)*

   b Calculate the centripetal force on a person of mass 61 kg at this position. *(1 mark)*

17.3 On the road
17.4 At the fairground

Specification reference: 3.6.1.1

When an object is in uniform circular motion, the forces acting on it give a resultant force which acts as the centripetal force. At A Level, many of the problems you will meet on uniform circular motion involve an object acted on by no more than three forces. The direction of one or more of the forces may change as the object moves round the circle.

Applying Newton's laws of motion

For a given position of an object on its circular path:

1 Identify the forces acting on the object and draw a 'free body diagram' of the object showing the forces acting on it.

2 Resolve the forces into components parallel and perpendicular to the line from the centre of the circle to the object.

3 Equate the resultant of the parallel components towards the centre of the circle to $\frac{mv^2}{r}$.

Using force diagrams

1 A vehicle of mass m moves at speed v on a road that passes over the top of a hill or over the top of a curved bridge.

At the top, the resultant force is directed towards the centre of the circle and is equal to the difference between the weight mg and the support force S of the road on the vehicle. Therefore

$$mg - S = \frac{mv^2}{r}$$

where r is the radius of curvature of the hill.

2 A fairground 'Big Dipper' takes its passengers at high speed through a big dip and 'pushes' them into their seats as they pass through the dip. At the bottom of the dip, the difference between the support force S on a passenger (acting upwards) and their weight mg acts as the centripetal force.

Therefore, for a speed v at the bottom of a dip of radius of curvature r,

$$S - mg = \frac{mv^2}{r}$$

3 A small object O of mass m on the end of a string is swung round in a horizontal circle of radius r at constant speed v. The object is acted on by its weight mg and the tension T in the string (see Figure 3).

The string must be at a non-zero angle θ to the vertical so that the vertical component of T is equal and opposite to the weight. The components of tension parallel and perpendicular to the horizontal are $T\sin\theta$ and $T\cos\theta$ respectively. Therefore:

• the vertical component of tension $T\cos\theta = mg$
• the horizontal component of tension $T\sin\theta = \frac{mv^2}{r}$

Therefore, $\tan\theta = \dfrac{\sin\theta}{\cos\theta} = \dfrac{\left(\frac{mv^2}{r}\right)}{mg} = \dfrac{v^2}{gr}$

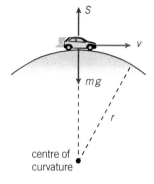

▲ **Figure 1** *Over the top*

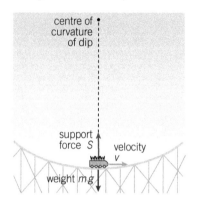

▲ **Figure 2** *In a dip*

> **Revision tip**
> In general, where there are just two forces acting on an object in uniform circular motion, work out the magnitude and direction of the resultant force and equate the magnitude to $\frac{mv^2}{r}$.

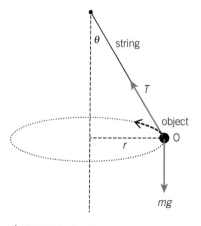

▲ **Figure 3** *The forces acting on an object*

Synoptic link

The horizontal and vertical components of a force F that is at an angle θ to the vertical must be $F \sin \theta$ and $F \cos \theta$, respectively. See Topic 6.1, Vectors and scalars.

Summary questions

$g = 9.81 \, \text{m s}^{-2}$

1 A vehicle of mass 1200 kg passes through a dip in the road at a speed of 22 m s^{-1}. The radius of curvature of the bottom of the dip is 82 m. Calculate:

 a the centripetal acceleration of the vehicle at the bottom of the dip *(1 mark)*

 b the magnitude and direction of the support force on the vehicle at the bottom of the dip. *(2 marks)*

2 Figure 4 shows a fairground wheel that rotates at a constant speed in a vertical plane with passengers strapped to the inside of the rim. Each passenger experiences a downward support force when they reach the highest position in the wheel. The wheel takes 7.0 s for each full rotation and it has a radius of 16 m. Calculate the support force on a passenger of mass 48 kg at the highest position. *(4 marks)*

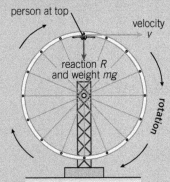

▲ **Figure 4**

3 A child of mass 23 kg on a swing of length 3.3 m passes through the lowest position of the swing at a speed of 2.8 m s^{-1}. Calculate the support force on the child from the swing when she passes through the lowest position. *(3 marks)*

Chapter 17 Practice questions

1   The planet Jupiter orbits the Sun once every twelve years on a circular orbit that is about five times larger than the Earth's orbit.

Which one of the estimates **A–D** gives the following ratio?

$\dfrac{\text{the speed of Jupiter on its orbit}}{\text{the speed of the Earth on its orbit}}$

  A 0.4   **B** 0.8   **C** 2.4   **D** 5   *(1 mark)*

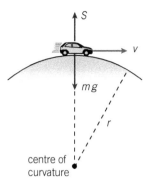

▲ Figure 1

2   An object O of mass m moves in uniform circular motion at angular speed ω on a circular path of radius r.

Which one of the expressions **A–D** gives the following ratio?

$\dfrac{\text{the kinetic energy of the object}}{\text{its centripetal acceleration}}$

  A $\dfrac{m}{2\omega r}$ **B** $\dfrac{m}{2r}$ **C** $\dfrac{mr}{2}$ **D** $\dfrac{m\omega}{2r}$ *(1 mark)*

3   A vehicle of mass m travels at constant speed v over a bridge that has a radius of curvature r.

Which one of the following expressions **A–D** gives the magnitude of the support force on the vehicle when it is at the top of the bridge?

  A $mg - \dfrac{mv^2}{r}$ **B** $mg + \dfrac{mv^2}{r}$ **C** mg **D** $\dfrac{mv^2}{r} - mg$ *(1 mark)*

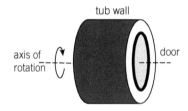

▲ Figure 2

4   A spin drier tub of diameter 0.45 m rotates about a horizontal axis at 1400 revolutions per minute (see Figure 2).

  a   Calculate:

    i   the angular speed of the tub   *(1 mark)*

    ii   the speed of a point on the wall of the tub   *(1 mark)*

    iii the centripetal acceleration of a point on the wall.   *(1 mark)*

  b   Wet clothing in the tub is forced against the wall of the tub when it spins. The tub is designed with many small holes in the tube wall. Explain why water from the wet clothing leaves the tub through these holes when the tub spins.   *(2 marks)*

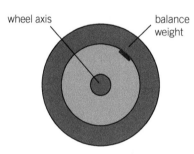

▲ Figure 3

5   Vehicle wheels are fitted with balance weights to prevent the wheels wobbling at high speed. A balance weight on a wheel is positioned to make the wheel perfectly balanced. A wheel of diameter 0.60 m is fitted with a 25 g balance weight at a distance of 0.21 m from the wheel axis as shown in Figure 3.

  a   Calculate the frequency of rotation of the wheel when the wheel is on a vehicle travelling at a speed of 25 m s⁻¹.   *(1 mark)*

  b   **i**   Calculate the centripetal force F on the balance weight when the wheel is travelling at this speed.   *(2 marks)*

    ii Draw a graph to show how the centripetal force F varies with the speed v of the vehicle.   *(3 marks)*

6   A cyclist travels at a speed of 8.2 m s⁻¹ on a bicycle that has wheels of diameter 50 cm.

Calculate:

  a   the frequency of rotation of each wheel   *(1 mark)*

  b   the centripetal acceleration of a point on the tyre.   *(1 mark)*

7 A microwave oven has a turntable that rotates at a constant angular speed when the oven is switched on.

   a Describe how you would measure the speed of a point on the edge of the turntable. *(2 marks)*

   b The turntable of a microwave oven has a diameter of 270 mm and it turns at an angular speed of 0.86 rad s⁻¹.

     i Calculate the number of turns made by the turntable in 180 s. *(2 marks)*

     ii Calculate the centripetal acceleration of a small piece of food at the edge of the turntable. *(1 mark)*

8 On a fairground train ride, a train at the highest point of the track travels at a speed of 1.2 m s⁻¹, then descends on the track, and then travels through a dip and onto a level section of track.

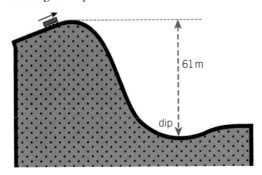

▲ **Figure 4**

   a The train descends through a vertical height of 61 m before it reaches the bottom of the dip. The total mass of the train and its passengers is 2200 kg.

     i Calculate the loss of potential energy of the train and its passengers in the descent. *(1 mark)*

     ii Calculate the maximum possible speed of the train at the bottom of the descent. *(2 marks)*

   b The dip has a radius of curvature of 52 m. Calculate the force on the train due to the track when the train is at its lowest point and travelling at its maximum possible speed. *(2 marks)*

18.1 Oscillations
18.2 The principles of simple harmonic motion

Specification reference: 3.6.1.2

The **time period**, T, of the oscillating motion is the time for one complete cycle of oscillation. One full cycle after passing through any position, the object passes through that same position in the same direction.

The **angular frequency** ω of the oscillating motion is defined as $\frac{2\pi}{T}(= 2\pi f)$.

The unit of ω is the radian per second (rad s^{-1}).

Simple harmonic motion is defined as oscillating motion in which the acceleration ***a*** is:

1 proportional to the displacement ***x***

2 always in the opposite direction to the displacement ***x***.

The defining equation for simple harmonic motion is

$$\textbf{acceleration, } a = -\omega^2 x$$

Phase difference

In general, for two objects oscillating at the same frequency,

$$\textbf{their phase difference in radians} = \frac{2\pi\Delta t}{T}$$

where Δt is the time between successive instants when the two objects are at maximum displacement in the same direction.

For any object undergoing simple harmonic motion, Figure 1 shows how its displacement, x, velocity, v, and acceleration, a, vary with time. All three graphs have the same sinusoidal (sine wave) shape but the velocity is $\frac{1}{4}$ of a cycle ahead of the displacement and the acceleration is $\frac{1}{4}$ of a cycle ahead of the velocity. You can see this from the position of the three peaks at $\frac{T}{2}$, $\frac{3T}{2}$, and T.

▲ **Figure 1** *Displacement (x), velocity, (v) and acceleration (a) against time graphs*

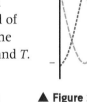

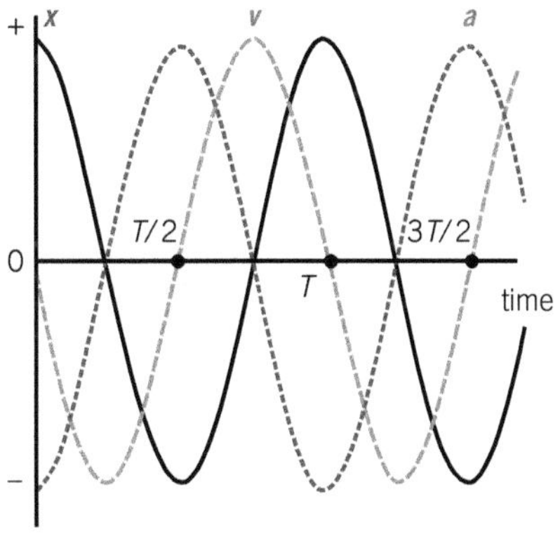

The variation of velocity with time is given by the gradient of the displacement–time graph.

The variation of acceleration with time is given by the gradient of the velocity–time graph.

Summary questions

1 An object suspended from the lower end of a vertical spring is displaced downwards from equilibrium and the time taken for 20 oscillations was measured three times as 15.2 s, 15.6 s, and 15.5 s. Calculate:

a **i** its time period *(1 mark)*

ii its frequency of oscillation *(1 mark)*

b the percentage uncertainty in its frequency. *(3 marks)*

2 For the oscillations in **Q1**, calculate the acceleration of the object when its displacement is:

a +25 mm *(2 marks)* **b** 0 *(1 mark)* **c** −25 mm. *(1 mark)*

Key terms

The **displacement** of the object (i.e., distance and direction) from equilibrium continually changes during the motion.

The **amplitude** of the oscillations is the maximum displacement of the oscillating object from equilibrium.

The **frequency**, f, of oscillations is the number of cycles per second made by an oscillating object.

The unit of frequency is the hertz (Hz), which is one cycle per second.

For oscillations of frequency f, the time period $T = \frac{1}{f}$.

Revision tip

Note that 2π radians = 360°.
So the phase difference in degrees is $360 \times \frac{\Delta t}{T}$.

Revision tip

1 The time period is independent of the amplitude of the oscillations.

2 Maximum displacement $x_{max} = \pm A$, where A is the amplitude of the oscillations. Therefore, when $x_{max} = +A$, the acceleration $a = -\omega^2 A$; $x_{max} = -A$, the acceleration $a = +\omega^2 A$

3 The acceleration equation may also be written as $a = -(2\pi f)^2 x$, where frequency $f = \frac{1}{T}$

18.3 More about sine waves

Specification reference: 3.6.1.2

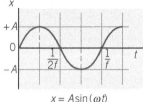

$x = A\sin(\omega t)$

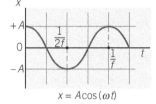

$x = A\cos(\omega t)$

▲ **Figure 1** *Graphical solutions*

Revision tip

The time period T does *not* depend on the amplitude of the oscillating motion. For example, the time period of an object oscillating on a spring is the same, regardless of whether the amplitude is large or small.

Revision tip

Remember that ωt must be in radians. So make sure your calculator is in radian mode when you use the equation $x = A\cos(\omega t)$

In this example the time period is 2.0 s so after 0.6 s, the object must have gone through 0.3 of a cycle. So it would be between $\frac{1}{4}$ and $\frac{1}{2}$ of a cycle at 0.6 s. Look at Figure 1 and you will see its displacement is therefore on the opposite side of the equilibrium position to its initial displacement which is why the answer is negative.

For any object oscillating at frequency f in simple harmonic motion, its acceleration a at displacement x is given by

$$a = -\omega^2 x$$

where $\omega = 2\pi f$

The variation of displacement with time depends on the initial displacement and the initial velocity (i.e., the displacement and velocity at time $t = 0$).

If $x = +A$ when $t = 0$ and the object has zero velocity at that instant, then its displacement at a later time t is given by

$$x = A\cos(\omega t)$$

If $x = 0$ when $t = 0$ and the object has zero velocity at that instant, then its displacement at a later time t is given by

$$x = A\sin(\omega t)$$

Question and model answer

A pendulum bob oscillates in simple harmonic motion after being displaced from rest then released. Its initial displacement was 32 mm and its frequency of oscillation is 0.50 Hz. Calculate:

a the angular frequency of the oscillations b the displacement of the object after 1.60 s

a Angular frequency $\omega = 2\pi f = 3.14$ rad s^{-1}

b Its displacement at time t after release, $x = A\cos(\omega t)$ where $A = 32$ mm

Therefore at $t = 1.60$ s, $x = A\cos(\omega t) = 32$ mm $\times \cos(3.14$ rad s$^{-1} \times 1.60$ s$)$

$= 32$ mm $\times \cos(1.88$ rad$) = -9.7$ mm

Summary questions

1 An object on a spring oscillates along a vertical line with a time period of 2.3 s and a maximum acceleration of 7.2 m s^{-2}. Calculate:
   a its angular frequency   *(1 mark)*
   b its amplitude.   *(2 marks)*

2 The displacement x from equilibrium of the object in **Q1** is given by $x = A\sin(\omega t)$ where t is the time in seconds from when the object was moving vertically upwards through its equilibrium position.
   a Sketch a graph to show how displacement x varies with t for 1 complete cycle.   *(2 marks)*
   b Calculate the displacement at $t = 2.0$ s.   *(1 mark)*

3 The bob of a simple pendulum was displaced from rest by a distance of 55 mm and then released. The pendulum oscillated in simple harmonic motion with a time period of 0.27 s.
   a Calculate the magnitude of the acceleration of the object half a cycle after it was released.   *(4 marks)*
   b Calculate its displacement and its direction of motion 0.20 s after it was released.   *(1 mark)*

18.4 Applications of simple harmonic motion

Specification reference: 3.6.1.3

For any oscillating object, the resultant force acting on the object acts towards the equilibrium position. The resultant force is described as a *restoring force* because it always acts towards equilibrium. As long as the restoring force is proportional to the displacement from equilibrium, the acceleration is proportional to the displacement and always acts towards equilibrium. Therefore, the object oscillates with simple harmonic motion.

The oscillations of a mass–spring system

For a small object of mass m attached to a spring which obeys Hooke's Law:

- the time period T of its oscillations is given by $T = 2\pi\sqrt{\frac{m}{k}}$ where k is the spring constant

- its acceleration a at displacement x is given by $a = -\omega^2 x$, where $\omega^2 = \frac{k}{m}$.

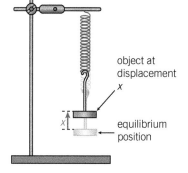

▲ **Figure 1** *The oscillations of a loaded spring*

Worked example

$g = 9.8\,\mathrm{m\,s^{-2}}$

A spring of natural length 300 mm hangs vertically with its upper end attached to a fixed point. When a small object of mass 0.20 kg is suspended from the lower end of the spring in equilibrium, the spring is stretched to a length of 379 mm. Calculate:

a **i** the extension of the spring at equilibrium
  ii the spring constant.
b the time period of oscillations that the mass on the spring would have if the mass was to be displaced downwards slightly then released.

a **i** Extension of spring at equilibrium is $\Delta L_0 = 79\,\mathrm{mm} = 0.079\,\mathrm{m}$

  ii Spring constant $k = \dfrac{mg}{\Delta L_0} = \dfrac{0.20 \times 9.8}{0.079} = 25\,\mathrm{N\,m^{-1}}$

b $T = 2\pi\sqrt{\dfrac{0.20}{25}} = 0.56\,\mathrm{s}$

The theory of the simple pendulum

For a simple pendulum that consists of a bob of mass m attached to a thread of length L, as shown in Figure 2, provided angle θ between the thread and the vertical does not exceed approximately 10°:

- the time period T of its oscillations is given by $T = 2\pi\sqrt{\frac{L}{g}}$

- its acceleration a at displacement s is given by $a = -\omega^2 s$, where $\omega^2 = \frac{g}{L}$.

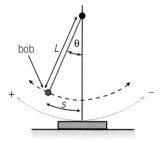

▲ **Figure 2** *The simple pendulum*

Data analysis

Squaring both sides of the equation $T = 2\pi\sqrt{\frac{m}{k}}$ gives $T^2 = \frac{4\pi^2 m}{k}$ so plotting a graph of T^2 on the y-axis against m on the x-axis should give a straight line through the origin with a gradient of $\frac{4\pi^2}{k}$ in accordance with the general equation for a straight-line graph $y = mx + c$.

You get a similar graph if you square both sides of $T = 2\pi\sqrt{\frac{L}{g}}$ except the gradient of the graph of T^2 against L gives a straight line through the origin with a gradient of $\frac{4\pi^2}{g}$.

Revision tip

1 The time period of the oscillations $T = \dfrac{1}{f} = 2\pi\sqrt{\dfrac{m}{k}}$
 The time period does *not* depend on g. A mass–spring system on the Moon would have the same time period as it would on Earth.

2 The tension in the spring varies from $mg + kA$ (at maximum displacement downwards) to $mg - kA$ (*at* maximum displacement upwards) where A = amplitude.

Revision tip

1 Note that the time period is independent of the mass of the pendulum bob.

2 Rearranging the time period formula gives frequency $f = \dfrac{1}{2\pi}\sqrt{\dfrac{g}{L}}$

Summary questions

$g = 9.81 \text{ m s}^{-2}$

1 A spring of natural length 300 mm hangs vertically with its upper end attached to a fixed point.
 When a small object of mass 0.20 kg is suspended from the lower end of the spring in equilibrium, the spring is stretched to a length of 379 mm. Calculate:
 a i the extension of the spring at equilibrium (1 mark)
  ii the spring constant (1 mark)
 b the time period of its oscillations if the object was displaced downwards then released. (1 mark)

2 In the arrangement described in **Q1**, the object was replaced by an object of different mass. When the second object was oscillating vertically, its acceleration a at displacement x was given by $a = -250x$.
 a Calculate the time period of the oscillations. (2 marks)
 b Calculate the mass of the second object. (3 marks)

3 a Calculate the time period of a simple pendulum of length 250 mm. (1 mark)

 b The length was measured to an accuracy of + 5 mm. Estimate the percentage uncertainty in the calculated value of the time period. (2 marks)

18.5 Energy and simple harmonic motion

Specification reference: 3.6.1.3

Free oscillations

A freely oscillating object oscillates with a constant amplitude because there is no friction acting on it. The only forces acting on it combine to provide the restoring force. If friction was present, the amplitude of oscillations would gradually decrease, and the oscillations would eventually cease.

The energy of a freely oscillating system changes from kinetic energy to potential energy and back again every half-cycle after passing though equilibrium. As long as friction is absent, the total energy of the system is constant and is equal to its maximum potential energy.

For a system consisting of an object of mass m oscillating on a spring:

- the potential energy E_P at displacement x from equilibrium $= \frac{1}{2}kx^2$, where k is the spring constant of the spring
- the total energy E_T of the system $= \frac{1}{2}kA^2$, where A is the amplitude of the oscillations
- the kinetic energy of the oscillating mass, $E_K = E_T - E_P = \frac{1}{2}k(A^2 - x^2)$.

Energy–displacement graphs

1 The potential energy curve is parabolic in shape, given by $E_P = \frac{1}{2}kx^2$.
2 The kinetic energy curve is an inverted parabola, given by $E_K = E_T - E_P = \frac{1}{2}k(A^2 - x^2)$.
3 The total energy $= \frac{1}{2}kA^2$, which is the potential energy at maximum displacement. The potential energy and kinetic energy curves add together to give a horizontal line for the total energy due to the oscillations.

The simple harmonic motion speed equation

Using $E_K = \frac{1}{2}mv^2$ gives $\frac{1}{2}mv^2 = \frac{1}{2}k(A^2 - x^2)$, where v is the speed of the object at displacement x.

Because $\omega^2 = \frac{k}{m}$, the above equation can be written as $v^2 = \omega^2(A^2 - x^2)$ which gives the speed equation:

$$v = \pm \omega \sqrt{A^2 - x^2}$$

Damped oscillations

In any oscillating system where friction or air resistance is present, the amplitude decreases. The forces dissipate the energy of the system to the surroundings as thermal energy. The motion is said to be **damped** if dissipative forces are present.

Light damping occurs when the time period is independent of the amplitude so each cycle takes the same length of time as the oscillations die away. See Figure 2.

Critical damping is just enough to stop the system oscillating after it has been displaced from equilibrium and released. The oscillating system returns to equilibrium in the shortest possible time without overshooting if the damping is critical.

Heavy damping occurs when the damping is so strong that the displaced object returns to equilibrium much more slowly than if the system is critically damped. No oscillating motion occurs.

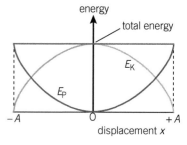

▲ **Figure 1** *Energy variation with displacement*

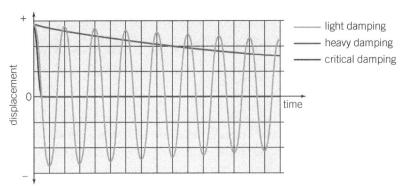

▲ **Figure 2** *Damping*

Summary questions

$g = 9.81\,\mathrm{m\,s^{-2}}$

1 A small object of mass 0.150 kg is attached to the lower end of a spring which hangs vertically from a fixed point. The object is displaced vertically downwards from its equilibrium position by a distance of 30 mm and then released. It takes 22.0 s to make 10 complete cycles of oscillation. Calculate:
   a the angular frequency of the oscillations   (*1 mark*)
   b the spring constant of the spring   (*2 marks*)
   c the maximum speed of the object.   (*1 mark*)

2 a Calculate the total energy of the oscillating object in **Q1**. (*1 mark*)
   b Sketch a graph to show how the kinetic energy and the potential energy of the oscillating object in **Q1** varies with displacement.
                                                                    (*4 marks*)

3 A simple pendulum consists of a small metal sphere of mass 0.30 kg attached to a thread. The sphere is displaced from its equilibrium position through a height of 10.0 mm with the thread taut then released. It takes 15.0 s to make 10 complete cycles of oscillation.
   a Calculate the speed of the pendulum bob as it moves through its equilibrium position.   (*2 marks*)
   b i Calculate the amplitude *A* of the oscillations.
      ii Calculate its speed when its displacement is 0.5*A*.   (*3 marks*)

18.6 Forced vibrations and resonance

Specification reference: 3.6.1.4

Forced vibrations

The response of an oscillating system to a periodic force depends on the frequency of the periodic force (the applied frequency), see Figure 1.

If the applied frequency is increased gradually from zero:

- **the amplitude of oscillations** of the system increases until it reaches a maximum amplitude at a particular frequency, and then the amplitude decreases again
- **the phase difference between the displacement and the periodic force** increases from zero to $\frac{1}{2}\pi$ at the maximum amplitude, then from $\frac{1}{2}\pi$ to π as the frequency increases further.

Resonance

At the maximum amplitude, the phase difference between the displacement and the periodic force is $\frac{1}{2}\pi$. The periodic force is then exactly in phase with the velocity of the oscillating system, and the system is in **resonance**. The frequency at the maximum amplitude is called the **resonant frequency**.

The lighter the damping:

- the larger the maximum amplitude is at resonance, and
- the closer the resonant frequency is to the natural frequency of the system.

As the applied frequency becomes increasingly larger than the resonant frequency of the oscillating system the amplitude of oscillations decreases more and more.

For an oscillating system with little or no damping, at resonance,

the applied frequency of the periodic force = the natural frequency of the system.

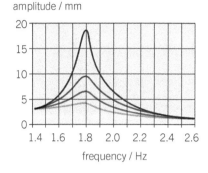

▲ **Figure 1** *Resonance curves*

amplitude / mm

frequency / Hz

Summary questions

1 A small object of mass m_0 suspended on a vertical spring has a natural frequency of oscillation f_0. The system is subjected to a periodic force of variable frequency f.
   a Explain why the amplitude of the oscillations is very large when $f = f_0$.
   (2 marks)
   b When $f = f_0$, the object oscillates with constant energy. Explain why the energy of the object is constant even though a force acts on it.
   (2 marks)

2 With reference to the mass–spring system described in **Q1**, state and explain what the effect would be of replacing:
   a the object by a small object of mass $4m_0$ *(2 marks)*
   b replacing the spring with a stiffer spring without changing the object. *(2 marks)*

3 A small object of mass m suspended on a vertical spring is made to resonate by applying a periodic force of frequency 1.8 Hz to it. The spring constant of the spring is $25\,\text{N}\,\text{m}^{-1}$.
   a Calculate the mass, m, of the object. *(3 marks)*
   b An identical spring is fitted to the system in parallel with the first spring. Determine the frequency at which the system with the same mass would resonate. *(3 marks)*

1 A spring fixed at its upper end hangs vertically and supports a small object of mass m attached to its lower end. The spring constant of the spring is k.

Which of the following expressions **A–D** gives the maximum speed of the object when it oscillates vertically with amplitude A? *(1 mark)*

A $\dfrac{Ak}{m}$ **B** $\dfrac{A\sqrt{k}}{m}$ **C** $\dfrac{Ak}{\sqrt{m}}$ **D** $\dfrac{A\sqrt{k}}{\sqrt{m}}$

2 Which **one** of the following statements **A–D** about damped oscillations is a correct statement? *(1 mark)*

A The time period of a lightly damped system increases as the amplitude decreases.

B A critically damped system never returns to equilibrium after it has been displaced from equilibrium and released.

C A heavily damped system never oscillates.

D A damped system subjected to a periodic force can never be made to resonate.

3 Table 1 represents the energy changes of a freely oscillating system during one cycle. Which one of the statements **A–D** is correct? *(1 mark)*

4 A simple pendulum consists of a small weight on the end of a thread. The weight is displaced from equilibrium and released. It oscillates with an amplitude of 22 mm, taking 35 s to execute 20 oscillations.

Calculate the magnitude of its maximum acceleration. *(3 marks)*

5 The displacement of an object oscillating in simple harmonic motion varies with time according to the equation $x\,(\text{mm}) = 40\cos 5t$, where t is the time in seconds after the object's displacement was at its maximum positive value.

a Determine:

  i the amplitude *(1 mark)*

  ii the time period. *(1 mark)*

b Calculate the displacement of the object at $t = 0.90$ s. *(1 mark)*

6 The upper end of a vertical spring of natural length 250 mm is attached to a fixed point. When a small object of mass 0.15 kg is attached to the lower end of the spring, the spring stretches to an equilibrium length of 320 mm. The object is then displaced vertically from its equilibrium position and released. Calculate the time period of its oscillations. *(3 marks)*

7 a Explain with the aid of a suitable graph what is meant by lightly damped oscillations. *(2 marks)*

b The amplitude of an oscillating pendulum decreases to 45% of its initial amplitude over five cycles. Estimate its percentage loss of energy per cycle. *(3 marks)*

8 A two-wheeled car trailer has a carrier box of mass 150 kg mounted on two identical springs, one near each wheel (see Figure 1).

a When the box carries a load of 60 kg, it oscillates in resonance when the trailer travels at a speed of 3.0 m s⁻¹ over speed bumps spaced 12 m apart.

  Calculate the spring constant of each spring. *(5 marks)*

b Discuss whether or not this effect will happen when the trailer goes over the same speed bumps after it has been unloaded. *(3 marks)*

▼ **Table 1**

	x	E_{K}	E_{P}
A	$0 \rightarrow +A$	increases	increases
B	$+A \rightarrow 0$	increases	decreases
C	$0 \rightarrow -A$	decreases	decreases
D	$-A \rightarrow 0$	decreases	increases

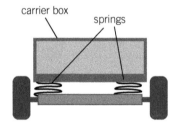

▲ **Figure 1** *Rear view of the trailer*

carrier box
springs

19.1 Internal energy and temperature

Specification references: 3.6.2.1; 3.6.2.2

The internal energy of an object is the energy of its atoms and molecules due to their individual movements and positions. The internal energy of an object due to its temperature is sometimes called **thermal energy**. An object's internal energy can change as a result of:

- energy transfer by heating between the object and its surroundings
- work done on the object by a force (including an electrical force).

The first law of thermodynamics

In general, when work is done on or by an object and/or energy is transferred by heating,

> **the change of internal energy of the object =**
> **the total energy transfer due to work done and heating**

About molecules

In a solid, electrical forces between the atoms and molecules hold them together in an arrangement in which they vibrate randomly about fixed positions.

- The energy supplied to raise the temperature of a solid increases the kinetic energy of the molecules.
- The energy supplied to melt a solid raises the potential energy of the molecules because they break free from each other.

In a liquid, the molecules move about at random in contact with each other. The forces between the molecules are not strong enough to hold the molecules in fixed positions.

- The energy supplied to a liquid to raise its temperature increases the average kinetic energy of the liquid molecules.
- Continued supply of energy gives the molecules enough kinetic energy to break free and move away from each other. The potential energy of the molecules increases because they use the kinetic energy they gained to do work and break away from each other.

In a gas or vapour, the molecules also move about randomly but much further apart on average than in a liquid. Heating a gas or a vapour makes the molecules speed up and so gain kinetic energy.

> **The internal energy of an object is the sum of the random**
> **distribution of the kinetic and potential energies of its molecules.**

Temperature and temperature scales

The **Celsius scale** of temperature, in units of °C, is defined in terms of ice point, 0 °C, and steam point, 100 °C.

The **absolute scale** of temperature, in units of kelvins (K), starts at **absolute zero**, 0 K, which is the lowest possible temperature.

> **temperature in °C = absolute temperature in kelvins − 273.15**

Revision tip

When the 1st Law of Thermodynamics is applied to an object, the directions of the energy transfers (i.e., *to* or *from* the object) are very important because the directions determine whether the overall internal energy of the object increases or decreases.

Summary questions

1   a   Explain what is meant by the internal energy of an object. (*1 mark*)

    b   When a certain gas is compressed, 400 J of work is done on it and 300 J of energy is transferred by heating from it. What change of its internal energy occurs? (*1 mark*)

2   Describe how the average kinetic energy and the average potential energy of the molecules of a liquid changes when:

    a   the liquid is cooled without changing its state (*1 mark*)

    b   the liquid solidifies at its melting point. (*2 marks*)

3   When a stretched rubber band is suddenly released, it becomes warmer for a short time. Describe how its internal energy changes and explain why it becomes warmer for a short time. (*4 marks*)

Revision tip

A temperature change is the same in °C as it is in K. If you are given the initial and final temperatures in °C, just calculate the temperature difference in °C.

Synoptic link

If the volume flow rate is given, you need to know the density of the fluid to calculate the rate of flow of mass $\left(\dfrac{m}{t}\right)$. See Topic 11.1, Density.

Summary questions

Use the data in Table 1 for the following calculations.

▼ **Table 1** Some specific heat capacities

Substance	Specific heat capacity/ J kg⁻¹ K⁻¹
aluminium	900
copper	390
water	4200

1 Calculate the energy needed to heat an insulated copper tank of mass 20 kg containing 430 kg of water from 15 °C to 60 °C. (2 marks)

2 Estimate the maximum possible temperature change of an insulated aluminium object that falls to Earth from a height of 8000 m. (2 marks)

3 A 5.0 kW electric shower is used to heat the water flowing through it when the flow rate is $3.8 \times 10^{-5}\,\mathrm{m^3\,s^{-1}}$. Calculate the maximum possible temperature rise of the water. Density of water = 1000 kg m⁻³. (3 marks)

The **specific heat capacity, c, of a substance is the energy needed to raise the temperature of unit mass of the substance by 1 K without change of state.** The unit of c is J kg⁻¹K⁻¹.

To raise the temperature of mass m of a substance from temperature T_1 to temperature T_2, the energy needed

$$\Delta Q = mc(T_2 - T_1)$$

The **heat capacity** of an object is the heat supplied to raise the temperature of the object by 1 K.

Measurement of specific heat capacity

1 **For a liquid**, a known mass of the liquid m_1 in an insulated calorimeter of known mass m_{cal} and known specific heat capacity c_{cal} is heated for a measured time t by a 12 V electrical heater in the liquid as shown in Figure 1. A variable resistor is used as shown in Figure 2 to keep the current constant. The heater current I and pd V are measured using the ammeter and voltmeter. If the temperature rise is ΔT, assuming no heat loss to the surroundings,

the electrical energy supplied =
(IVt)

$$\underset{(m_1 c_1 \Delta T)}{\substack{\text{the energy needed to} \\ \text{heat the liquid}}} + \underset{(m_{cal} c_{cal} \Delta T)}{\substack{\text{the energy needed to} \\ \text{heat the calorimeter}}}$$

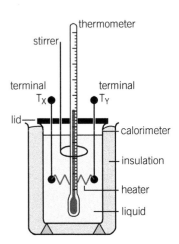

▲ **Figure 1** Measurement of the specific heat capacity of a liquid

The specific heat capacity of the liquid c_1 can be calculated from this equation.

2 **For a solid metal**, a block of the metal of known mass m in an insulated container is used. A 12 V electrical heater is inserted into a hole drilled in the metal and used to heat the metal as explained above. A thermometer inserted into a second hole drilled in the metal is used to measure the temperature rise ΔT. Assuming no heat loss to the surroundings,

$$mc\Delta T = IVt$$

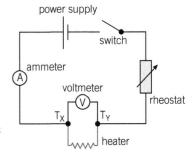

▲ **Figure 2** Circuit diagram

The specific heat capacity of the solid c can be calculated from this equation.

Continuous flow heating

In an electric shower, water passing steadily through copper coils is heated by an electrical heater. For mass m of liquid, passing through the heater in time t at a steady flow rate, when the outflow is at a constant temperature, assuming no heat loss to the surroundings:

the electrical energy supplied per second, $IV = \dfrac{mc\Delta T}{t}$

where ΔT is the temperature rise of the water and c is its specific heat capacity.

19.3 Change of state

The three physical states of a substance, solid, liquid, and vapour, have different physical properties. For example:

- The density of a gas is much less than the density of the same substance in the liquid or the solid state.

- Liquids and gases can flow, but solids can't.

Latent heat

When a solid or a liquid is heated so that its temperature increases, its molecules gain kinetic energy. In a solid, the atoms vibrate mainly about their mean positions. In a liquid, the molecules move about faster, still keeping in contact with each other, but free to move about.

1 **When a solid is heated at its melting point**, its atoms vibrate so much that they break free from each other. The solid therefore becomes a liquid due to energy being supplied at the melting point. The energy needed to melt a solid at its melting point is called **latent heat of fusion**.

2 **When a liquid is heated at its boiling point**, the molecules gain enough kinetic energy to overcome the bonds that hold them close together. The molecules therefore break away from each other to form bubbles of vapour in the liquid. The energy needed to vaporise a liquid is called **latent heat of vaporisation**.

Some solids vaporise directly when heated. This process is called **sublimation**.

For a pure substance, change of state is at constant temperature.

The **specific latent heat of fusion**, l_f, of a substance is the energy needed to change the state of unit mass of the substance from solid to liquid without change of temperature.

The **specific latent heat of vaporisation** of a substance is the energy needed to change the state of unit mass of the substance from liquid to vapour without change of temperature.

So, the energy Q needed to change the state of mass m of a substance from solid to liquid (or liquid to vapour) without change of temperature is given by

$$Q = ml$$

where l is the specific latent heat of fusion (or the specific latent heat of vaporisation). The unit of specific latent heat is $J\,kg^{-1}$.

> **Synoptic link**
>
> If the volume of a substance is given, you need to know its density to find its mass. See Topic 11.1, Density.

Maths skill: Temperature–time graphs

If a pure solid is heated to its melting point and beyond, its temperature–time graph will be as shown in Figure 1.

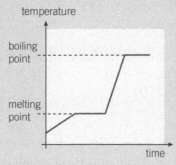

▲ **Figure 1** *Temperature against time for a solid being heated*

Assuming that no heat loss occurs during heating, and assuming that energy is transferred to the substance at a constant rate P (i.e., power supplied), then:

- Before the solid melts, $P = mc_s\left(\dfrac{\Delta T}{\Delta t}\right)_s$, where $\left(\dfrac{\Delta T}{\Delta t}\right)_s$ is the rise of temperature per second and c_s is the specific heat capacity of the solid.

  So the rise of temperature per second of the solid is $\left(\dfrac{\Delta T}{\Delta t}\right)_s = \dfrac{P}{mc_s}$.

- After the solid melts, $P = mc_L\left(\dfrac{\Delta T}{\Delta t}\right)_L$, where $\left(\dfrac{\Delta T}{\Delta t}\right)_L$ is the rise of temperature per second and c_L is the specific heat capacity of the liquid.

  So the rise of temperature per second of the liquid is $\left(\dfrac{\Delta T}{\Delta t}\right)_L = \dfrac{P}{mc_L}$

If the solid has a larger specific heat capacity than the liquid, the rate of temperature rise of the solid is less than that of the liquid. In other words, the liquid heats up faster than the solid.

Summary questions

1 Explain why energy is released when a solid freezes. *(3 marks)*

2 A plastic beaker containing 0.095 kg of water at 12 °C took 600 s to freeze completely. Calculate the energy per second transferred from the water.
specific heat capacity of water = 4200 J kg⁻¹ K⁻¹
specific latent heat of fusion of water = 3.40×10^5 J kg⁻¹ *(3 marks)*

3 a In a 3.0 kW electric kettle, 0.015 kg of water is boiled away in 20 s. Calculate the percentage of energy supplied to the kettle that is used to boil the water.
The specific latent heat of vaporisation of water is 2.25 MJ kg⁻¹. *(2 marks)*

  b A jet of steam at 100 °C was directed into an insulated plastic beaker containing 0.11 kg of water at 20 °C until the water temperature was 55 °C. Calculate the mass of steam that condensed in the beaker. *(5 marks)*

Chapter 19 Practice questions

1 When the outlet of a bicycle pump is blocked and the air in the pump is compressed rapidly, the air becomes warmer for a short time before it returns to its initial temperature. Which **one** of the following changes **A–D** correctly describes the overall energy changes that take place?

	Work done on the air	Energy transfer by heating to surroundings	Internal energy change
A	Yes	No	Increase
B	Yes	Yes	No change
C	Yes	Yes	Decrease
D	No	Yes	No change

(1 mark)

2 Figure 1 shows how the temperature of a pure substance changes as it is heated in an insulated container from below its melting point to above its melting point. The heater transfers energy to the substance at a constant rate.

Which **one** of the following statements **A–D** about the substance is correct?

A No energy was transferred to the substance when it was melting.

B The specific heat capacity of the substance in the solid state is less than in the liquid state.

C The rate of change of temperature in the solid state is less than in the liquid state.

D The internal energy of the substance decreased when it was melting.

(1 mark)

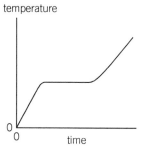

▲ Figure 1

3 Which one of the following combinations of SI base units has the same combination of base units as specific latent heat?

A $kg\,m\,s^{-1}$ B $kg\,m\,s^{-2}$ C $m\,s^{-2}$ D $m^2\,s^{-2}$ *(1 mark)*

4 In an experiment to measure the specific heat capacity of an insulated metal block using the method described in Topic 19.2, Specific heat capacity the following measurements were made:

mass of block = 1.210 ± 0.005 kg
current = 2.60 ± 0.05 A
potential difference = 12.2 ± 0.10 V
heating time = 200 ± 0.40 s
initial temperature = 18.0 ± 0.5 °C
final temperature = 33.5 ± 0.5 °C

a Use this data to calculate the specific heat capacity of the metal.

(3 marks)

b Calculate the percentage uncertainty in:

i the energy supplied *(2 marks)*

ii the temperature change *(1 mark)*

iii the specific heat capacity of the metal. *(1 mark)*

c Discuss how you might test the assumption that no energy is transferred to the surroundings from the metal block during the heating time. *(2 marks)*

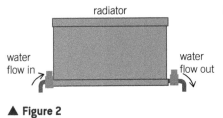

▲ Figure 2

5 An ice cube at 0 °C is placed in an insulated copper can of mass 0.080 kg containing 0.120 kg of water at 15 °C. After the ice cube has melted and the water has been stirred, the temperature of the water is 12 °C. Calculate the mass of the ice cube.

specific heat capacities: copper = 390 J kg⁻¹ K⁻¹, water = 4200 J kg⁻¹ K⁻¹

specific latent heat of fusion of water = 3.40×10^5 J kg⁻¹ *(6 marks)*

6 In a heating system, hot water is pumped through a steel radiator at a flow rate of 0.042 kg s⁻¹.

a The water enters the radiator at a temperature of 50 °C and leaves it at a temperature of 44 °C. Calculate the energy transfer from the radiator to the surroundings, assuming the radiator is at a constant temperature.

The specific heat capacity of water is 4200 J kg⁻¹ K⁻¹ *(2 marks)*

b The radiator is made of steel and has a mass of 28 kg. Estimate the time taken for the radiator to warm up from 15 °C, assuming energy is transferred to it at the same rate as in part **a**. In making your estimate, state any further assumptions you have made.

The specific heat capacity of steel = 560 J kg⁻¹ K⁻¹ *(4 marks)*

7 In an experiment, a solid was heated by a 50 W electric heater placed inside a hole drilled into the solid. The heater was switched on for seven minutes. The temperature of the solid was measured at the start of every minute. Figure 3 shows how the temperature changed when the solid was heated.

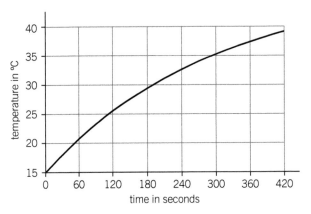

▲ Figure 3

a Explain why the temperature did not increase at a constant rate.

(3 marks)

b The mass of the solid was 0.58 kg. Estimate the specific heat capacity of the solid. *(4 marks)*

c Describe and explain how the graph would have differed if the solid had been heated for five minutes only. *(2 marks)*

8 When a plastic beaker containing 70 g of water at 10 °C was placed in a freezer, the temperature of the water decreased steadily to 0 °C in 140 s before it started to freeze.

a Calculate the average rate at which energy was removed from the water.

The specific heat capacity of water is 4200 J kg K⁻¹ *(2 marks)*

b i Estimate the time taken to freeze the water.

The specific latent heat of fusion of water is 340 kJ kg⁻¹ *(2 marks)*

ii Discuss the assumptions you made in your estimate and their validity. *(4 marks)*

20.1 The experimental gas laws

Specification reference: 3.6.2.2

The molecules of a gas move at random with different speeds. The pressure of a gas on a surface is due to the gas molecules hitting the surface. Each impact causes a tiny force on the surface. Because there are a very large number of impacts each second, the overall result is that the gas exerts a measurable pressure on the surface.

- For a fixed mass of gas at constant temperature, pV = **constant**

  where p = gas pressure and V = gas volume.

  A gas that obeys this equation (Boyle's Law) is called an **ideal gas**.

  Figure 1 shows a graph of gas pressure p against volume V for a fixed mass of gas for different constant temperatures.

Maths skill: Data analysis

A graph of pressure on the y-axis against $\dfrac{1}{\text{volume}}$ on the x-axis for a fixed mass of gas at constant temperature gives a straight line through the origin. This is because Boyle's law can be written as $p = \text{constant} \times \dfrac{1}{V}$, which represents the equation $y = mx$ for a straight-line graph through the origin.

- For a fixed mass of an ideal gas at constant pressure, $\dfrac{V}{T}$ = **constant**
  where V is the gas volume and T is the absolute temperature of the gas in kelvins. A graph of V against T is a straight line through the origin.

- For a fixed mass of gas at constant volume, $\dfrac{p}{T}$ = **constant** where p is its pressure and T is its temperature in kelvins.

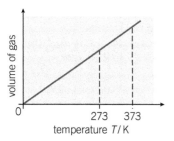

▲ **Figure 3** *A graph of V against T for an ideal gas*

Key terms

Pressure is defined as the force per unit area exerted normally (i.e., at right angles) on a surface. Pressure is measured in pascals (Pa), where

$$1\,\text{Pa} = 1\,\text{N}\,\text{m}^{-2}$$

Absolute zero is the lowest possible temperature.

Absolute temperature T in kelvins (K) = temperature in °C + 273.15

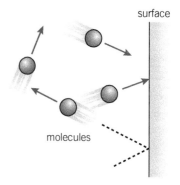

▲ **Figure 1** *Molecules in motion*

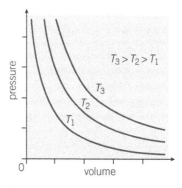

▲ **Figure 2** *Boyle's law*

Summary questions

1 Air at a pressure of 120 kPa is trapped in a piston and compressed until its volume has decreased from $8.0 \times 10^{-4}\,\text{m}^3$ to $5.0 \times 10^{-4}\,\text{m}^3$ at constant temperature. The gas is then cooled at constant volume from a temperature of 50 °C to 20 °C. Calculate its final pressure.   (*3 marks*)

2 A fixed mass of ideal gas at an initial pressure of 100 kPa in a sealed container is cooled from a temperature of 100 °C to −100 °C.
   a Calculate the pressure of the gas at −100 °C.   (*2 marks*)
   b Sketch a graph to show how the pressure of the gas varied with its temperature in kelvins.   (*3 marks*)

3 The volume of a fixed mass of an ideal gas at 15 °C was 0.085 m³. The gas was then heated to 55 °C without change of pressure. Calculate the new volume of this gas.   (*2 marks*)

Key terms

An **isothermal** change is any change at constant temperature.

An **isobaric** change is any change at constant pressure.

Revision tip

For a volume change V of a gas at constant pressure p, the work done by the gas = pV.

20.2 The ideal gas law

Specification reference: 3.6.2.2; 3.6.2.3

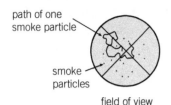

▲ **Figure 1** *Brownian motion*

Synoptic link

Randomness occurs in radioactive decay as well as in Brownian motion of molecules. You can't predict when a random event will take place. See Topic 26.5, Radioactive decay.

Key terms

The Avogadro constant, N_A, is defined as the number of atoms in exactly 12 g of the carbon isotope $^{12}_{6}C$. The value of N_A is 6.023×10^{23} (to four significant figures). So the mass of an atom of $^{12}_{6}C$ is

$$1.993 \times 10^{-23}\,g = \left(\frac{12\,g}{6.023 \times 10^{23}}\right)$$

One atomic mass unit (u) is $\frac{1}{12}$th of the mass of a $^{12}_{6}C$ atom. The mass of a carbon atom is 1.993×10^{-26} kg, so $1\,u = 1.661 \times 10^{-27}$ kg.

Revision tip

Don't mix up molar mass and molecular mass.

Revision tip

The unit of R is the joule per mole per kelvin ($J\,mol^{-1}\,K^{-1}$), which is the same as the unit of

$$\frac{pressure \times volume}{absolute\ temperature \times number\ of\ moles}$$

This is because the unit of pressure ($Pa = N\,m^{-2}$) × the unit of volume (m^3) is the joule [$= 1\,N\,m$].

Molecules in a gas

Molecules are too small to see individually. You can see the effect of individual molecules in a gas if you observe smoke particles with a microscope. The smoke particles appear as tiny specks of light wriggling about unpredictably. This type of motion is called **Brownian motion**. Each smoke particle is bombarded unevenly and randomly by individual molecules so it experiences forces due to these impacts, which change its magnitude and direction of motion at random.

Molar mass

One **mole** of a substance consisting of identical particles is defined as the quantity of substance that contains N_A particles. The **molar mass** of a substance is the mass of 1 mol of the substance. The unit of molarity (the number of moles in a given quantity of a substance) is the mol. The unit of molar mass is $kg\,mol^{-1}$.

Therefore, the number of moles n in mass M_S of a substance $= \frac{M_S}{M}$, where M is the molar mass of the substance. For example, because the molar mass of carbon dioxide is 0.044 kg, then n moles of carbon dioxide has a mass of $0.044n$ kg and contains nN_A molecules.

The ideal gas equation

The three experimental gas laws can be combined to give the equation

$$\frac{pV}{T} = \text{constant, for a fixed mass of ideal gas}$$

where p is the pressure, V is the volume, and T is the absolute temperature.

For 1 mole of any ideal gas, the value of $\frac{pV}{T}$ is equal to $8.31\,J\,mol^{-1}\,K^{-1}$ $\left(= \frac{pV}{T} = \frac{101 \times 10^3\,Pa \times 0.0224\,m^3}{273\,K}\right)$. This value is called the **molar gas constant, R**.

Therefore for n moles of ideal gas,

$$pV = nRT$$

where V is the volume of the gas at pressure p and temperature T in kelvins.

This equation is called the **ideal gas equation**.

Note: In the equation $pV = nRT$, substituting the number of moles $n = \frac{N}{N_A}$ gives

$$pV = NkT$$

where the **Boltzmann constant** $k = \frac{R}{N_A}$, and N is the number of molecules. Prove for yourself that $k = 1.38 \times 10^{-23}\,J\,K^{-1}$. You will meet k again in the next topic.

Worked example

$R = 8.31\,\text{J}\,\text{mol}^{-1}\text{K}^{-1}$

Calculate the number of moles and the mass of air in a 3 litre plastic bottle in which the air pressure 100 kPa, and the temperature of the air in the balloon is 17 °C.

molar mass of air = 0.029 kg mol^{-1}, 1 litre = 0.001 m^3

$T = 273 + 17 = 290\,\text{K}$

Using $pV = nRT$ gives $n = \dfrac{pV}{RT} = \dfrac{100 \times 10^3\,\text{Pa} \times 0.003\,\text{m}^3}{8.31\,\text{J}\,\text{mol}^{-1}\text{K}^{-1} \times 290\,\text{K}} = 0.124\,\text{mol}$

Mass of air = number of moles × molar mass = 0.124 mol × 0.029 kg mol^{-1} = 3.6×10^{-3} kg

Summary questions

$N_A = 6.02 \times 10^{23}\,\text{mol}^{-1}$, $R = 8.31\,\text{J}\,\text{mol}^{-1}\text{K}^{-1}$

1 A cylinder of volume 0.020 m^3 contains nitrogen gas at a pressure of 120 kPa and at a temperature of 20 °C.

   a The molar mass of nitrogen = 0.028 kg. Calculate the mass of gas in the cylinder. *(2 marks)*

   b Calculate the pressure of the gas if the temperature of the gas decreases to 5 °C. *(1 mark)*

2 In a chemistry experiment, 2.2×10^{-5} m^3 of oxygen gas is collected at a pressure of 101 kPa and a temperature of 12 °C. Calculate the mass of gas collected. *(2 marks)*
   The molar mass of oxygen is 0.032 kg mol^{-1}

3 a Estimate the average volume per molecule of air molecules at a pressure of 100 kPa and a temperature of 300 K. *(3 marks)*

   b Use your estimate to make a further estimate of the average spacing between the air molecules in part **a**. *(2 marks)*

20.3 The kinetic theory of gases

Specification reference: 3.6.2.3

The gas laws were explained by assuming that a gas consists of point molecules moving about at random, continually colliding with the container walls. Each impact causes a force on the container. The force of many impacts is the cause of the pressure of the gas on the container walls.

The pressure of a gas can be increased:

- at constant temperature by reducing its volume so the gas molecules travel less distance on average between impacts at the walls. Therefore, there are more impacts per second, and so the pressure is greater.
- at constant volume by increasing its temperature. This raises the kinetic energy and therefore the speed of the molecules. Therefore, the impacts of the molecules on the container walls are harder and more frequent so the pressure is raised.

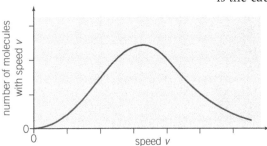

▲ **Figure 1** *Distribution of molecular speeds*

Revision tip

If the temperature of a gas is raised, the distribution curve becomes flatter and broader because the greater the temperature, the more molecules there are moving at higher speeds.

Molecular speeds

The molecules in an ideal gas have a continuous spread of speeds, as shown in Figure 1.

The **root mean square speed** of the molecules,

$$c_{rms} = \left[\frac{c_1^2 + c_2^2 + \ldots + c_N^2}{N} \right]^{\frac{1}{2}}$$

where $c_1, c_2, c_3, \ldots c_N$ represent the speeds of the individual molecules, and N is the number of molecules in the gas.

The kinetic theory equation

For an ideal gas consisting of N identical molecules, each of mass m, in a container of volume V, the pressure p of the gas is given by the equation

$$pV = \frac{1}{3}Nmc_{rms}^{\ 2}$$

where c_{rms} is the root mean square speed of the gas molecules.

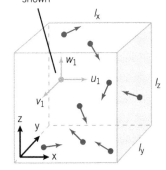

chosen molecule with its velocity components shown

▲ **Figure 2** *Molecules in a box*

Proof

1 Consider one molecule of mass m in a rectangular box of dimensions l_x, l_y, and l_z as shown in Figure 2. Let u_1, v_1, and w_1 represent its velocity components in the x, y, and z directions, respectively.

For each impact of the molecule with the shaded face of the box, its change of momentum = final momentum – initial momentum = $(-mu_1) - (mu_1) = -2mu_1$.

The time, t, between successive impacts on this face is given by $t = \frac{2l_x}{u_1}$.

Revision tip

To derive the kinetic theory equation, assume the molecules are point molecules, do not attract each other, move about in continual random motion, collide elastically with each other and with the container surface (i.e., no overall loss of kinetic energy in a collision), and they undergo collisions of much shorter duration than the time between impacts.

Therefore, the force on the molecule = $\dfrac{\text{change of momentum}}{\text{time taken}}$

$$= \frac{-2mu_1}{\frac{2l_x}{u_1}} = \frac{-mu_1^2}{l_x}$$

The force F_1 of the impact on the surface is $F_1 = \dfrac{+mu_1^2}{l_x}$ (as it is equal and opposite to the force on the molecule).

The pressure p_1 of the molecule on the surface = $\dfrac{\text{force } F_1}{\text{area } (l_y l_z)} = \dfrac{mu_1^2}{l_x l_y l_z} = \dfrac{mu_1^2}{V}$ where V = the volume of the box.

2 Consider N identical molecules in the box.

The total pressure of all the molecules on the surface, $p = p_1 + p_2 + p_3 + \ldots + p_N$ where each subscript refers to each molecule.

Therefore $p = \dfrac{mu_1^2}{V} + \dfrac{mu_2^2}{V} + \dfrac{mu_3^2}{V} + \ldots + \dfrac{mu_N^2}{V} = \dfrac{m}{V}(u_1^2 + u_2^2 + u_3^2 + \ldots + u_N^2) = \dfrac{Nm\bar{u}^2}{V}$

where $\bar{u}^2 = \dfrac{u_1^2 + u_2^2 + u_3^2 + \ldots + u_N^2}{N}$

Because the motion of the molecules is random, there is no preferred direction of motion. Therefore, in terms of the y-components of velocity $v_1, v_2, v_3, \ldots v_N$, or the z-components of velocity $w_1, w_2, w_3, \ldots w_N$:

$p = \dfrac{Nm\bar{v}^2}{V}$ where $\bar{v}^2 = \dfrac{v_1^2 + v_2^2 + v_3^2 + \ldots + v_N^2}{N}$

$p = \dfrac{Nm\bar{w}^2}{V}$ where $\bar{w}^2 = \dfrac{w_1^2 + w_2^2 + w_3^2 + \ldots + w_N^2}{N}$

Therefore, $p = \dfrac{Nm}{3V}(\bar{u}^2 + \bar{v}^2 + \bar{w}^2) = \dfrac{Nm}{3V}c_{\text{rms}}^2$ (because $c_{\text{rms}}^2 = \bar{u}_{\text{rms}}^2 + \bar{v}_{\text{rms}}^2 + \bar{w}_{\text{rms}}^2$)

So $pV = \dfrac{1}{3}Nmc_{\text{rms}}^2$

Molecules and kinetic energy

For an ideal gas, its internal energy is due only to the kinetic energy of the molecules of the gas.

The mean **kinetic energy of a molecule of a gas** =

$\dfrac{\text{total kinetic energy of all the molecules}}{\text{total number of molecules}} = \dfrac{1}{2}mc_{\text{rms}}^2$

Since $(pV =)$ $\dfrac{1}{3}Nm(c_{\text{rms}})^2 = nRT$

$\dfrac{1}{2}m(c_{\text{rms}})^2 = \dfrac{3}{2}\left(\dfrac{nR}{N}\right)T = \dfrac{3}{2}kT$ where $k = \dfrac{nR}{N} = \dfrac{R}{N_A}$

The mean kinetic energy of a molecule of an ideal gas $= \dfrac{3}{2}kT$

The total kinetic energy of n moles of an ideal gas $= N \times \dfrac{3}{2}kT = \dfrac{3}{2}nRT$

and since an ideal gas has no potential energy its

$$\textbf{internal energy} = \dfrac{3}{2}nRT$$

Synoptic link

Recall that the constant k is called the Boltzmann constant. Its value $\left(\dfrac{R}{N_A}\right)$ is $1.38 \times 10^{-23}\,\text{J K}^{-1}$. See Topic 20.2, The ideal gas law.

Worked example

$k = 1.38 \times 10^{-23}\,\text{J mol}^{-1}\,\text{K}^{-1}$

Calculate the mean kinetic energy of an oxygen molecule in oxygen gas at 0 °C.

$T = 273\,\text{K}$

Mean kinetic energy $= \dfrac{3}{2}kT$

$= 1.5 \times 1.38 \times 10^{-23}\,\text{J K}^{-1} \times 273\,\text{K}$

$= 5.7 \times 10^{-21}\,\text{J}$

Summary questions

$N_A = 6.02 \times 10^{23}\,\text{mol}^{-1}$, $R = 8.31\,\text{J mol}^{-1}\text{K}^{-1}$, $k = 1.38 \times 10^{-23}\,\text{J K}^{-1}$

1 Explain in molecular terms why the pressure of a gas in a container increases when its volume is decreased. (*2 marks*)

2 Calculate the root mean square speed of a hydrogen molecule at 20 °C. The molar mass of hydrogen is 0.002 kg mol⁻¹ (*3 marks*)

3 A cylinder of volume 0.018 m³ contains nitrogen gas at a pressure of 120 kPa and a temperature of 323 K. The molar mass of nitrogen is 0.028 kg mol⁻¹. Calculate:
   a the number of moles of nitrogen in the cylinder (*1 mark*)
   b the internal energy of the gas in the container. (*2 marks*)

4 A sample of carbon dioxide is mixed with nitrogen. Calculate the ratio of the root mean square speed of a carbon dioxide molecule to that of a nitrogen molecule in the sample.
   nitrogen = 0.028 kg mol⁻¹, carbon dioxide = 0.044 kg mol⁻¹ (*4 marks*)

5 An ideal gas of molar mass 0.032 kg mol⁻¹ is in a container of volume 0.037 m³ at a pressure of 100 kPa and a temperature of 300 K. Calculate:
   a the number of moles (*1 mark*)
   b the root mean square speed of the molecules (*3 marks*)
   c the density of the gas. (*2 marks*)

6 The molar mass of air is 0.029 kg m⁻³.
   Calculate the density of air at 15 °C and 101 kPa pressure. (*3 marks*)

1   A gas cylinder P contains n moles of an ideal gas at pressure p and absolute temperature T. An identical cylinder Q contains the same type of gas at pressure $0.60p$ and a temperature of $0.75T$. Which one of the following options, **A–D**, gives the number of moles of gas in Q?

    A $0.45n$ **B** $0.60n$ **C** $0.75n$ **D** $0.80n$ *(1 mark)*

2   Which one of the following expressions, **A–D**, is a correct expression for the density of an ideal gas of molar mass M at pressure p and absolute temperature T?

    A $\dfrac{p}{MRT}$ **B** $\dfrac{pM}{RT}$ **C** $\dfrac{pRT}{M}$ **D** $\dfrac{MR}{pT}$ *(1 mark)*

3   The root mean square speed of the molecules of a certain gas X of molar mass M at a certain temperature is $800\,\mathrm{m\,s^{-1}}$. The root mean square speed of the molecules of a different gas Y at the same temperature is $1600\,\mathrm{m\,s^{-1}}$. Which one of the following alternatives, **A–D**, gives the molar mass of Y?

    A $0.25M$ **B** $0.50M$ **C** $2.0M$ **D** $4.0M$ *(1 mark)*

4   **a** Sketch a graph of pressure against $\dfrac{1}{\text{volume}}$ for a fixed mass of ideal gas at constant temperature. Label this graph O. *(2 marks)*

    b On the *same axes* sketch **two** additional curves labelled P and Q, if the following changes are made:

        i The same mass of gas at a higher constant temperature (P).

        ii A smaller mass of gas at the original constant temperature (Q). *(2 marks)*

5   **a** A cylinder of volume $0.020\,\mathrm{m^3}$ contains an ideal gas at a pressure of $140\,\mathrm{kPa}$ and a temperature of $285\,\mathrm{K}$.

    Calculate:

        i the amount of gas, in moles, in the cylinder

        ii the total kinetic energy of the molecules of gas in the cylinder

        iii the average kinetic energy of a gas molecule in the cylinder. *(4 marks)*

    b The cylinder is heated and the temperature of the gas increases to $320\,\mathrm{K}$. Calculate the increase of internal energy of the gas. *(2 marks)*

6   **a** State what is meant by the root mean square speed of the molecules of a gas. *(1 mark)*

    b **i** Sketch a graph on the axes in Figure 1 to show the distribution of speeds of the molecules of an ideal gas at constant temperature T_1. *(2 marks)*

        ii Sketch a further graph on the same axes to show the distribution of speeds at a higher temperature T_2. *(2 marks)*

    c Calculate the root mean square speed of nitrogen molecules in air at a temperature of $273\,\mathrm{K}$. The molar mass of nitrogen is $0.028\,\mathrm{kg}$. *(3 marks)*

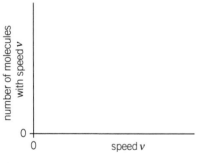

▲ **Figure 1**

7   **a** The molecules in a gas move about at random and collide elastically. Explain what is meant by:

        i random motion   *(1 mark)*   **ii** elastic collisions.   *(1 mark)*

    b A gas molecule of mass $5.3 \times 10^{-26}\,\mathrm{kg}$ moving at a speed of $1200\,\mathrm{m\,s^{-1}}$ collides normally with the wall of the gas container and rebounds without loss of speed.

        i Calculate the change of momentum of this molecule.   *(3 marks)*

        ii The pressure of the gas in the container is $120\,\mathrm{kPa}$. Estimate how many such collisions would need to occur each second on each square millimetre of the wall to produce this pressure.   *(4 marks)*

21.1 Gravitational field strength

Specification reference: 3.7.2.2

Any two masses exert a gravitational pull on each other. The mass of an object creates a force field around itself. Any other mass placed in the field is attracted towards the object. The force field round a mass is called a **gravitational field**.

The path which a small mass would follow is called a **field line** or a **line of force**. Figure 1 shows the field lines near a planet. The lines are directed to the centre of the planet because a small object released near the planet would fall towards its centre.

The strength of a gravitational field, g, is the force per unit mass on a small test mass placed in the field.

If a small test mass, m, at a particular position in a gravitational field is acted on by a gravitational force F,

$$g = \frac{F}{m}$$

The unit of gravitational field strength is the newton per kilogram ($N\,kg^{-1}$).

Free fall in a gravitational field

The weight of an object is the force of gravity on it. If an object of mass m is in a gravitational field, the gravitational force on the object is $F = mg$, where g is the gravitational field strength at the object's position. If the object is not acted on by any other force, it accelerates with

$$\text{acceleration } a = \frac{\text{force}}{\text{mass}} = \frac{mg}{m} = g$$

The object therefore falls freely with acceleration g. So, g may also be described as the acceleration of a freely falling object.

Field patterns

1   A **radial field** is where the field lines are always directed towards the same point. See Figure 1.

2   A **uniform field** is where the gravitational field strength has the same magnitude and direction throughout the field. The field lines are therefore parallel to one another and equally spaced.

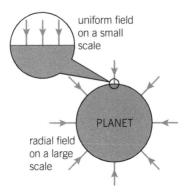

uniform field on a small scale

PLANET

radial field on a large scale

▲ **Figure 1** *Field patterns*

Revision tip

Is the Earth's gravitational field uniform or radial? The force of gravity due to the Earth on a small mass is directed towards the centre of the Earth so the field is therefore radial. However, over distances that are small compared with the Earth's radius, the Earth's gravitational field strength hardly changes so the field may be considered to be uniform.

Summary questions

$g = 9.81\,N\,kg^{-1}$

1   An object of mass 23 kg has a weight of 37 N on the Moon's surface.

   a   Calculate the gravitational field strength of the Moon at its surface.
        (*1 mark*)

   b   Calculate the weight of space vehicle of mass 250 kg on the Moon's surface.
        (*1 mark*)

2   a   Explain what is meant by a uniform gravitational field.   (*1 mark*)

   b   The gravitational field strength of the Earth is $9.78\,N\,kg^{-1}$ at the equator and $9.83\,N\,kg^{-1}$ at the poles. Calculate the difference in the weight of a 70 kg person at the equator and the same person at the poles.
        (*1 mark*)

3   The gravitational field strength on the surface of Mars is $3.70\,N\,kg^{-1}$. Calculate the weight on Mars of an astronaut whose weight on the Earth is 650 N.
        (*2 marks*)

21.2 Gravitational potential

Gravitational potential energy (gpe) is the energy of an object due to its position in a gravitational field. The position for zero gpe is at infinity. In other words, the object would be so far away that the gravitational force on it is negligible. When the object is in the gravitational field, its potential energy is therefore negative.

The gravitational potential, V, at a point in a gravitational field is the work done per unit mass to move a small object from infinity to that point.

Therefore, for a small object of mass m at a position where the gravitational potential is V, the work W that must be done to enable it to escape completely is given by $W = mV$. Rearranging this gives

$$V = \frac{W}{m}$$

The unit of gravitational potential is $J\,kg^{-1}$

Potential gradients

Equipotentials are surfaces of constant potential. Because of this, no work needs to be done to move along an equipotential surface. The equipotentials near the Earth are circles as shown in Figure 1.

At increasing distance from the surface, the equipotentials for equal increases of potential are spaced further apart.

Near the surface over a small region, the equipotentials are horizontal (i.e., parallel to the ground). This is because the gravitational field over a small region is uniform.

The **potential gradient** at a point in a gravitational field is the change of potential per unit distance at that point.

In general, for a change of potential ΔV over a small distance Δr,

$$\text{the potential gradient} = \frac{\Delta V}{\Delta r}$$

Consider a test mass m being moved away from a planet as shown in Figure 2. To move m a small distance Δr in the opposite direction to the gravitational force F_{grav} on it, its gravitational potential energy must be increased:

- by an equal and opposite force F acting through the distance Δr
- by an amount of energy equal to the work done by F, that is, $\Delta W = F\Delta r$.

For the test mass, the change of potential ΔV of the test mass $= \frac{\Delta W}{m}$, so

$\Delta V = \frac{F\Delta r}{m}$ and $F = \frac{m\Delta V}{\Delta r}$. As $F_{grav} = -F$, then $F_{grav} = -\frac{m\Delta V}{\Delta r}$. Therefore the

gravitational field strength $g = \frac{F_{grav}}{m} = -\frac{\Delta V}{\Delta r}$.

Gravitational field strength $g = -\dfrac{\Delta V}{\Delta r}$ (equals the negative of the potential gradient) where the minus sign here shows you that g acts in the opposite direction to the potential gradient.

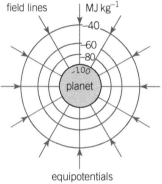

▲ **Figure 1** *Equipotentials near a planet*

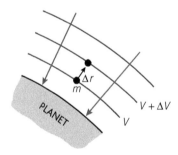

▲ **Figure 2** *Potential gradients*

Summary questions

$g = 9.81\,\text{N}\,\text{kg}^{-1}$

1 An object of mass 15 kg is lowered through a vertical distance of 50 m.
   a Calculate its change of gravitational potential energy (*1 mark*)
   b Calculate its change of gravitational potential. (*1 mark*)

2 **a** Define gravitational potential. (*1 mark*)
   b Calculate the change of gravitational potential between a point on the Earth's surface and a point 20.0 m above the surface. (*2 marks*)

3 A rocket of mass 300 kg launched from the Earth's surface reaches a height of 500 km above the surface. After its engine stops it then falls to the ground. The gravitational potential at the Earth's surface is $-62.5\,\text{MJ}\,\text{kg}^{-1}$ and $-57.9\,\text{MJ}\,\text{kg}^{-1}$ at a height of 500 km above the surface.
   a Estimate the mean gravitational field strength between the surface and a point 500 km above it. (*3 marks*)
   b Calculate the change of gravitational potential energy of the rocket when it moves from the surface to 500 km above the surface. (*2 marks*)
   c Calculate the maximum possible speed of the rocket on impact if it falls to the ground from this height. (*3 marks*)

21.3 Newton's law of gravitation

Specification reference: 3.7.2.1

Newton's law of gravitation assumes that the gravitational force between any two *point* objects is:

- always an attractive force
- proportional to the mass of each object
- proportional to $\frac{1}{r^2}$, where r is their distance apart.

In mathematical terms,

$$\text{gravitational force } F = \frac{Gm_1m_2}{r^2}$$

where m_1 and m_2 are masses of the two objects and G, is the **universal constant of gravitation** $(6.67 \times 10^{-11}\,\text{Nm}^2\,\text{kg}^{-2})$.

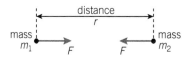

▲ Figure 1

Revision tip

The distance r is the distance between their centres of mass.

Worked example

$G = 6.67 \times 10^{-11}\,\text{Nm}^2\,\text{kg}^{-2}$

The distance from the centre of the Moon to the centre of the Earth is $3.8 \times 10^8\,\text{m}$. The mass of the Moon is $7.4 \times 10^{22}\,\text{kg}$, and the mass of the Earth is $6.0 \times 10^{24}\,\text{kg}$.

a Calculate the force of gravitational attraction between the Moon and the Earth.

b Calculate the gravitational field strength of the Moon's gravitational field at the Earth.

a $F = \dfrac{6.67 \times 10^{-11} \times 7.4 \times 10^{22} \times 6.0 \times 10^{24}}{(3.8 \times 10^8)^2} = 2.1 \times 10^{20}\,\text{N}$

b $g = \dfrac{F}{m} = \dfrac{21 \times 10^{20}\,\text{N}}{7.4 \times 10^{22}\,\text{kg}} = 2.8 \times 10^{-3}\,\text{Nkg}^{-1}$

Summary questions

$G = 6.67 \times 10^{-11}\,\text{N m}^2\,\text{kg}^{-2}$

1 Calculate the force of gravitational attraction between the Moon and a 2000 kg space vehicle in orbit about the Earth when the space vehicle is at a distance of $3.7 \times 10^8\,\text{m}$ from the centre of the Moon. The mass of the Moon $= 7.4 \times 10^{22}\,\text{kg}$. *(1 mark)*

2 The Earth orbits the Sun at a distance of $1.5 \times 10^{11}\,\text{m}$ from the Sun. Mars orbits the Sun at a distance of $2.3 \times 10^{11}\,\text{m}$ from the Sun. Calculate the force of gravitational attraction between the Earth and Mars when they are closest to each other. The mass of the Earth is $6.0 \times 10^{24}\,\text{kg}$, the mass of Mars is $6.4 \times 10^{23}\,\text{kg}$. *(2 marks)*

3 Estimate the ratio the gravitational force on the Earth due to the Sun and the gravitational force on the Earth due to the Moon. *(3 marks)* The mass of the Sun $= 2.0 \times 10^{30}\,\text{kg}$, the distance from the Earth to the Moon is $3.8 \times 10^8\,\text{m}$. See above for mass of the Moon and distance from Sun to Earth.

21.4 Planetary fields

Specification reference: 3.7.2.2; 3.7.2.4

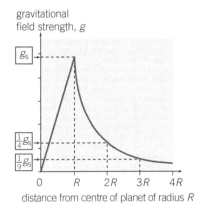

▲ **Figure 1** *Gravitational field strength*

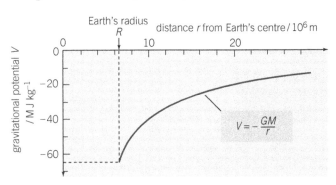

▲ **Figure 2** *Gravitational potential*

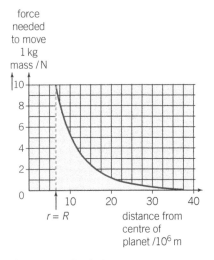

▲ **Figure 3** *Work done*

Gravitational field strength

The **magnitude of the gravitational field strength** $g = \dfrac{GM}{r^2}$ at distance r from a point object or from the centre of a sphere of mass M.

For a spherical planet of mass M and radius R, its surface gravitational field strength $g_s = \dfrac{GM}{R^2}$ so $GM = g_s R^2$

Therefore, for distance $r \geq R$, the gravitational field strength $g = \dfrac{GM}{r^2} = \dfrac{g_s R^2}{r^2}$

Gravitational potential near a spherical planet

At or beyond the surface of a spherical planet, the gravitational potential V at distance r from the centre of the planet of mass M is given by

$$V = -\frac{GM}{r}$$

The potential energy of an object of mass m at distance r from the centre of the planet $= mV = -\dfrac{GMm}{r}$

The **escape velocity** from a planet is the minimum velocity an object must be given to escape from the planet when projected vertically from the surface.

For an object of mass m to escape from the surface of a planet of radius R to infinity, its initial kinetic energy, $\dfrac{1}{2}mv^2 \geq$ its gain of potential energy $\dfrac{GMm}{R}$

Therefore, its escape velocity $v_{esc} = \sqrt{2gR}$ (as $GM = gR^2$)

Estimating work done

The area under the curve in Figure 3 represents the work done to move the 1 kg mass from infinity to the surface, and is therefore the magnitude of the gravitational potential at the surface. To estimate the work done from the area under the curve, count the number of blocks under the curve. Include blocks that are half full or over as full blocks but not blocks that are less than half full. Use the graph scales to calculate the amount of work corresponding to one full block. The total work done is given by multiplying the number of blocks you counted by the amount of work representing one full block.

21.5 Satellite motion

Specification reference: 3.7.2.4

For a satellite of mass m orbiting a planet (or a planet orbiting a star) of mass M in a circular orbit of radius r at speed v, the gravitational force $\dfrac{GMm}{r^2}$ = the centripetal force $\dfrac{mv^2}{r}$.

Therefore, the speed of the satellite is given by $v^2 = \dfrac{GM}{r}$.

Since speed $v = \dfrac{\text{circumference of the orbit}}{\text{time period}} = \dfrac{2\pi r}{T}$

$$\frac{(2\pi r)^2}{T^2} = \frac{GM}{r}$$

Rearranging gives $\qquad \dfrac{r^3}{T^2} = \dfrac{GM}{4\pi^2}$

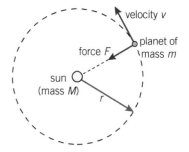

▲ **Figure 1** *Explanation of planetary motion*

The mass of the Sun

The Earth orbits the Sun once per year on a circular orbit of radius 1.5×10^{11} m. Given $G = 6.67 \times 10^{-11}$ Nm^{2}kg^{-2}, the above equation can be used to show that the mass of the Sun is 2.0×10^{30} kg.

Geostationary satellites

A geostationary satellite orbits the Earth directly above the equator and has a time period of exactly 24 hours. It therefore remains in a fixed position above the equator because it has exactly the same time period as the Earth's rotation. The radius of orbit of a geostationary satellite can be calculated using the equation $\dfrac{r^3}{T^2} = \dfrac{GM}{4\pi^2}$

The energy of an orbiting satellite

Consider a satellite of mass m moving at speed v in a circular orbit of radius r about a spherical planet or star of mass M.

- Its kinetic energy, $E_K = \dfrac{1}{2}mv^2 = \dfrac{1}{2}m \times \dfrac{GM}{r} = \dfrac{GMm}{2r}$

- Its gravitational potential energy, $E_P = mV = -\dfrac{GMm}{r}$

So, the total energy of the satellite $E = E_P + E_K = -\dfrac{GMm}{r} + \dfrac{GMm}{2r} = -\dfrac{GMm}{2r}$

Figure 2 shows how E, E_P, and E_K vary with r. Notice that all three curves are $\dfrac{1}{r}$ curves and that the total energy is always negative.

> **Revision tip**
> For planets orbiting the Sun, because $\dfrac{GM}{4\pi^2}$ is the same for all of the planets, then $\dfrac{r^3}{T^2}$ is the same for all of the planets.

> **Revision tip**
> A geosynchronous orbit is a 24 hour orbit inclined to the equator (i.e., it crosses the equator).

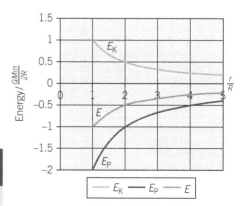

▲ **Figure 2** *Satellite energy*

Summary questions

$G = 6.67 \times 10^{-11}$ Nm^{2}kg^{-2}, $g = 9.81$ Nkg^{-1} at the Earth's surface.
The radius of the Earth is 6400 km and the mass of Earth is 6.0×10^{24} kg.

1   The International Space Station (ISS) orbits the Earth once every 90 minutes in a circular orbit. Calculate:
    a   the average height of the orbit above the Earth's surface   (*3 marks*)
    b   the speed of the satellite.   (*3 marks*)

2   A satellite orbits the Earth in a circular polar orbit at a height of 1200 km. Calculate:
    a   the time period of the orbit   (*2 marks*)
    b   the angular speed of the satellite.   (*1 mark*)

▼ Table 1

	gravitational field strength	gravitational potential
A	$\frac{1}{2}g_s$	$\frac{1}{2}g_s R$
B	$\frac{1}{2}g_s$	$g_s R$
C	$\frac{1}{4}g_s$	$\frac{1}{2}g_s R$
D	$\frac{1}{4}g_s$	$g_s R$

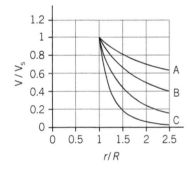

▲ Figure 1

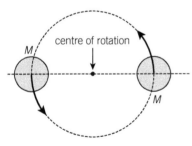

▲ Figure 2

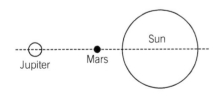

▲ Figure 3

$G = 6.67 \times 10^{-11} \text{Nm}^2\text{kg}^{-2}$, $g = 9.81 \text{Nkg}^{-1}$ at the Earth's surface.

The radius of the Earth is 6400 km and the mass of Earth is 6.0×10^{24} kg.

1 The gravitational field strength at the surface of a spherical planet of radius R is g_s.

Which one of the alternatives **A–D** in Table 1 gives the gravitational field strength and the gravitational potential at height R above the surface?

(1 mark)

2 Which one of the following alternatives **A–D** is the correct combination of SI base units for the potential gradient in a gravitational field?

A ms^{-2} **B** kgs^{-2} **C** $\text{m}^{-1}\text{s}^{-2}$ **D** kgms^{-2} *(1 mark)*

3 A spherical planet has a radius R and the gravitational potential at its surface is V_s. Which line **A–D** in the graph in Figure 1 shows how the gravitational potential V of its gravitational field varies from its surface with the distance r from the centre of the planet? *(1 mark)*

4 The Moon has a radius of 1740 km and its surface gravitational field strength is 1.62Nkg^{-1}.

a Use this data to calculate the mean density of the Moon. *(3 marks)*

b Calculate the escape velocity from the surface of the Moon. *(2 marks)*

5 Titan is the largest moon of the planet Saturn and it orbits Saturn once every 16.0 days in an orbit that has a mean radius of 1.22×10^9 m.

a Use this data to calculate the mass of Saturn. *(2 marks)*

b Saturn is a gaseous sphere which has a radius of 60 300 km. Calculate the mean density of Saturn. *(2 marks)*

c Titan has a mass of 1.35×10^{23} kg. Calculate the distance from the centre of Titan to the point at which the gravitational field strength of Titan is equal and opposite to the gravitational field strength of Saturn. *(4 marks)*

6 A certain binary star system consists of two identical stars, each of mass M. The two stars move in an orbit of diameter D about the midpoint between their centres (see Figure 2).

a **i** Show that for each star, its orbital speed, $v = \sqrt{\dfrac{GM}{2D}}$. *(2 marks)*

    ii Derive an expression for the time period T of each star. *(2 marks)*

b Two identical stars each of mass 2.0×10^{30} kg orbit each other with time period of 80 years. Calculate the distance between their centres in astronomical units (AU) where 1 AU (the mean distance between the Sun and the Earth) = 1.5×10^{11} m. *(3 marks)*

7 A steel spring has an unstretched length of 300 mm and a spring constant of 5.0Nm^{-1}. Its limit of proportionality is about 480 mm. Discuss whether or not the spring and a metre ruler could be used to measure the difference between g at the equator and g at the poles, which is about 0.05Nkg^{-1}. *(4 marks)*

8 Jupiter has a mass of 1.90×10^{27} kg and the mean radius of its orbit about the Sun is 7.8×10^{11} m. The mean radius of the orbit of Mars about the Sun is 2.3×10^{11} m.

a Estimate the gravitational field strength of Jupiter at Mars when Mars is closest to Jupiter, as shown in Figure 3 *(2 marks)*

b Calculate the resultant gravitational field strength at Mars due to the Sun and Jupiter when Mars is closest to Jupiter. *(2 marks)*

The mass of the Sun is 2.0×10^{30} kg.

22.1 Field patterns
22.2 Electric field strength

Specification reference: 3.7.3.2

Electrical conductors such as metals contain **free electrons**. These are electrons which move about inside the metal and are not attached to any one atom.

Electrically insulating materials do not contain free electrons. All the electrons in an insulator are attached to individual atoms. Some insulators, such as perspex or polythene, are easy to charge because their surface atoms easily gain or lose electrons.

Field lines and patterns

The path a free positive test charge (a small charged object) follows is called a **line of force** or a **field line**. The direction of an electric field line is the direction a positive test charge would move along.

Figure 1 shows the patterns of the field between two oppositely charged parallel plates. The field is *uniform* between the plates because the field lines are parallel to each other.

Electric field strength

The electric field strength, E, at a point in the field is defined as the force per unit charge on a positive test charge placed at that point.

The unit of E is the newton per coulomb (NC^{-1}). Electric field strength is a vector in the same direction as the force on a positive test charge.

If a positive test charge Q at a certain point in an electric field is acted on by force F due to the electric field, the electric field strength $E = \dfrac{F}{Q}$.

For two parallel plates with a potential difference (pd) V between the plates at separation d, $E = \dfrac{V}{d}$. This is because the work done moving Q between plates $= Fd = QV$ so $E = \dfrac{F}{Q} = \dfrac{V}{d}$.

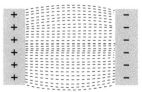

▲ **Figure 1** *A uniform electric field pattern*

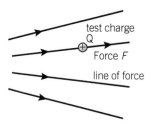

electric field strength, $E = \dfrac{F}{Q}$ (at Q)

▲ **Figure 2** *Electric field strength*

Summary questions

$e = 1.6 \times 10^{-19}\,C$

1 Figure 3 shows some lines of force of the electric field between a negatively charged point object Q and an earthed metal plate.
   a Copy Figure 3 and show the direction of the field lines. *(1 mark)*
   b State the direction of the force on an electron at:
      i point A   ii point B. *(2 marks)*

2 A charged metal sphere of radius 0.15 m has a charge of 2.5 nC.
   a The charge is distributed evenly on the surface of the sphere. Calculate the number of electrons per mm^2 on the surface. *(3 marks)*
   b When the sphere is discharged, the discharge current is about 0.2 mA. Estimate the discharge time. *(1 mark)*

3 a A beam of ions of charge $+3e$ is directed into a uniform electric field of strength 900 kV m^{-1}. Calculate the force on each ion. *(2 marks)*
   b The field is due to two oppositely charged parallel plates 50 mm apart. Calculate the pd between the plates. *(2 marks)*

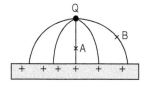

▲ **Figure 3**

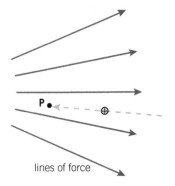

lines of force

▲ **Figure 1** *Electric potential energy*

Worked example 🖩

A + 0.5 µC test charge is moved into an electric field from infinity to reach a certain position P, where the electric potential is −3000 V. Calculate the electric potential energy of the test charge at P.

$E_P = QV = +0.5 \times 10^{-6} \, C \times (-3000 \, V)$
$= -1.5 \times 10^{-3} \, J$

The electric potential at a certain position in any electric field is the work done per unit positive charge on a positive test charge (i.e., a small positively charged object) when it is moved from infinity to that position. By definition, the position of zero potential energy is infinity

The unit of electric potential is the volt (V), equal to $1 \, J \, C^{-1}$. For a positive test charge Q placed at a position in an electric field where its electric potential energy is E_P, the electric potential V at this position is given by

$$V = \frac{E_P}{Q}$$

Note:

If a test charge $+Q$ is moved in an electric field from one position where the electric potential is V_1 to another where the electric potential is V_2, the work done $\Delta W = Q(V_2 - V_1)$.

Potential gradients

Equipotentials are surfaces of constant potential. The field lines cross the equipotentials at right angles to the surface.

The **potential gradient** $\frac{\Delta V}{\Delta x}$ at any position in an electric field is the change of potential per unit change of distance in a given direction.

At any point in an electric field, the **electric field strength $E = -\frac{\Delta V}{\Delta x}$**

The minus sign means that the electric field is in the opposite direction to the potential gradient. See Figure 2

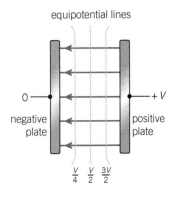

equipotential lines

negative plate — positive plate

$\frac{V}{4}$ $\frac{V}{2}$ $\frac{3V}{2}$

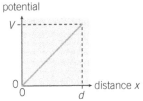

potential

▲ **Figure 2** *A uniform potential gradient*

Summary questions

$e = 1.6 \times 10^{-19} \, C$

1   An electron in a uniform electric field of strength $8.0 \, kV \, m^{-1}$ moves from X to Y through a distance of 28 mm in the opposite direction to the direction of the electric field.
   a   Calculate the work done by the electric field on the electron. *(2 marks)*
   b   The potential at X was + 400 V. Calculate the potential at Y. *(2 marks)*

2   An ion of charge $+2e$ was moved a distance of 36 mm from A where the potential was +1000 V to a point B at a higher potential of 6000 V.
   a   Calculate the work done by the ion when it moved from A to B. *(2 marks)*
   b   Estimate an average value for the potential gradient between A and B. *(2 marks)*

3   Two vertical parallel plates L and M are at a fixed separation of 80 mm. Plate L is at a potential of +4200 V and plate M is earthed.
   a   Sketch a graph to show how the potential between L and M varies along the shortest straight line between them. *(2 marks)*
   b   Calculate: i   the potential gradient between L and M
               ii   the electric field strength between the plates. *(3 marks)*

22.4 Coulomb's law
22.5 Point charges

Specification reference: 3.7.3.1; 3.7.3.3

Coulomb's law states that the force F between two point charges Q_1 and Q_2 is proportional to the amount of each charge and to the inverse of the square of the distance between the two charges. Coulomb's law can be expressed as the following equation $F = \dfrac{1}{4\pi\varepsilon_0}\dfrac{Q_1 Q_2}{r^2}$

where r is the distance between the two charges and $\dfrac{1}{4\pi\varepsilon_0}$ is the constant of proportionality. See Topic 22.5 of the Student Book if necessary.

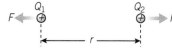

$$F = \frac{1}{4\pi\varepsilon_0}\frac{Q_1 Q_2}{r^2}$$

▲ **Figure 1** *Coulomb's law*

The electric field near a point charge

At distance r from a point charge Q,

$$\text{the electric field strength } E = \frac{1}{4\pi\varepsilon_0}\frac{Q}{r^2}$$

$$\text{the electric potential } V = \frac{1}{4\pi\varepsilon_0}\frac{Q}{r}$$

the potential energy of point charges q and Q (at separation r) $= qV = \dfrac{1}{4\pi\varepsilon_0}\dfrac{qQ}{r}$

Where two or more point charges are present, the resultant electric field strength is the vector sum of the electric field strengths of the point charges and the overall potential is the sum of the potentials of the point charges.

Graphs

Figure 3 shows how the electric field strength and the electric potential vary with distance from a point charge (or from the centre of a charged metal sphere).

Electric field strength = the negative of the gradient of a potential against distance graph.

Change of potential = area under an electric field strength against distance graph.

> **Revision tip**
> Remember to square r when calculating F.
> When substituting charges, get the powers of 10 correct: $\mu = 10^{-6}$, $n = 10^{-9}$, $p = 10^{-12}$.

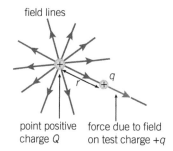

▲ **Figure 2** *Force near a point charge Q*

> **Revision tip**
> A negative E for the field of a point charge indicates the point charge is negative. But a negative V indicates a value less than zero: E is a vector, whilst V is a scalar.

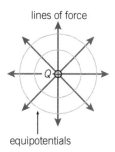

Summary questions

$\varepsilon_0 = 8.85 \times 10^{-12}\,\text{Fm}^{-1}$ $e = 1.6 \times 10^{-19}\,\text{C}$

1 Calculate the force between two identical ions, each of charge $+2e$, at a separation of $5.6 \times 10^{-7}\,\text{m}$. *(2 marks)*

2 Two charged point objects, X and Y, at a separation of $1.6 \times 10^{-2}\,\text{m}$ exert a force on each other of $5.1 \times 10^{-5}\,\text{N}$. When the charge of X is increased by 2.1 nC, the force increases to $7.8 \times 10^{-5}\,\text{N}$. Calculate the charge of each object. *(5 marks)*

3 A proton and an electron are at a separation of 0.11 nm.
   a Calculate their potential energy at this separation. *(2 marks)*
   b Calculate the electric field strength at the midpoint between the proton and electron. *(3 marks)*

4 The potential at a point P which is a distance of 80 mm from a point charge X is +6.2 kV.
   a Calculate: i the charge of X *(2 marks)*
              ii the potential at a distance of 1000 mm from X. *(2 marks)*
   b Calculate the magnitude of the electric field strength at P. *(2 marks)*

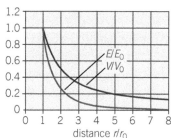

▲ **Figure 3** *The electric field and potential near a point charge*

22.6 Comparing electric fields and gravitational fields

Specification reference: 3.7.1

Revision tip

A force field is a region in which a body experiences a non-contact force.

Table 1 summarises the similarities and differences between electric and gravitational fields.

▼ **Table 1** *Similarities and differences between gravitational and electric fields*

	Gravitational fields	Electrostatic fields
	Similarities	
line of force or a field line	path of a free test mass in the field	path of a free positive test charge in the field
inverse-square law of force	Newton's law of gravitation $F = \dfrac{Gm_1m_2}{r^2}$	Coulomb's law of force $F = \dfrac{Q_1Q_2}{4\pi\varepsilon_0 r^2}$
field strength	force per unit mass, $g = \dfrac{F}{m}$	force per unit + charge, $E = \dfrac{F}{q}$
potential energy of two point mass or charges	$E_P = \dfrac{-Gm_1m_2}{r}$	$E_P = \dfrac{Q_1Q_2}{4\pi\varepsilon_0 r}$
potential	work done per unit mass moved from infinity	work done per unit + charge moved from infinity
uniform fields	Field strength is the same everywhere, field lines parallel and equally spaced	
radial fields	due to a point mass or a uniform spherical mass M, $g = \dfrac{GM}{r^2}$ $V = \dfrac{-GM}{r}$	due to a point charge Q, $E = \dfrac{Q}{4\pi\varepsilon_0 r^2}$ $V = \dfrac{Q}{4\pi\varepsilon_0 r}$

Summary questions

1 State two differences between a uniform field and a radial field.
(2 marks)

2 The equations for the field strength and potential of a point charge or a point mass can be written as $Z = kYr^n$ where Z is the field strength or potential. Complete the table below by inserting the correct symbols for Y and k and the correct value of n.
(2 marks)

	Z	Y	k	n
Gravitational field strength	g			
Electric field strength				
Gravitational potential	V			
Electric potential		Q		

3 a Express the unit of $\dfrac{1}{4\pi\varepsilon_0}$ in SI base units.
(2 marks)

   b The escape velocity from a planet of radius R is $\sqrt{(2gR)}$. Write down an equivalent equation for the escape velocity of a charged particle from an oppositely charged sphere of radius R.
(1 mark)

1 Two point charges S and T at separation *d* exert a force of attraction *F* on each other. The charge on T is then doubled without changing the charge on S then their separation is doubled.

Which one of the following alternatives, **A–D**, gives the force of attraction after the changes?

A $\frac{1}{4}F$ **B** $\frac{1}{2}F$ **C** $2F$ **D** $4F$ *(1 mark)*

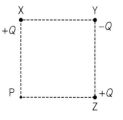
▲ **Figure 1**

2 Three point charges X, Y, and Z are placed at the vertices of a square with X and Z at opposite vertices as shown in Figure 1. Their charges are +*Q* for X and for Z and –*Q* for Y.

Point P in Figure 1 is the vertex of the square where there is no charge. Which one of the following alternatives, **A–D**, gives the direction of the resultant electric field strength at P?

A Parallel to PY away from Y

B Parallel to PY towards Y

C At 45° to PX to the left of PX

D At 45° to PZ below PZ *(1 mark)*

3 Which one of the following alternatives, **A–D**, gives a correct unit for $G\varepsilon_o$?

A $kg\,C^{-2}$ **B** $kg^2\,C^{-2}$ **C** $C kg^{-2}$ **D** $C^2 kg^{-2}$ *(1 mark)*

4 Figure 2 shows how the potential outside a positively charged metal sphere of radius 0.20 m varies with distance *r* from the centre of the sphere.

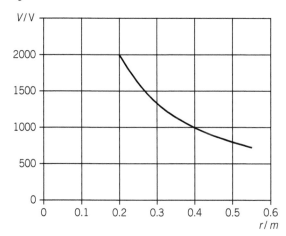

▲ **Figure 2**

a The potential at and outside the sphere is inversely proportional to *r*. Calculate the potential at a distance of 0.25 m from the centre of the sphere. *(2 marks)*

b The electric field at the surface and outside the sphere is the same as if the charge *Q* on the sphere was concentrated at its centre.

i Calculate the charge on the surface of the sphere. *(2 marks)*

ii Calculate the electric field strength at a point P which is a distance of 0.25 m from the centre of the sphere. *(2 marks)*

c An uncharged insulated metal rod XY of length 0.10 m is placed near the sphere along a radial line such that its nearest end X is at P.

i Explain why end X of the rod becomes negatively charged and end Y becomes positively charged. *(4 marks)*

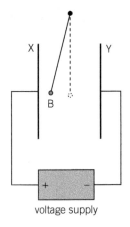

voltage supply

▲ **Figure 3**

ii Sketch a second curve on a copy of Figure 2 to show how the potential varies along the radial line from the surface of the sphere through X and through Y to a distance of 0.5 m from the centre of the sphere. *(4 marks)*

iii Discuss the effect the rod has had on the electric field strength at and near P. *(2 marks)*

5 A charged polystyrene ball B is suspended on the end of an insulating vertical thread between two vertical parallel plates as shown in Figure 3.

When a potential difference of 2800 V is applied between the two plates, the ball is attracted towards one of the plates and the thread is displaced through an angle θ of 5.5°.

a i Show that the force F on the ball due to the electric field is equal to $mg \tan \theta$, where m is the mass of the ball. *(2 marks)*

ii The mass of the ball is 0.82 g and the perpendicular distance d between the plates is 35 mm. Calculate the charge on the ball. *(4 marks)*

b When the distance d between the plates is increased, describe and explain the effect this change has on the ball. *(3 marks)*

6 Figure 4 shows two oppositely charged point charges A and B at a fixed separation of 0.25 m. Their total charge is + 4.0 nC.

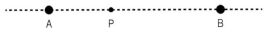

A   P   B

▲ **Figure 4**

a The total electric potential is zero at a point P which is 0.10 m from A along the line AB.

Calculate the charge of A and the charge of B. *(4 marks)*

b Copy Figure 4 and draw on your copy:

i three lines of force, each one from one charge to the other, including one through P *(1 mark)*

ii an equipotential line through P. *(2 marks)*

23.1 Capacitance
23.2 Energy stored in a charged capacitor
23.3 Charging and discharging a capacitor through a fixed resistor

Specification reference: 3.7.4.1; 3.7.4.3; 3.7.4.4

The capacitance C of a capacitor is defined as the charge stored per unit pd.

The unit of capacitance is the farad (F), which is equal to one coulomb per volt. Note that $1.0\,\mu F = 1.0 \times 10^{-6}\,F$.

For a capacitor that stores charge Q at pd V, its capacitance can be calculated using the equation

$$C = \frac{Q}{V}.$$

Charging a capacitor at constant current

In Figure 1, after the switch is closed, the variable resistor is continually adjusted to keep the microammeter reading constant and the capacitor pd is measured at regular intervals using a stopwatch and a high-resistance voltmeter.

The charge Q on the capacitor can be calculated using the equation $Q = It$, where I is the current and t is the time since the switch was closed.

A graph of charge stored, Q, on the y-axis against pd, V, on the x-axis gives a straight line passing through the origin. The gradient of the line is equal to the capacitance.

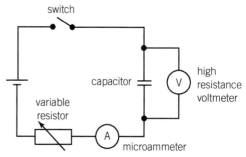

▲ **Figure 1** Investigating capacitors

Energy stored in a charged capacitor

The energy stored by the capacitor when its pd is V and it stores charge Q is given by

$$E = \frac{1}{2}QV.$$

> **Maths skill**
>
> 1 Using $Q = CV$ or $V = \frac{Q}{C}$ you can write the above equation as $E = \frac{1}{2}CV^2$ or $E = \frac{1}{2}\frac{Q^2}{C}$.
>
> 2 Figure 2 shows how the pd increases as the charge stored increases from zero. The total work done is given by the total area of the triangle under the line which $= \frac{1}{2} \times \text{base} \times \text{height} = \frac{1}{2}QV$.
>
> 3 In the charging process, the battery transfers energy QV to the circuit. Thus, 50% of the energy supplied by the battery $\left(= \frac{1}{2}QV\right)$ is stored in the capacitor. The other 50% is wasted due to resistance in the circuit as it is transferred to the surroundings when the charge flows in the circuit.

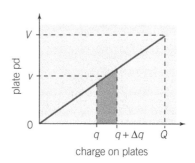

▲ **Figure 2** Energy stored in a capacitor

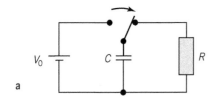

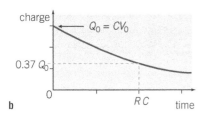

▲ Figure 3 *Capacitor discharging*

Capacitor discharge through a fixed resistor

When a capacitor discharges through a fixed resistor, the current and the charge decrease exponentially with time.

For capacitor discharge through a resistor of fixed resistance R, when the charge is Q and the current is $I\left(=\dfrac{dQ}{dt}\right)$,

the capacitor pd $\left(\dfrac{Q}{C}\right)$ + the resistor pd (IR) = 0.

So $I = -\dfrac{Q}{RC}$. Therefore, $\dfrac{dQ}{dt} = -\dfrac{Q}{CR}$.

The mathematical solution is where $Q = Q_{0}e^{-\frac{t}{RC}}$ where Q_{0} is the initial charge, and e is the exponential function (sometimes written 'exp').

> **Maths skill: Data analysis using natural logarithms**
>
> 1 The quantity RC is called the **time constant** for the circuit. At time $t = RC$ after the start of the discharge, the charge falls to 0.37 $(= e^{-1})$ of its initial value. Note the half-life of the discharge $= RC \ln 2$.
>
> 2 Taking natural logs of both sides of the above equation gives $\ln Q = \ln Q_{0} - \dfrac{t}{RC}$.
>
> A graph of $\ln Q$ on the y-axis against t on the x-axis gives a straight line with a negative gradient $\left(=-\dfrac{1}{RC}\right)$.

Revision tip

An exponential decrease graph starts at an intercept with the vertical axis and is asymptotic with the time axis (i.e., the curve approaches but never cuts the time axis).

Charging a capacitor through a fixed resistor

1 At any instant during the charging process,

the source pd V_{0} = the resistor pd + the capacitor pd.

Therefore, $V_{0} = IR + \dfrac{Q}{C}$ at any instant.

2 The initial current $I_{0} = \dfrac{V_{0}}{R}$, assuming that the capacitor is initially uncharged.

3 At time t after charging starts, $I = I_{0}e^{\frac{-t}{RC}}$

Rearranging this gives $Q = CV_{0}\left(1 - e^{\frac{-t}{RC}}\right)$

At time $t = 0$, $e^{\frac{-t}{RC}} = 1$, so $Q = 0$ and the capacitor pd $V = 0$.

As time $t \to \infty$, $e^{\frac{-t}{RC}} \to 0$ so $Q \to Q_{0}$ and $V \to V_{0}$.

Worked example 🖩

A 470 µF capacitor is charged to a pd of 6.0 V then discharged through a 220 kΩ resistor using a circuit as shown in Figure 1. Calculate the pd after 200 s.

When $t = 200\,\text{s}$,

$$\frac{t}{RC} = \frac{200\,\text{s}}{220\times10^{3}\,\Omega \times 470\times10^{-6}\,F}$$

$$= 1.93$$

Therefore $V = V_{0}e^{-\frac{t}{RC}} = 6.0\,\text{V} \times e^{-1.93}$

$= 0.87\,\text{V}$

> **Summary questions**
>
> 1 A 47 µF capacitor is charged by using a constant current of 1.5 µA to a pd of 6.0 V. Calculate:
>   a the charge stored on the capacitor at 6.0 V   *(1 mark)*
>   b the time taken.   *(1 mark)*
>
> 2 A 22 µF capacitor was charged by connecting it to a 3.0 V battery. The battery was then disconnected from the charged capacitor and a 10 µF capacitor which was initially uncharged was then connected across the charged capacitor.
>   a Calculate the charge stored on the 22 µF capacitor when it was connected to the battery.   *(1 mark)*

b The pd across the capacitors after they were connected together
was less than 3.0 V.

  i Explain why the pd was less than 3.0 V   *(2 marks)*

  ii Calculate the pd across the capacitors after they were
connected together.   *(4 marks)*

3 An uncharged 10 μF capacitor is connected to a 6.0 V battery. Calculate:

  a **i** the charge and energy stored in the capacitor   *(2 marks)*

     ii the energy supplied by the battery.   *(1 mark)*

  b Explain the difference between the energy supplied by the battery
and the energy stored in the capacitor.   *(2 marks)*

4 A 47 μF capacitor is charged by connecting it to a 6.0 V battery then
discharged through a 200 kΩ resistor.

  a **i** Calculate the time constant of the circuit.   *(1 mark)*

     ii Estimate how long the capacitor would take to
discharge to 2 V.   *(2 marks)*

  b Calculate the pd and the discharge current 5.0 s after the
discharge started.   *(3 marks)*

5 An uncharged 68 μF capacitor is charged to a pd of 5.0 V through a
20 kΩ resistor. Calculate:

  a **i** the initial charging current   *(1 mark)*

     ii the current when the pd across the capacitor is 3.0 V.   *(2 marks)*

  b the time taken for the pd to increase from zero to 3.0 V.   *(3 marks)*

a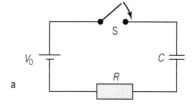

b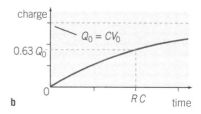

c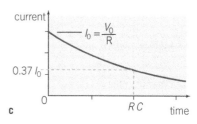

▲ **Figure 4** *Capacitor charging*

23.4 Dielectrics

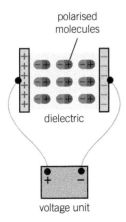

polarised molecules

dielectric

voltage unit

▲ **Figure 1** *Dielectric action*

Dielectrics are insulators used to increase the capacitance of a capacitor. When a dielectric is placed between two oppositely charged parallel plates connected to a battery, each molecule of the dielectric becomes **polarised** and forms **a dipole**. This means that its electrons are pulled slightly towards the positive plate as shown in Figure 1. So the surface of the dielectric facing the positive plate becomes negatively charged and the other surface of the dielectric becomes positively charged. As a result, the positive side of the dielectric attracts more electrons from the battery onto the negative plate. and the negative side of the dielectric pushes electrons back to the battery from the positive plate. So more charge is stored on the plates. In some dielectric substances, the molecules are already polarised. These molecules are called polar molecules. The electric field makes them turn so their electrons become nearer the positive plate. As a result, more charge is stored on the plates as explained above.

Relative permittivity

The **relative permittivity** (or dielectric constant) of a dielectric substance,

$$\varepsilon_r = \frac{\text{charge stored with dielectric present, } Q}{\text{charge stored without dielectric, } Q_0}$$

Capacitor design

For a parallel-plate capacitor with dielectric between the plates, its capacitance $C = \dfrac{A\varepsilon_0\varepsilon_r}{d}$ where A is the surface area of each plate and d is the spacing between the plates. A large capacitance can be achieved by increasing A, decreasing d and using a dielectric which has a relative permittivity as large as possible.

Dielectrics and alternating potential differences

In an alternating electric field, at low frequencies, the dipoles alternate in phase with the field regardless of the mass of the dipoles. As the frequency increases, the mass of the particles being moved by the field determines if they respond. The greater the mass of each particle, the greater its inertia is, and so the frequency at which it ceases to respond is lower.

Summary questions

$\varepsilon_0 = 8.85 \times 10^{-12}\,\text{Fm}^{-1}$

1   A parallel-plate capacitor consists of two insulated metal plates separated by an air gap. A battery in series with a switch is connected to the plates. The capacitor is charged by closing the switch to charge the capacitor to a pd V.
   a   When the separation of the plates is increased, state and explain the change that takes place to:
      i   the capacitance C of the capacitor   (2 marks)
      ii the pd between the plates.   (2 marks)
   b   State and explain the change that takes place to the energy stored by the capacitor when the plates are moved further apart.   (2 marks)

2   Explain why a dielectric material in a capacitor increases its capacitance compared with the same capacitor with the same plate spacing and without dielectric.   (4 marks)

Chapter 23 Practice questions

1 A capacitor of capacitance C is charged to a potential difference V then discharged through a variable resistor which is adjusted to keep the current at a constant value I for as long as possible. Which one of the following alternatives, **A–D**, gives the time taken for the discharge?

   A $\dfrac{V}{CI}$ **B** $\dfrac{CV}{I}$ **C** $\dfrac{I}{CV}$ **D** $\dfrac{IC}{V}$ *(1 mark)*

2 A uncharged capacitor is connected in series with a resistor, an open switch, and a battery. When the switch is closed, the capacitor charges to the same pd as the battery.
   Which one of the following alternatives, **A–D**, describes how the charge and the current changes with time?   *(1 mark)*

	charge	current
A	Increases at a decreasing rate	Increases at a decreasing rate
B	Increases at an increasing rate	Increases at a decreasing rate
C	Increases at a decreasing rate	Decreases at a decreasing rate
D	Increases at an increasing rate	Decreases at an increasing rate

3 A capacitor charges through a resistor R from a battery to a potential difference V_B. The potential difference across the capacitor after time T from the start of the charging process is $0.40V_B$. Which one of the following alternatives, **A–D**, gives the potential difference across the capacitor after a further time T?

   A $0.50V_B$ **B** $0.60V_B$ **C** $0.64V_B$ **D** $0.80V_B$ *(1 mark)*

4 The circuit in Figure 1 is used to charge a capacitor using a constant current.

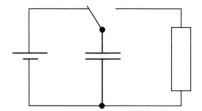

▲ **Figure 1**

   a The current is maintained at $30\,\mu\text{A}$ and the voltmeter reading is measured every 15 seconds until the capacitor is fully charged. The measurements are shown in the table below.

Time t / s	0	15	30	45	60	75	90
Capacitor pd V/ V	0	0.51	1.02	1.50	1.98	2.52	3.01
Charge stored Q / mC	0	0.45					

   i Complete the last row of the table by calculating Q at time t.

              (1 mark)

   ii Plot a graph of Q on the y-axis against V on the x-axis.   *(3 marks)*

   iii Use your graph to determine the capacitance of the capacitor.

              (2 marks)

   b Discuss how the voltmeter reading would have changed with time if the variable resistor had not been adjusted.   *(2 marks)*

5 A 2200 μF capacitor is charged to a potential difference of 5.5 V.

    a Calculate the energy stored in the capacitor. *(2 marks)*

    b **i** The charged capacitor is to be connected to a 5.0 V electric motor which has a thread attached to its axis supporting a small object of mass 3.8 g at its lower end.
Estimate the maximum possible height the object could be raised when the capacitor is discharged through the motor. *(3 marks)*

       ii Give two physical reasons why the height gain of the object would be less than your estimate above. *(2 marks)*

6 The potential difference across a capacitor was measured as it discharged through a 47 kΩ resistor. The measurements are shown in the table below:

Time t / s	0	30	60	90	120	150	180
Capacitor pd V/ V	5.95	4.55	3.60	2.70	2.15	1.70	1.35
ln V	1.78	1.52					

    a Use the equation $V = V_0 e^{-\frac{t}{RC}}$, where V_0 is the initial pd across the capacitor to show that a graph of ln V against time t should be straight line. *(3 marks)*

    b **i** Complete the last row of the table and use the results in the table to plot a graph of ln V against time t. *(5 marks)*

       ii Use your graph to determine the capacitance of the capacitor. *(4 marks)*

      iii Calculate the time constant of the circuit. *(1 mark)*

7 **a** A dielectric substance is inserted into the space between the plates of a parallel-plate capacitor which is connected to a battery.

      i Describe the effect on the dielectric of the electric field between the capacitor plates. *(3 marks)*

      ii Explain why the charge stored by the capacitor increases when the dielectric is inserted. *(3 marks)*

    b The battery is disconnected from the capacitor without altering the charge stored. When the dielectric is removed from the capacitor:

      i describe the effect on the capacitance of the capacitor *(1 mark)*

      ii state and explain the effect on the potential difference and the energy stored by the capacitor. *(5 marks)*

    c A 47 μF capacitor has an effective surface area of 1.2 m² and it contains a dielectric which has a relative permittivity of 7. Calculate the spacing between the capacitor plates, assuming the dielectric fills the entire space between the plates. *(2 marks)*

8 A 68 μF capacitor is charged to a pd of 6.0 V and then discharged through a 50 kΩ resistor.

    a Calculate the time taken after the discharge started for the pd to decrease to 1.5 V. *(3 marks)*

    b Estimate the resistance of the resistor that you would use in place of the 50 kΩ resistor if the discharge to 1.5 V is to take about 20 s. *(2 marks)*

24.1 Current-carrying conductors in a magnetic field

Specification reference: 3.7.5.1

Magnetic field patterns

A magnetic field is a force field surrounding a magnet or current-carrying wire which acts on any other magnet or current-carrying wire placed in the field. The magnetic field of a bar magnet is strongest at its ends which are referred to as **north-seeking** and **south-seeking** 'poles'. A **line of force** (or magnetic field line) of a magnetic field is a line along which a north pole would move in the field.

The force on a current-carrying wire in a magnetic field

A current-carrying wire placed at a non-zero angle to the lines of force of an external magnetic field experiences a force due to the field. This effect is known as the **motor effect**. The force is perpendicular to the wire and to the lines of force.

The motor effect can be tested using the simple arrangement shown in Figure 3. The wire is placed between opposite poles of a U-shaped magnet so it is at right angles to the lines of force of the magnetic field. When a current flows, the section of the wire in the magnetic field experiences a force that pushes it out of the field. The force is:

- greatest when the wire is at right angles to the magnetic field
- zero when the wire is parallel to the magnetic field.

The direction of the force can be related to the direction of the field and to the direction of the current using **Fleming's left-hand rule** shown in Figure 3. If the current is reversed or if the magnetic field is reversed, the direction of the force is reversed.

The magnitude of the force on the wire is proportional to:

1 the current I
2 the length l of the wire.

The **magnetic flux density** B of the magnetic field, sometimes referred to as the strength of the magnetic field, is defined as the force per unit length per unit current on a current-carrying conductor at right angles to the magnetic field lines.

Therefore, for a wire of length l carrying a current I in a uniform magnetic field B at 90° to the field lines, the force F on the wire is given by

$$F = BIl$$

1 The unit of B is the tesla (T), equal to $1\,\text{Nm}^{-1}\text{A}^{-1}$.
2 The direction of the force is given by Fleming's left-hand rule. See Figure 3.

Applications

1 The electric motor

The simple electric motor consists of a coil of insulated wire which spins between the poles of a U-shaped magnet.

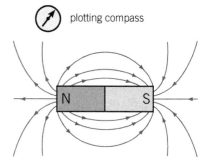

▲ **Figure 1** *The magnetic field near a bar magnet*

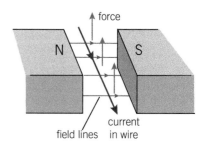

▲ **Figure 2** *The motor effect*

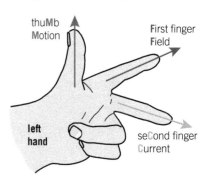

▲ **Figure 3** *Fleming's left-hand rule*

Revision tip

$F = BIl$ applies only when B and I are at right angles.

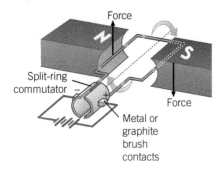

▲ **Figure 4** *In an electric motor*

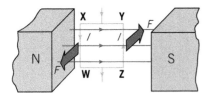

a *top view of coil parallel to field*

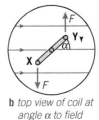

b *top view of coil at angle α to field*

▲ **Figure 5** *Couple on a coil*

When a direct current passes round the coil:

- the wires at opposite edges of the coil are acted on by forces in opposite directions
- the force on each edge makes the coil spin about its axis.

Current is supplied to the coil via a split-ring commutator. The direction of the current round the coil is reversed by the split-ring commutator each time the coil rotates through half a turn. This ensures the current along an edge changes direction when it moves from one pole face to the other. As shown in Figure 4, the result is that the force on each edge continues to turn the coil in the same direction.

2 The couple on a coil in a magnetic field

Consider a rectangular current-carrying coil of length l and width w in a uniform horizontal magnetic field, as shown in Figure 5. The coil has n turns of wire and can rotate about a vertical axis. When the current in the coil is I:

- Each long side therefore experiences a horizontal force $F = (BIl)n$ in opposite directions at right angles to the field lines.
- The pair of forces acting on the long sides form a couple. If the plane of the coil is at angle $α$ to the field lines, the torque of the couple = $Fw\cos α = BIlnw\cos α = BIAn\cos α$, where the coil area $A = lw$.

If $α = 0$ (i.e., the coil is parallel to the field), the torque = $BIAn$ as $\cos 0 = 1$.

If $α = 90°$ (i.e., the coil is perpendicular to the field), the torque = 0 as $\cos 90° = 0$.

Summary questions

1 A straight vertical wire of length 80 mm carrying a constant current of 4.3 A downwards is in a uniform horizontal magnetic field which acts due North. The wire experiences a force of 29 mN.
   a Determine the magnitude of the magnetic flux density of the magnetic field (*2 marks*)
   b State the direction of the force. (*1 mark*)

2 A square coil of length 60 mm has 50 turns. The coil is in a uniform horizontal magnetic field of flux density 55 mT which is perpendicular to the plane of the coil. A current of 5.9 A is in the coil. Sketch the arrangement and determine the magnitude and direction of the force on each side of the coil. (*4 marks*)

3 A horizontal cable suspended between two pylons is aligned in a north-south plane. The Earth's magnetic field at this position has a vertical component of 66 μT and is in a downwards direction. Determine the magnitude of the force per unit length on the cable when the current in it is 65 A. (*2 marks*)

24.2 Moving charges in a magnetic field
24.3 Charged particles in circular orbits
Specification reference: 3.7.5.1; 3.7.5.2

Force on a moving charge in a magnetic field

For a particle of charge Q moving through a uniform magnetic field at speed v in a perpendicular direction to the field, the force on the particle is given by

$$F = BQv$$

The velocity selector

This apparatus has a magnetic field and an electric field perpendicular to each other. A charged particle entering the fields perpendicular to both fields is undeflected if its speed is such that the magnetic field force on it equals the electric field force on it.

$$B_s Qv = \frac{QV_P}{d}$$

In a velocity selector, ions moving at different speeds in a narrow beam are directed into the arrangement. The ions moving at speed $v = \frac{V_P}{B_s d}$ (from the above equation) pass through undeflected and can therefore pass through a collimator slit that stops all the other ions in the beam.

Charged particles in circular orbit

The magnetic force on a charged particle moving in a magnetic field is always perpendicular to the velocity at any point along the path. The particle therefore moves in a circular path with the force always acting towards the centre of curvature of the circular path.

Figure 2 shows a beam of electrons in a uniform magnetic field on a circular path. At any point on the path, the magnetic force BQv = the centripetal force $\frac{mv^2}{r}$ where r is the radius of the circular path.

Rearranging this equation gives $r = \frac{mv}{BQ}$.

Applications

The **cyclotron** consists of two hollow D-shaped electrodes (referred to as 'dees') in a vacuum chamber. A uniform magnetic field and an alternating pd are applied to the dees as shown in Figure 3.

Charged particles entering one of the dees near the centre of the cyclotron are forced on a semi-circle by the magnetic field, causing them to emerge from the dee. As they cross into the other dee, the alternating voltage reverses so they are accelerated into the other dee where they are once again forced by the magnetic field on a semi-circle with a larger radius. On emerging from this dee, the process is repeated until the particles leave one of the dees through a gap at the edge.

The particles are accelerated each time they cross from one dee to the other because the time taken t to complete each semi-circle equals $\frac{\pi r}{v}$ where v is their speed and r is the radius of the semi-circle.

The radius $r = \frac{mv}{BQ}$ therefore $t = \frac{m\pi}{BQ}$ which is independent of the particle's speed and is equal to the time taken for half a cycle of the alternating voltage.

> **Revision tip**
> The equation $F = BQv\sin\theta$ (when the direction of motion of a charged particle in a magnetic field is at angle θ to the lines of the field) is not required for this specification.

> **Revision tip**
> The direction of the magnetic force on a moving electron is given by Fleming's left-hand rule. For negative particles, remember that the current direction is opposite to the direction in which the particles move.

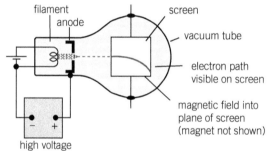

▲ **Figure 1** *Electrons in a magnetic field*

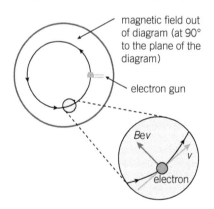

▲ **Figure 2** *A circular orbit in a magnetic field*

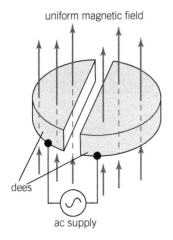

uniform magnetic field

dees

ac supply

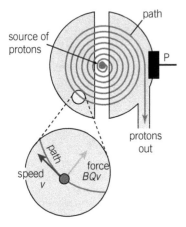

source of protons

path

P

protons out

speed v

path

force BQv

▲ **Figure 3** *The cyclotron*

The mass spectrometer

In a mass spectrometer, the atoms of the sample are ionised and directed in a narrow beam at the same velocity (after passing through a velocity selector) into a uniform magnetic field. Each ion is deflected in a semi-circle by the magnetic field onto a detector. The radius of curvature of the path of each ion depends on the specific charge $\frac{Q}{m}$ of the ion in accordance with the equation $r = \frac{mv}{BQ}$. Each type of ion is deflected by a different amount onto the detector.

The detector is linked to a computer which is programmed to show the relative abundance of each type of ion in the sample.

Summary questions

$e = 1.6 \times 10^{-19}\,\text{C}$

1 Calculate the force on an electron that enters a uniform magnetic field of flux density 150 mT at a velocity of $8.0 \times 10^6\,\text{ms}^{-1}$ at an angle of:

   a 90°   *(1 mark)*

   b 0° to the field.   *(1 mark)*

2 A beam of protons at a speed of $3.2 \times 10^7\,\text{ms}^{-1}$ is directed into a uniform magnetic field of flux density 85 mT in a direction perpendicular to the field lines. The protons move on a circular path in the field.

   mass of proton $= 1.67 \times 10^{-27}\,\text{kg}$

   a Calculate the radius of curvature of the path.   *(3 marks)*

   b The flux density is halved and the kinetic energy of the protons is halved. Determine the effect of this change on the radius of curvature of the path.   *(3 marks)*

3 In a mass spectrometer, a beam of ions was directed in a velocity selector and the ions that emerged had a speed of $9.5 \times 10^5\,\text{ms}^{-1}$.

   a The electric field strength in the velocity selector was 82 kVm⁻¹. Calculate the magnetic flux density in the velocity selector.   *(2 marks)*

   b These ions were then directed into a uniform magnetic field of flux density 910 mT where they were deflected in a semi-circular path of diameter 370 mm on to the detector. Calculate the specific charge of these ions.   *(3 marks)*

1 A velocity selector is an arrangement in which a uniform electric field is perpendicular to a uniform magnetic field. The magnetic flux density is B and the speed of the charged particles that pass through undeflected is v. Which one of the following alternatives, **A–D**, gives the electric field strength in this situation?

A vB **B** $\dfrac{B}{v}$ **C** $\dfrac{v}{B}$ **D** B *(1 mark)*

2 In an electron tube, electrons of mass m are accelerated from rest through a potential difference V before they enter a uniform magnetic field in a direction perpendicular to the field lines. The electrons move on a circular path of radius r in the field when the magnetic flux density is B. Which one of the following alternatives, **A–B**, gives the equation for r in terms of V and B?

A $\sqrt{\dfrac{2mV}{eB}}$

B $\sqrt{\dfrac{2mV}{eB^2}}$

C $\sqrt{\dfrac{2meV}{B^2}}$

D $\sqrt{\dfrac{2meV}{eB}}$ *(1 mark)*

3 In a high-energy collision in a magnetic field, several particles and antiparticles emerge from the collision, each on a different curved track. One of the tracks P is created by a proton. A second track X curves in the same direction as P with a smaller radius of curvature than P. Which one of the following statements about the particle that created track X is correct?

A X could be a proton with a greater momentum than P

B X could be an antiproton with less momentum that P

C X could be a π^+ meson with less momentum than P

D X could be a π^- meson with greater momentum than P   *(1 mark)*

4 **a** A charged particle in a uniform magnetic field is moving along a circular path at a constant speed. Explain why the particle moves a constant speed even though it is acted on by a force due to the magnetic field.   *(3 marks)*

   b The most abundant isotope of neon is $^{20}_{10}\text{Ne}$. A sample of neon gas is analysed in the spectrometer. Each neon ion carries a single charge of $+1.6 \times 10^{-19}\,\text{C}$. The neon ions enter the magnetic field of the spectrometer at a velocity of $7.3 \times 10^5\,\text{m s}^{-1}$ and they are deflected by the magnetic field through $180°$ onto a detector.

   i The ions strike the detector $0.31\,\text{m}$ from the point **P** at which the ion beam enters the magnetic field. Calculate the magnetic flux density of the magnetic field.   *(3 marks)*

   ii Neon ions are also detected at a distance of $0.34\,\text{m}$ from point **P**. Calculate the relative atomic mass of these neon ions.   *(2 marks)*

   iii Explain why these ions have a different mass from the $^{20}_{10}\text{Ne}$ ions.   *(2 marks)*

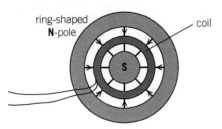

▲ **Figure 1**

5 In an earphone, a circular coil of insulated wire is in a radial magnetic field as shown in Figure 1. The coil is supported on a thin flexible disc attached to the ring-shaped pole.

   a The coil has 60 turns and a diameter of 8.2 mm. The magnetic field between the poles has a magnetic flux density of 120 mT. Calculate the force on the coil due to the magnetic field when the current in the coil is 0.19 A. *(2 marks)*

   b i Explain why the coil vibrates when there is an alternating current in the coil. *(2 marks)*

     ii Discuss how the amplitude of the vibrations of the coil varies with the rms current in the coil. *(2 marks)*

6 A current balance consists of a rectangular frame ABCD balanced on pivots midway along AB and CD as shown in Figure 2. A small piece of wire W of mass 0.67 g is positioned along AB to balance the frame horizontally. End AD is positioned at the end of the solenoid at right angles to the solenoid axis.

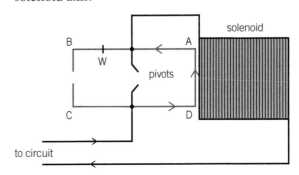

▲ **Figure 2**

When a constant current of 4.5 A is passed through the solenoid and along end AD of the solenoid in series with one another, W is moved 32 mm towards the pivot to regain the balance. Given AB = CD = 240 mm and AD = BC = 95 mm, calculate:

   a the force on end AD due to the magnetic field of the solenoid *(4 marks)*

   b the magnetic flux density at the end of the solenoid. *(2 marks)*

7 In a direct-current electric motor, the coil has 160 turns wound on a rectangular frame of length 35 mm and width 28 mm. The magnetic field in which the coil spins has a magnetic flux density of 110 mT.

   a The current in the coil is 0.67 A. Calculate:

     i the force on each long side of the coil when the current in the coil is 0.67 A *(1 mark)*

     ii the maximum torque on the coil due to the magnetic field. *(2 marks)*

   b A pulley of diameter 5.4 mm fixed to the motor axle is used to raise a 1.4 N weight W attached to a thread wrapped round the pulley. Estimate the least current in the coil that would be necessary to raise the weight. *(4 marks)*

Specification reference: 3.7.5.4

Investigating electromagnetic induction

When a magnet is moved near a wire connected to a sensitive meter, an electromotive force (emf) is **induced** in the wire and a current passes through the meter. See Figure 1. The **induced emf** forces electrons round the circuit. The induced emf can be increased by:

- moving the wire faster
- using a stronger magnet
- making the wire into a coil and pushing the magnet in or out of the coil.

Energy changes

The rate of transfer of energy from the source of emf to the other components of the circuit is equal to the product of the induced emf and the current. This is because the induced emf × the current = energy transferred per unit charge from the source × the charge flow per second = energy transferred per second from the source.

The dynamo rule

Look at Figure 2. The magnetic field is into the plane of the diagram and the motion of the conductor is to the right. The electrons in the rod experience a force downwards due to the magnetic field so the top end becomes positive and the lower end becomes negative so an emf is induced in the rod. If the rod is part of a complete circuit, an induced current passes round the circuit. The direction of the induced current in the rod is given by **Fleming's right-hand rule**, as shown in Figure 3. The direction of the induced current is, in accordance with the current convention, opposite to the direction of the flow of electrons in the conductor.

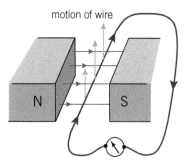

▲ **Figure 1** *Generating an electric current*

Revision tip

An emf is induced whenever there is relative motion between the coil and the magnet. The wire must cut across the lines of the magnetic field for an emf to be induced in the wire.

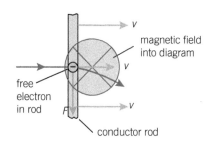

▲ **Figure 2** *Deflection of electrons in a magnetic field*

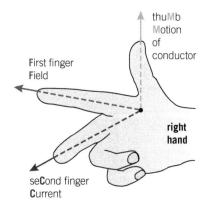

▲ **Figure 3** *Fleming's right-hand rule*

Summary questions

1 A straight section XY of insulated wire is connected to a sensitive meter.
   a Explain why the meter shows a brief reading when XY is moved between the poles of a U-shaped magnet. *(2 marks)*
   b State one way in which the meter reading could be made larger. *(1 mark)*

2 In Q1, XY is vertical with X above Y. The magnetic field lines are horizontal in a direction from east to west. State and explain the direction of the induced current in the wire when it is moved in a direction from south to north without changing its vertical alignment. *(2 marks)*

3 A flat circular coil of wire is connected to a sensitive centre-reading meter. The coil is placed between the poles of a U-shaped magnet with its plane perpendicular to the field lines. The coil is then removed from between the poles then returned to its original position between the poles. Describe and explain how the meter reading changes if it is removed quickly and returned slowly. *(4 marks)*

25.2 The laws of electromagnetic induction

Specification reference: 3.7.5.3; 3.7.5.4

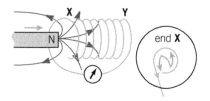

▲ **Figure 2** *Lenz's law*

Revision tip

The induced current could never be in a direction to help the change that causes it – that would mean producing electrical energy from nowhere, which is forbidden by conservation of energy.

Summary questions

1 A horizontal metal rod of length 0.063 m is moved at a speed of 12 mms⁻¹ across a vertical magnetic field of flux density 72 mT in a direction perpendicular to its length. Calculate:

   a the magnetic flux swept out by the rod each second   (*3 marks*)

   b the induced emf in the rod as it cuts across the field lines.   (*1 mark*)

2 A solenoid is connected to a centre-reading meter. When a bar magnet is brought near end X of a solenoid, the meter deflects briefly.

   a Explain why the meter deflects briefly.   (*3 marks*)

   b State and explain how the meter reading changes when the magnet is moved away from end X of the solenoid. (*3 marks*)

The solenoid rule

Figure 1 shows the magnetic field pattern of a current-carrying solenoid. If each end in turn is viewed from outside the solenoid:

- current passes a**N**ticlockwise (or cou**N**terclockwise) round the '**N**orth pole' end
- current passes clockwise round the 'south pole' end.

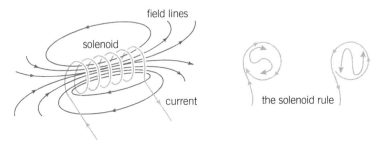

▲ **Figure 1** *The magnetic field near a solenoid*

Lenz's law

Consider the north pole of a bar magnet approaching end X of a coil, as shown in Figure 2. The induced current in the circuit creates a magnetic field in the coil. The induced polarity of end X must be a N-pole so as to repel the incoming N-pole. **Lenz's law states that the direction of the induced current is always such as to oppose the change that causes the current.**

Magnetic flux and flux linkage

Consider a uniform magnetic field of magnetic flux density B which is perpendicular to an area A. The **magnetic flux ϕ through this area $A = BA$.** The unit of magnetic flux is the **weber** (Wb), equal to 1 Tm². The **magnetic flux linkage** through a coil of N turns and area $A = N\phi = NBA$ When the magnetic field is at angle θ to the normal at the coil face, the flux linkage through the coil $= BAN\cos\theta$

Faraday's law of electromagnetic induction

Faraday's law of electromagnetic induction states that the induced emf in a circuit is equal to the rate of change of flux linkage through the circuit.

$$\text{Induced emf } \varepsilon = -N\frac{\Delta\phi}{\Delta t}$$

where $N\dfrac{\Delta\phi}{\Delta t}$ is the change of flux linkage per second. The minus sign represents the fact that the induced emf acts in such a direction as to oppose the change that causes it (as per Lenz's law).

The unit of flux change per second, the weber per second, is the same as the volt. Therefore, the weber is equal to 1 volt second. When a wire of length L moving at speed v cuts across the lines of a magnetic field of flux density B, the induced emf $= BLv$ (because the area swept out by the rod $= Lv$ so the flux swept out per second $= BLv$).

25.3 The alternating current generator
25.4 Alternating current and power
Specification reference: 3.7.5.4; 3.7.5.5

The alternating current generator

A simple ac generator is shown in Figure 1. When the coil spins at a steady rate, the flux linkage changes continuously. At an instant when the normal to the plane of the coil is at angle θ to the field lines, **the flux linkage through the coil equals BANcos θ** where B is the magnetic flux density, A is the coil area, and N is the number of turns on the coil. For a coil spinning at a steady frequency, f, $\theta = 2\pi ft$ at time t after $\theta = 0$. See Figure 2. The flux linkage = $BAN\cos 2\pi ft$. The induced emf $\varepsilon = \varepsilon_o \sin 2\pi ft = \varepsilon_o \sin \omega t$ where $\omega = 2\pi f$ and the peak emf, $\varepsilon_o = BAN\omega$. The gradient of the graph is the change of flux linkage per second $\dfrac{N\Delta\phi}{\Delta t}$, so is equal to $-\varepsilon$.

Back emf

A back emf, ε, is induced in the spinning coil of an electric motor because the flux linkage through the coil changes. The induced emf acts against the pd V applied to the motor in accordance with Lenz's law. At any instant, $V - \varepsilon = IR$, where I is the current through the motor coil and R is the circuit resistance.

Because the induced emf is proportional to the speed of rotation of the motor at low speed, the current is high because the induced emf is small, and at high speed, the current is low because the induced emf is high.

Alternating current and power

The **peak value** of an alternating current I_o (or pd V_o) is the maximum current (or pd) in either direction. Power ($= I^2R$) supplied to a resistor by an alternating current varies with time. At peak current I_o, maximum power is supplied equal to I_o^2R. The mean power $= \dfrac{1}{2}I_o^2R$.

The root mean square value of an alternating current I_{rms} is the value of direct current that would give the same heating effect as the alternating current in the same resistor.

Therefore $\quad I_{rms}^2R = \dfrac{1}{2}I_o^2R \quad$ so $\quad I_{rms} = \dfrac{1}{\sqrt{2}}I_o \quad$ and $\quad V_{rms} = \dfrac{1}{\sqrt{2}}V_o$

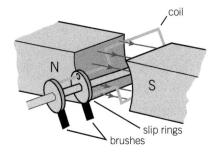

▲ **Figure 1** *The ac generator*

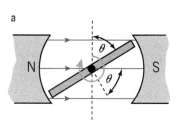

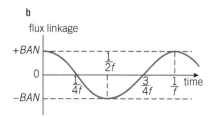

▲ **Figure 2** *Flux linkage in a spinning coil*

> **Revision tip**
>
> The induced emf is zero when the sides of the coil move parallel to the field lines and maximum when the sides of the coil cut the field lines at right angles.

> **Revision tip**
>
> **The peak-to-peak value** of an alternating current or pd (from the top to the bottom of the wave) = 2 × the peak value. An oscilloscope may be used to measure the peak-to peak value of an alternating pd. See Topic 4.7, Using an oscilloscope.

Summary questions

1   The coil of an ac generator has 120 turns, a length of 22 mm, and a width of 18 mm. It spins in a uniform magnetic field of flux density 110 mT at a constant frequency of 50 Hz.
    a   Calculate the maximum flux linkage through the coil.   *(2 marks)*
    b   Show that the peak voltage is 1.6 V.   *(2 marks)*

2   An electric motor is used to raise a weight. The motor is connected in series with a battery and an ammeter. Explain why the motor current increases when the load on it is increased.   *(3 marks)*

3   A mains heater gives a mean output power of 1.0 kW when the rms pd across it is 230 V. Calculate:
    a   the rms current   *(1 mark)*
    b   the peak current   *(1 mark)*
    c   the peak power.   *(2 marks)*

25.5 Transformers

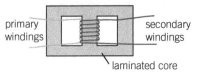

a *practical arrangement*

b *transformer symbol*

▲ **Figure 1** *The transformer*

A transformer consists of two coils (the primary coil and the secondary coil.) wound on the same iron core. The symbol for the transformer is shown in Figure 1.

The **transformer rule**

$$\frac{V_s}{V_p} = \frac{N_s}{N_p}$$

where V_s is the secondary voltage, V_p is the primary voltage, N_s is the number of turns on the secondary coil, and N_p is the number of turns on the primary coil.

- A **step-up transformer** has more turns on the secondary coil than on the primary coil. So the secondary voltage is greater than the primary voltage.

- A **step-down transformer** has fewer turns on the secondary coil than on the primary coil. So the secondary voltage is less than the primary voltage.

Transformer efficiency

Transformers are almost 100% efficient because they have:

1 **low-resistance windings** to reduce power wasted due to the heating effect of the current

2 **a laminated core** (layers of iron separated by layers of insulator). Eddy currents induced in the core by the changing magnetic field in the core are reduced in this way, so their heating effect is reduced

3 **a core of 'soft iron'** which is easily magnetised and demagnetised. This reduces power wasted through repeated magnetisation and demagnetisation of the core.

The efficiency of a transformer = $\dfrac{\textbf{power delivered by the secondary coil}}{\textbf{power supplied to the primary coil}}$

$$= \frac{I_s V_s}{I_p V_p} \times 100\%$$

The efficiency of most transformers is almost 100%, therefore $I_s V_s = I_p V_p$ assuming the transformer is 100% efficient.

So the current ratio $\dfrac{I_s}{I_p} = \dfrac{V_p}{V_s} = \dfrac{N_p}{N_s}$.

> ### Worked example
>
>
> A transformer is used to step down 230 V mains to 12 V. When a 12 V, 36 W lamp is connected to the transformer's secondary coil, the lamp lights normally. When the lamp is on, the primary current in the transformer is 0.16 A. Calculate the efficiency of the transformer.
>
> $$\text{Efficiency} = \frac{I_s V_s}{I_p V_p} = \frac{36\,\text{W}}{230\,\text{V} \times 0.16\,\text{A}} = 0.98\ [= 98\%]$$

The grid system

Transmission of electrical power is more efficient at high voltage than at low voltage. This is because the current needed to deliver a certain amount of power is reduced if the voltage is increased. So the power wasted due to the heating effect of the current through the cables is reduced.

> ### Revision tip
> In a step-up transformer, the voltage is stepped up and the current is stepped down. In a step-down transformer, the voltage is stepped down and the current is stepped up.

1   A rectangular coil which has *n* turns and an area *A* is in a uniform magnetic field which is perpendicular to the plane of the coil. When the coil is removed from the field in a time *t*, a potential difference *V* is induced in the coil. Which one of the following alternatives, **A–D**, gives a correct expression for the magnetic flux density of the magnetic field?

   A $\dfrac{VAt}{n}$ **B** $\dfrac{At}{Vn}$ **C** $\dfrac{nA}{Vt}$ **D** $\dfrac{Vt}{nA}$ *(1 mark)*

2   Which one of the following combination of units has the same SI base units as the unit of magnetic flux ?

   A $T\,A^{-1}$ **B** Tm^{-2} **C** Vs **D** VA^{-1} *(1 mark)*

3   A metal rod of length *L* moving at a constant velocity *v* cuts across a uniform magnetic field which has a magnetic flux density *B*. Which one of the lines **A–D** in the graph in Figure 1 shows how the induced emf *V* varies with time *t* during the time the rod is in the field?   *(1 mark)*

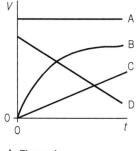

▲ Figure 1

4   A circular flat coil of radius 25 mm with 75 turns is connected to a data recorder. The coil is placed between the poles of a U-shaped magnet with the plane of the coil parallel to the pole faces. The coil is then rapidly removed from the field and the data recorder is used to measure the induced emf at 10 ms intervals. Figure 1 shows a graph of the measurements.

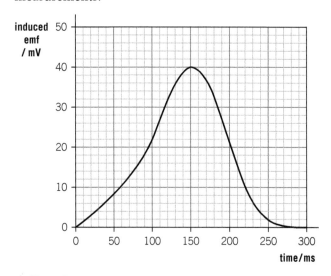

▲ Figure 2

   a Explain why the area under the line in Figure 2 represents the change of flux linkage through the coil.   *(2 marks)*

   b From the graph, estimate the change of flux linkage when the coil was removed from the magnetic field.   *(3 marks)*

   c i Calculate the magnetic flux density between the poles of the magnet.   *(3 marks)*

   ii The diameter of the coil was estimated to within + 0.5 mm. Estimate the uncertainty in the magnetic flux density calculated in part **c i**.   *(3 marks)*

5   An alternating current generator has a rectangular coil of area $2.3 \times 10^{-3}\,m^2$ with 180 turns. The coil rotates at a constant frequency of 50 Hz in a uniform magnetic field of flux density 92 mT with the coil axis perpendicular to the field lines.

   a i Calculate the maximum flux linkage in the coil.   *(2 marks)*

   ii Calculate the peak emf S induced in the coil.   *(2 marks)*

b Sketch a graph to show how the induced emf varies with time over two cycles from when the flux linkage in the coil is a maximum. *(3 marks)*

c Describe how the graph would differ if the frequency had been increased to 90 Hz. *(2 marks)*

6 A transformer consists of a primary coil with 1800 turns and a secondary coil with 94 turns. A 12 V, 50 W lamp is connected to the secondary coil.

a i Calculate the rms potential difference of the alternating pd that must be applied to the primary coil to operate the lamp at its normal brightness. *(2 marks)*

ii Calculate the peak value of the pd applied to the primary coil. *(1 mark)*

b When the lamp is at normal brightness, the primary current has an rms value of 0.23 A. Calculate the efficiency of the transformer. *(2 marks)*

c State and explain two possible reasons why the transformer is not 100% efficient. *(2 marks)*

7 A speedometer in a vehicle consists of a metal disc that rotates between the poles of a U-shaped magnet when the vehicle is moving. A voltmeter is connected between the rim and the centre of the disc as shown in Figure 3. The frequency of rotation of the disc is proportional to the speed of the vehicle.

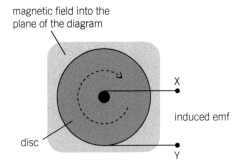

▲ **Figure 3**

a i Explain why an emf is induced between the rim and the centre of the disc when the disc is rotating. *(3 marks)*

ii Determine the polarity of the induced emf at X when the disc rotates clockwise. *(1 mark)*

b When the disc rotates at constant frequency, each radial element of the disc sweeps out an area equal to the area *A* of the disc in each complete rotation. Show that the induced emf is equal to *BAf* where *B* is the magnetic flux density of the magnetic field. *(3 marks)*

c The radius of the disc is 15 mm and the magnetic flux density is 270 mT. Calculate the induced emf when the frequency of rotation of the disc is 46 Hz. *(2 marks)*

8 A square coil of length 18 mm has 55 turns. The coil is placed in a uniform magnetic field of flux density 0.13 T with its plane perpendicular to the field lines. The coil was reversed in a time of 92 ms.

Calculate the magnitude of the induced emf. *(5 marks)*

26.1 The discovery of the nucleus

Specification reference: 3.8.1.1

Figure 1 shows an outline of the arrangement Ernest Rutherford used in 1911 to probe the structure of the atom.

A thin metal foil in the path of a narrow beam of α particles scattered the particles. When the particles hit the atoms of a fluorescent screen in front of a microscope, each impact created a pinpoint of light. The number of α particles reaching the screen per minute was measured for different angles of deflection from zero to almost 180°. The measurements showed that:

1   most α particles passed straight through the foil with little or no deflection – about 1 in 2000 were deflected

2   about 1 in 10 000 particles were deflected through angles of more than 90°.

Rutherford concluded that:

* Most of the atom's mass is concentrated in a small region, the nucleus, at the centre of the atom.

* The nucleus is positively charged because it repels α particles (which carry positive charge) that approach it too closely.

* By testing foils of different metal elements, the magnitude of the charge of a nucleus is +Ze, where e is the charge of the electron and Z is the atomic number of the element.

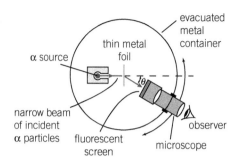

▲ **Figure 1** *Rutherford's α-scattering apparatus*

Estimate of nuclear diameter

In a head-on impact, the α particle stops momentarily at the least distance of approach, d. At this point, the potential energy of the α particle in the electric field of the nucleus is equal to the initial kinetic energy, E_K, of the α particle. Therefore, $E_K = \dfrac{Q_\alpha Q_N}{4\pi\varepsilon_0 d}$, where Q_α is the charge of the α particle (+2e) and Q_N is the charge of the nucleus (= +Ze). For an α particle with kinetic energy of 8×10^{-13} J directed at an aluminium nucleus ($Z = 13$), using this formula gives a least distance of approach of about 7×10^{-15} m.

Synoptic link

A more accurate method of determining the radius of a nucleus is explained in Topic 26.1, The discovery of the nucleus.

Revision tip

Nuclear diameter is of the order of 10^{-15} m.

Summary questions

1   **a**   In the Rutherford α particle scattering experiment, only a small fraction of the α particles were deflected through large angles. What did Rutherford deduce about the atom from this discovery? *(3 marks)*

    b   Discuss how the results would have differed if the α particles in the beam had a wide range of speeds. *(3 marks)*

2   An α particle collided with a nucleus and was deflected by 90°.

    a   Sketch the path of the α particle as it approaches the nucleus and then moves away from it. Label the nucleus N and label the point P on the path at which the α particle is nearest to the nucleus. *(3 marks)*

    b   Describe how the kinetic energy and the potential energy of the α particle change as the particle approaches the nucleus then moves away from it. *(3 marks)*

3   Calculate the least distance of approach of an α particle to an aluminium nucleus ($Z = 13$) if the initial kinetic energy of the α particle is 5.1 MeV. *(3 marks)*

$\varepsilon_0 = 8.85 \times 10^{-12}$ Fm^{-1}, $e = 1.6 \times 10^{-19}$ C

26.2 The properties of α, β, and γ radiation
26.3 More about α, β, and γ radiation

Specification reference: 3.8.1.2

Revision tip
See Topic 26.3, in the Student Book for the experimental verification of the inverse square law for gamma radiation.

Summary questions

1 a A certain isotope emits β^+ radiation. Describe in terms of protons and neutrons the changes that take place when a nucleus of this isotope emits a β^+ particle.
(2 marks)

b A point source of γ radiation is placed 75 mm from the end of a Geiger tube. The corrected count rate was measured at 25.2 counts per second. Calculate the corrected count rate if the source was moved to a distance of 160 mm from the tube. (3 marks)

2 The astatine isotope $^{210}_{85}\text{At}$ decays by electron capture to form an unstable isotope of polonium (Po) which then decays by emitting an α particle to form a stable isotope of lead (Pb). Write down the equation for:

a the decay of $^{210}_{85}\text{At}$
(2 marks)

b the decay of the polonium nucleus.
(1 mark)

Summary of the properties of α, β, and γ radiation

▼ **Table 1** *Nature and properties of α, β, and γ radiation*

	α radiation	β radiation	γ radiation
Nature	2 protons + 2 neutrons	β^- = electron (β^+ = positron)	photon of energy of the order of MeV
Range in air	fixed range, depends on energy, up to $\approx$100 mm	range up to about 1 m	follows the inverse-square law
Deflection in a magnetic field	deflected	opposite direction to α particles; more easily deflected	not deflected
Absorption	stopped by paper or thin metal foil	stopped by approx 5 mm of aluminium	stopped or significantly reduced by several cm of lead
Ionisation	$\approx 10^4$ ions per mm in air at standard pressure	$\approx$100 ions per mm in air at standard pressure	very weak ionising effect
Energy of each particle/photon	constant for a given source	varies up to a maximum for a given source	constant for a given source
Equations	$^A_Z\text{X} \rightarrow ^4_2\alpha + ^{A-4}_{Z-2}\text{Y}$	$^A_Z\text{X} \rightarrow ^0_{-1}\beta + ^A_{Z+1}\text{Y} + \bar{v}_e$	no change to A and Z

β^+ emission

A positron (i.e., a β^+ particle) is emitted from a proton-rich nucleus when a proton in the nucleus changes into a neutron. An electron neutrino v_e is emitted at the same time as the β^+ particle is created.

$$^A_Z\text{X} \rightarrow ^0_{+1}\beta + ^A_{Z-1}\text{Y} + v_e$$

Electron capture

Some proton-rich nuclides can capture an inner-shell electron. In this event, a proton in the nucleus changes to a neutron and an electron neutrino is emitted. The inner-shell vacancy is filled by an outer-shell electron with the emission of an X-ray photon.

$$^A_Z\text{X} + ^0_{-1}e \rightarrow + ^A_{Z-1}\text{Y} + v_e$$

The inverse-square law for γ radiation

The **intensity** I of the radiation is the radiation energy per second passing normally through unit area.

- For a point source that emits n γ photons per second, each of energy hf, the radiation energy per second from the source $= nhf$.
- At distance r from the source, the intensity $I = \dfrac{\text{radiation energy per second}}{\text{total area}}$ $= \dfrac{k}{r^2}$, where $k = \dfrac{nhf}{4\pi}$.

26.4 The dangers of radioactivity

Specification reference: 3.8.1.2

The hazards of ionising radiation

Ionising radiation includes X-rays, protons, and neutrons as well as α, β, and γ radiation. High doses of ionising radiation kill living cells. Ionising radiation affects living cells because:

- it can destroy cell membranes which causes cells to die

- it can damage vital molecules such as DNA directly or indirectly by creating 'free radical' ions which react with vital molecules. Damaged DNA may cause cells to divide and grow uncontrollably, causing a tumour which may be cancerous. Damaged DNA in a sex cell (i.e., an egg or a sperm) may cause a mutation which may be passed on to future generations.

The biological effect of ionising radiation depends on the dose received and the type of radiation. The dose is measured in terms of the energy absorbed per unit mass of matter from the radiation. The same dose of different types of ionising radiation has different effects. For example, α radiation produces far more ions per millimetre than γ radiation in the same substance so it is far more damaging. However, α radiation from a source outside the body cannot penetrate the skin's outer layer of dead cells so is much less damaging than if the source were inside the body.

Background radiation

Background radiation occurs naturally due to various sources such as cosmic radiation and from radioactive materials in rocks, soil, and in the air. See Figure 1.

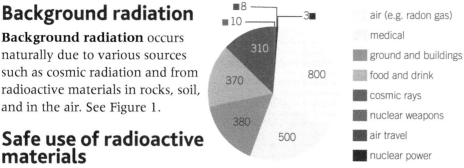

air (e.g. radon gas)
medical
ground and buildings
food and drink
cosmic rays
nuclear weapons
air travel
nuclear power

▲ **Figure 1** *Sources of background radiation in the UK in terms of the average radiation dose per person per year in µSv*

Safe use of radioactive materials

Because radioactive materials produce ionising radiation, they must be stored and used with care. In addition, disposal of a radioactive substance must be carried out in accordance with specific regulations.

1 **Storage of radioactive materials** should be in lead-lined containers. The lead lining of a container must be thick enough to reduce the γ radiation from the sources in the container to about the background level.

2 **When using radioactive materials**, established rules and regulations must be followed. No source should be allowed to come into contact with the skin.

- Sources should be transferred using handling tools such as long-handled tongs or using robots with the material as far from the user as possible.

- Liquid and gas sources and solids in powder form should be in sealed containers. This is to ensure radioactive gas cannot be breathed in and radioactive liquid cannot be splashed on the skin or drunk.

- Radioactive sources should not be used for longer than is necessary. The longer a person is exposed to ionising radiation, the greater the dose of radiation received.

Summary questions

1 Explain why a source of α radiation is dangerous in the body but not as dangerous outside the body. *(3 marks)*

2 Explain why a radioactive source should be:
 a kept in a lead-lined storage box when not in use *(2 marks)*
 b out of its storage box for as little time as possible. *(2 marks)*

3 Describe two effects that ionising radiation has on living cells. *(2 marks)*

26.5 Radioactive decay
26.6 The theory of radioactive decay

Specification reference: 3.8.1.3

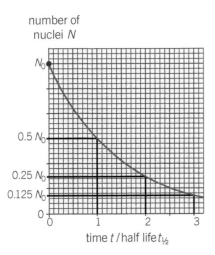

▲ **Figure 2** $N = N_0 e^{-\lambda t}$

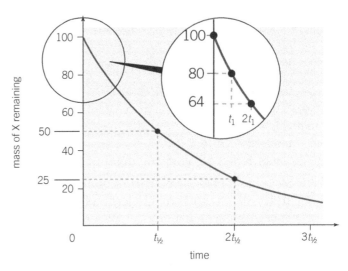

▲ **Figure 1** *A radioactive decay curve*

The half-life, $T_{\frac{1}{2}}$, of a radioactive isotope is the time taken for the mass of the isotope to decrease to half the initial mass. This is the same as the time taken for the number of nuclei or activity of the isotope to decrease to half the initial value.

The activity A of a radioactive isotope is the number of nuclei of the isotope that disintegrate per second. The unit of activity is the **becquerel (Bq),** where $1\,Bq = 1$ disintegration per second.

For a radioactive source of activity A that emits particles (or photons) of the same energy E, the energy transfer per second from the source = AE

Consider a sample of a radioactive isotope X that initially contains N_0 nuclei of the isotope. Let N represent the number of nuclei of X remaining at time t after the start. Suppose in time Δt, the number of nuclei that disintegrate is ΔN.

The rate of disintegration $\dfrac{\Delta N}{\Delta t} = -\lambda N$ where λ is a constant referred to as the **decay constant**. The minus sign is because ΔN is a decrease.

The solution of the equation is $N = N_0 e^{-\lambda t}$, where e^x is the exponential function.

Note:

1   For a given radioactive isotope, its activity $A = (-)\dfrac{\Delta N}{\Delta t} = \lambda N$.

2   For a sample of a radioactive isotope, the mass m, the activity A, and the corrected count rate C of a detector at a fixed distance from the sample decreases from the initial values m_0 or A_0 or C_0 in accordance with the equation $X = X_0 e^{-\lambda t}$ where X represents m, A, or C (because these are all proportional to the number of nuclei, N, of the isotope).

Maths skill: Data analysis using natural logarithms

From the equation $N = N_0 e^{-\lambda t}$, taking natural logs of both sides gives $\ln N = \ln N_0 - \lambda t$. Therefore, a graph of $\ln N$ (or $\ln m$ or $\ln A$) against time t is a straight line with a negative gradient $-\lambda$ and a y-intercept equal to $\ln N_0$ (or $\ln m_0$ or $\ln A_0$).

The decay constant

The decay constant λ is the probability of an individual nucleus decaying per second $= \left(\dfrac{\Delta N}{N\Delta t}\right)$.

The half-life $T_{\frac{1}{2}}$ is related to the decay constant λ according to the equation

$$T_{\frac{1}{2}} = \frac{\ln 2}{\lambda}$$

The unit of decay constant is s^{-1}. A large value of λ means fast decay and short half-life.

Summary questions

$N_A = 6.02 \times 10^{23}\,mol^{-1}$, $1\,MeV = 1.6 \times 10^{-13}\,J$

1 A freshly prepared sample of a radioactive isotope X contains 1.8×10^{15} atoms of the isotope. The half-life of the isotope is 2.0 hours. Calculate the number of atoms of this isotope remaining after:

  a 8.0 h   *(2 marks)*

  b 3.0 h.   *(2 marks)*

2 **a** Explain what is meant by the decay constant of a radioactive isotope. *(1 mark)*

  b The isotope $^{226}_{88}\mathrm{Ra}$ has a half-life of 1620 years. For an initial mass of 1.0 kg of this isotope, calculate:

    i the initial activity of this mass *(3 marks)*

    ii the mass of this isotope remaining after 3000 years *(4 marks)*

    iii how many atoms of the isotope will remain after 3000 years. *(1 mark)*

3 A fresh sample of a radioactive isotope has an initial activity of 40 kBq. After 24 h, its activity has decreased to 36 kBq. Calculate:

  a the decay constant of this isotope *(3 marks)*

  b its half-life *(2 marks)*

  c its activity after 120 h. *(3 marks)*

26.7 Radioactive isotopes in use

Radioactive dating

A radioactive isotope in an ancient object can be used to determine the age of the object by comparing:

- the activity per unit mass of the isotope with the activity per unit mass of the same isotope in a recently-formed object, or

- the ratio of the number of atoms of the radioactive isotope to the number of atoms of the isotope it decays into, provided the second isotope is stable.

Radioactive tracers

The radioactive isotope(s) in the tracer should:

- have a half-life which is long enough for the necessary measurements to be made and short enough to decay quickly after use

- emit β radiation or γ radiation so it can be detected outside the flow path.

Examples of industrial uses of radioactivity

Thickness monitoring

Figure 1 shows how metal foil is made using a continuous production line. The source used is a β⁻ emitter with a long half-life. α radiation would be absorbed completely by the foil and γ radiation would pass straight through without absorption.

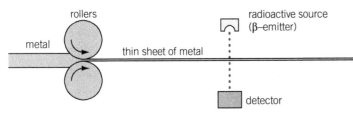

▲ **Figure 1** *The manufacture of metal foil*

Power sources for remote devices

Remote devices can be powered using a radioactive isotope. The isotope needs to have a reasonably long half-life but a very long half-life may require too much mass to generate the necessary power.

Revision tip

The choice of a radioactive isotope for a particular application is determined by:

- the half-life and the type of radiation needed

- whether or not a stable product is needed.

Summary questions

$N_A = 6.02 \times 10^{23} \, mol^{-1}$, $1 \, MeV = 1.6 \times 10^{-13} \, J$

1. The uranium isotope $^{238}_{92}U$ decays in several stages with an effective half-life of 4500 million years to form the lead isotope $^{206}_{82}Pb$. An ancient rock was found to contain 6.4 times as many $^{206}_{82}Pb$ nuclei as $^{238}_{92}U$ nuclei.

   a. Estimate the proportion of $^{238}_{92}U$ in the rock now to when it formed.

   (1 mark)

   b. Use your estimate to determine the age of the rock. *(4 marks)*

2. A cardiac pacemaker is a device used to ensure that a faulty heart beats at a suitable rate. The electrical energy in a cardiac pacemaker is obtained from the plutonium isotope $^{238}_{94}Pu$ which has a half-life of 88 years and emits α particles of energy 5.6 MeV. The initial mass of the isotope in the pacemaker is 0.20 grams.

   a. Calculate the initial activity of the radioactive source. *(3 marks)*

   b. Calculate the energy per second released from the source. *(2 marks)*

26.8 More about decay modes

The N–Z graph

Figure 1 shows the N–Z graph of the neutron number N against the proton number Z for all known isotopes. Stable nuclei lie along a belt curving upwards with an increasing neutron–proton ratio from the origin to $N = 120$, $Z = 80$ approximately.

- For light isotopes (Z from 0 to no more than 20), the stable nuclei follow the straight line $N = Z$.

- As Z increases beyond about 20, stable nuclei have more neutrons than protons. The neutron/proton ratio increases. The extra neutrons help to bind the nucleons together without introducing repulsive electrostatic forces as more protons would do.

- α **emitters** occur beyond about $Z = 60$, most of them with more than 80 protons and 120 neutrons. The largest stable nuclide is $^{209}_{83}\text{Bi}$. Most α-emitting nuclides are above this.

- β⁻ emitters occur to the left of the stability belt where the isotopes are neutron-rich compared to stable isotopes.

- β⁺ emitters occur to the right of the stability belt where the isotopes are proton-rich compared to stable isotopes.

The change that takes place when an unstable nucleus becomes stable or less unstable can be represented on the N–Z graph as shown in Figure 2.

Radioactive series

An unstable nucleus, before it becomes stable, may undergo a series of changes in which each change involves an emission of an α particle or a β particle. Any radioactive series may be represented on the N–Z graph by a sequence of 'decay arrows'.

Nuclear energy levels

After an unstable nucleus emits an α or a β particle or undergoes electron capture, it might emit a γ photon in order to lose energy. This happens if the 'daughter' nucleus is formed in an **excited state** after it emits an α or a β particle or undergoes electron capture. The excited nucleus moves to its lowest energy state, its **ground state**, either directly or via one or more lower-energy excited states.

The technetium generator

Some radioactive isotopes such as the technetium isotope $^{99}_{43}\text{Tc}$ stay in the excited state long enough to be separated from the parent isotope. Such a long-lived excited state is said to be a **metastable state**. The technetium isotope $^{99}_{43}\text{Tc}$ in a metastable state (represented by the symbol $^{99}_{43}\text{Tc}^{\text{m}}$) forms after β⁻ emission from nuclei of the molybdenum isotope $^{99}_{42}\text{Mo}$ which has a half-life of 67 h. See Figure 3.

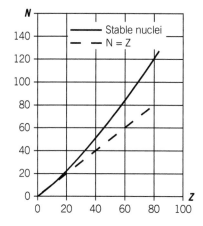

▲ **Figure 1** The N–Z graph

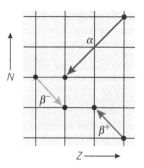

▲ **Figure 2** N–Z changes

Synoptic link

β⁻ emitters can be manufactured by bombarding stable isotopes with neutrons. β⁺ emitters can only be produced by bombarding stable isotopes with protons. The protons need to have sufficient kinetic energy to overcome coulomb repulsion from the nucleus.

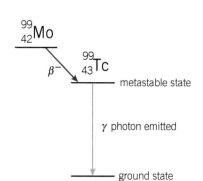

▲ **Figure 3** The metastable state of technetium $^{99}_{43}\text{Tc}$

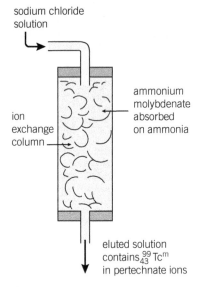

sodium chloride
solution

ion
exchange
column

ammonium
molybdenate
absorbed
on ammonia

eluted solution
contains $^{99}_{43}$Tc^m
in pertechnate ions

▲ **Figure 4** *The technetium generator.
The ion exchange column contains
ammonium molybdenate exposed to
neutron radiation several days earlier to
make a significant number of the molybdenum nuclei unstable. When a solution
of sodium chloride is passed through
the column, some of the chlorine ions
exchange with pertechnate ions but not
with molybdenate ions so the solution
that emerges contains $^{99}_{43}$Tc^m nuclei*

$^{99}_{43}$Tc^m has a half-life of 6 h and decays to the ground state by γ emission. The ground state of $^{99}_{43}$Tc is a β$^-$ emitter with a half-life of 500 000 years and it forms a stable product. Therefore, a sample of $^{99}_{43}$Tc^m with no molybdenum present effectively emits only γ photons.

$^{99}_{43}$Tc^m is used in medical diagnosis, for example:

- **monitoring blood flow** through the brain after a small quantity of sodium pertechnate solution is administered intravenously

- the **γ camera** is designed to 'image' internal organs and bones by detecting γ radiation from sites in the body where a γ emitting isotope such as $^{99}_{43}$Tc^m nuclei is located. For example, bone deposits can be located using a phosphate tracer labelled with $^{99}_{43}$Tc^m.

Summary questions

1 a Sketch an *N–Z* graph to show how *N* varies with *Z* for the known stable isotopes. *(2 marks)*

   b Show on your diagram possible locations of an isotope that is:
      i   an α-emitter
      ii a β$^-$ emitter
      iii a β$^+$ emitter. *(3 marks)*

2 A nucleus of uranium $^{238}_{92}$U decays to form a stable nucleus of lead $^{206}_{82}$Pb by emitting 8 alpha particles and a number of β$^-$ particles.

   a Determine the number of β$^-$ particles it emits. *(2 marks)*

   b Use your *N-Z* chart to explain why no β$^+$particles are emitted. *(3 marks)*

3 The neon isotope $^{24}_{10}$Ne releases 2.45 MeV of energy when it decays to form an isotope of sodium (Na). In the process, it releases β$^-$ particles of maximum energy 1.98 MeV and 1.10 MeV and γ photons of energies 0.47 MeV and 0.88 MeV.

   a Construct an energy level diagram to show the possible ways in which a $^{24}_{10}$Ne nucleus could decay. *(4 marks)*

   b Write the equation to represent the β$^-$ decay of $^{24}_{10}$Ne. *(2 marks)*

26.9 Nuclear radius

Specification reference: 3.8.1.5

High-energy electron diffraction

When a narrow beam of high-energy electrons is directed at a thin solid sample of an element, the incident electrons are diffracted by the nuclei of the atoms in the foil. Measurements of the number of electrons per second diffracted through different angles show that as the angle of diffraction θ is increased, the number of electrons per second diffracted into the detector decreases then increases slightly then decreases again. Scattering of the beam electrons by the nuclei due to their charge causes the intensity to decrease as θ increases. Diffraction of the beam electrons by each nucleus causes intensity maxima and minima to be superimposed on the effect above. The angle of the first minimum from the centre, θ_{min}, is measured and used to calculate the radius of the nucleus, provided the wavelength of the incident electrons is known.

Dependence of nuclear radius on nucleon number

Using samples of different elements, the radius R of different nuclides can be measured. It can be shown graphically that R depends on mass number A according to $R = r_0 A^{\frac{1}{3}}$, where the constant $r_0 = 1.05\,\text{fm}$. A graph of $\ln R$ against $\ln A$ gives a straight line with a gradient of $\frac{1}{3}$ which confirms $R = r_0 A^{\frac{1}{3}}$ and gives a y-intercept equal to $\ln r_0$. A graph of R against $A^{\frac{1}{3}}$ gives a straight line through the origin with a gradient equal to r_0. Plotting this graph gives an accurate value of r_0.

Nuclear density

Assuming the nucleus is spherical:

* its volume $V = \frac{4}{3}\pi R^3 = \frac{4}{3}\pi (r_0 A^{\frac{1}{3}})^3 = \frac{4}{3}\pi r_0^3 A$
* its mass in atomic mass units = Au where 1u = 1 atomic mass unit = $1.661 \times 10^{-27}\,\text{kg}$.

Therefore, its density $\rho = \dfrac{A\text{u}}{\frac{4}{3}\pi r_0^3 A} = \dfrac{1\text{u}}{\frac{4}{3}\pi r_0^3}$ which is independent of A.

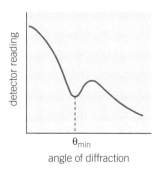

▲ **Figure 1** *High-energy electron diffraction*

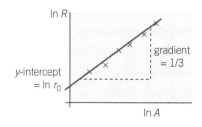

▲ **Figure 2** *Nuclear radius graphs*

Summary questions

$r_0 = 1.05\,\text{fm}$, 1u = $1.661 \times 10^{-27}\,\text{kg}$

1 A beam of high-energy electrons directed at a target is diffracted by the nuclei in the target.

  a The angle of diffraction of the first minimum was 39° when the energy of the electrons was 310 MeV. Calculate the de Broglie wavelength of the electrons and use $R\sin\theta_{min} = 0.61\lambda$ to calculate the radius of the target nuclei. *(3 marks)*

2 In the high-energy electron diffraction experiment in **Q1**, the angle of diffraction was measured to within ± 2.0°. Estimate the uncertainty in the radius of the target nucleus due to this measurement. *(2 marks)*

3 Use the equation $R = r_0 A^{\frac{1}{3}}$ to calculate the radius of a $^{28}_{14}\text{Si}$ nucleus and hence determine its density. *(3 marks)*

1 A polonium nucleus $^{218}_{84}$Po undergoes a series of α and β decays before it forms a stable lead nucleus $^{206}_{82}$Pb. Which one of the following alternatives, **A–D**, gives the total number of decays ?

   A 4   **B** 5   **C** 6   **D** 7   (*1 mark*)

2 The half-life of a radioactive isotope X is 10 hours. A pure sample of X is prepared. How long will it take for the sample to contain three times as many decayed atoms as undecayed atoms of X?

   A 10 hours   **B** 20 hours   **C** 30 hours   **D** 40 hours   (*1 mark*)

3 Two radioactive isotopes P and Q have half-lives of 50 s and 100 s respectively. Separate samples of the two isotopes are prepared at the same time with the same initial activity. The isotopes decay into stable isotopes R and S. Which one of the following alternatives, **A–D**, gives the ratio $\dfrac{\text{the number of atoms of Q}}{\text{the number of atoms of P}}$ after 200 s?

   A 4   **B** 8   **C** 12   **D** 16   (*1 mark*)

4 The following count rate measurements were made of the radiation from a radioactive isotope at a fixed distance from a Geiger tube connected to a counter. The average background count was 25 counts per minute.

Time t / s	0	30	60	90	120	150	180
Count rate /s	11.7	8.4	6.6	5.2	4.2	3.0	2.5
Corrected count rate C /s							

   a Calculate the corrected count rate C and copy and complete the last row of the table.   (*1 mark*)

   b Plot a graph of ln C on the y-axis against time t on the x-axis.   (*3 marks*)

   c i Use your graph to determine the half-life of the radioactive isotope.   (*3 marks*)

   ii Explain why the Geiger tube needed to be at the same distance from the isotope throughout the measurements.   (*2 marks*)

5 **a** A beam of monoenergetic α particles is directed normally at a thin metal foil in an α-scattering experiment.

   i Explain why a small proportion of the α particles are deflected through angles of more than 90°.   (*3 marks*)

   ii Explain why multiple scattering of an individual α particle is unlikely.   (*3 marks*)

   b A narrow beam of 4.5 MeV α particles is directed normally at a thin copper foil. Estimate the least distance of approach of an α particle to a copper $^{63}_{29}$Cu nucleus.   (*3 marks*)

6 A small source of γ radiation inside a sealed container was placed at different distances d from a Geiger tube and the count rate was measured at each distance. The results corrected for background radiation are shown in the table below.

Distance d / mm	102	136	170	202	245	305	385
Corrected count rate C / s^{-1}	218	126	73	57	40	27	15

Assuming the source in the container is at a distance d_o from the front of the container, the distance from the source to the Geiger tube is $d + d_o$, so the corrected count rate C is inversely proportional to $(d + d_o)^2$ in accordance with the inverse-square law.

a Show that $d = \dfrac{a}{\sqrt{C}} + b$, where a and b are constants. (*2 marks*)

b i Tabulate the values of $\dfrac{1}{\sqrt{C}}$ and d. (*2 marks*)

ii Plot a suitable graph to find out if the inverse-square law applies to these results. (*3 marks*)

c Use your graph to determine the distance d_o. (*1 mark*)

7 A sample of the stable copper isotope ${}^{63}_{29}\mathrm{Cu}$ is placed in the core of a nuclear reactor where the copper atoms are bombarded by neutrons. As a result, the sample becomes radioactive due to some of its nuclei each absorbing a neutron.

a Write down an equation to represent a copper nucleus absorbing a neutron. (*1 mark*)

b i What type of radiation is emitted by the copper nucleus after it has absorbed a neutron? Give a reason for your answer. (*3 marks*)

ii After absorbing a neutron, the copper nucleus emits radiation and forms a stable zinc (Zn) nucleus. Write down an equation to represent the change that takes place as a result of the emission of the radiation in part **b i**. (*2 marks*)

c Discuss whether or not the sample would become radioactive if it were to be bombarded by protons instead of neutrons. (*2 marks*)

8 When a radioactive source near a Geiger tube was moved gradually away from the tube, the count rate due to the source suddenly dropped significantly.

a Explain why the count rate suddenly dropped. (*2 marks*)

b The count rate did not drop to zero. Describe a further test you would do to determine the type(s) of radiation from the source still being detected. (*4 marks*)

27.1 Energy and mass

Specification reference: 3.8.1.6

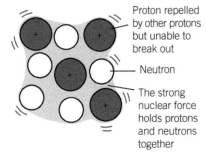

Proton repelled by other protons but unable to break out

Neutron

The strong nuclear force holds protons and neutrons together

▲ **Figure 1** *The strong nuclear force*

Revision tips

1 When calculating Q in beta decay, assume the mass of the neutrino is negligible.

2 To calculate the energy released in a reaction in MeV, calculate the mass difference in u then multiply by 931.5 MeV/u.

Revision tip

If the mass of each atom is given instead of the mass of its nucleus, calculate the mass of each nucleus by subtracting the mass of the electrons $(= Zm_e)$ in the atom from the mass of each atom.

Revision tip

Einstein showed that the mass m of an object increases (or decreases) when it gains (or loses) energy E in accordance with the equation $E = m c^2$. This equation applies to all energy changes.

Energy and the strong nuclear force

- The strong nuclear force acts between any two protons or neutrons when they are closer than about 3 to 4 femtometres. The force is attractive at distances greater than 0.5 fm (10^{-15} m) and repulsive at shorter distances.

- The strength of the strong nuclear force is sufficient to overcome the Coulomb repulsion between 2 protons at a separation of about 1 fm which is about 200 N. So the strong nuclear force is at least 200 N.

- Nuclear changes involve energies of the order of MeV ($\approx 200\,\text{N} \times 3 \times 10^{-15}\,\text{m} = 6 \times 10^{-13}\,\text{J} \approx 3\text{–}4\,\text{MeV}$ as $1\,\text{MeV} = 1.6 \times 10^{-13}\,\text{J}$).

Energy changes in reactions

Reactions on a nuclear or sub-nuclear scale involve measurable changes of mass. For a difference Δm in the total mass of all the particles involved before and after a reaction, the energy transferred

$$Q = \Delta m c^2$$

1 In **α decay**, the energy released is shared as kinetic energy between the α particle and the nucleus. The two particles move in opposite directions with equal and opposite momentum. So the energy released is shared between the α particle and the nucleus in inverse proportion to their masses.

2 In **β decay**, the energy released is shared in variable proportions between the β particle, the nucleus, and the neutrino or antineutrino released in the decay. The maximum kinetic energy of the β particle is very slightly less than the energy released in the decay because of recoil of the nucleus.

3 In **electron capture**, the nucleus emits a neutrino which causes the nucleus to recoil. The energy released is shared between the nucleus and the neutrino which carries away most of the energy released. The atom also emits an X-ray photon when the inner-shell vacancy due to the electron capture is filled.

Worked example

The radon isotope $^{220}_{86}$Rn emits α particles and decays to form the polonium isotope $^{220}_{86}$Rn.

Write down an equation to represent this process and calculate the energy released when a $^{220}_{86}$Rn nucleus emits an α particle.

mass of $^{220}_{86}$Rn nucleus = 219.964 18 u

mass of $^{216}_{84}$Po nucleus = 215.955 76 u

mass of α particle = 4.001 50 u

1u is equivalent to 931.5 MeV

$^{220}_{86}$Rn $\rightarrow$ $^{4}_{2}$α + $^{216}_{84}$Po (+ energy released Q)

mass difference = total initial mass − total final mass = 219.964 18 − (215.955 76 + 4.001 50) = 6.92×10^{-3} u

energy released Q = mass difference in u × 931.5 MeV/u = 6.45 MeV

Summary questions

1 u = 931.5 MeV   **g = 9.81 m s^{-2}**

1 A carbon $^{14}_{6}C$ nucleus decays to form a stable nucleus of nitrogen (N) and emits a β^- particle and an antineutrino in the process.

   a Write down an equation to represent this process.   *(2 marks)*

   b Calculate the energy released.

   mass of $^{14}_{6}C$ nucleus = 13.999 95 u,

   mass of the nitrogen

   nucleus = 13.999 23 u

   mass of β^- particle = 0.000 55 u   *(3 marks)*

2 The argon isotope $^{37}_{18}Ar$ decays through electron capture to form the stable isotope of chlorine (Cl).

   a Write down an equation to represent this process.   *(4 marks)*

   b Calculate the energy released.

   mass of $^{37}_{18}Ar$ nucleus = 36.956 89 u,

   mass of chlorine nucleus

   nucleus formed = 36.956 57 u

   mass of electron = 0.000 55 u   *(3 marks)*

3 The americium isotope $^{241}_{95}Am$ emits α particles and decays to form an isotope of neptunium (Np).

   a Write down an equation to represent this process.   *(2 marks)*

   b Calculate the energy released.   *(4 marks)*

   atomic mass of $^{241}_{95}Am$ = 241.056 69 u,

   atomic mass of neptunium

   isotope = 237.048 03 u

   mass of α particle = 4.001 50 u, mass of electron = 0.000 55 u

27.2 Binding energy

Specification reference: 3.8.1.6

Revision tip

The energy corresponding to a mass of 1 u is $1.661 \times 10^{-27} \times (3.00 \times 10^8)^2 \, J = 931.5 \, MeV$. If you are given the mass of the nucleus in kilograms, you can convert this to atomic mass units and use the method above to calculate the mass defect. The values of the mass of the proton and the neutron are given in atomic mass units in the data sheet.

Note: The atomic mass unit, $1 \, u = 1.661 \times 10^{-27} \, kg$, is defined as $\frac{1}{12}$th of the mass of an atom of the carbon isotope $^{12}_{6}C$.

Binding energy

When a nucleus forms from separate neutrons and protons, energy is released as the strong nuclear force does work pulling the nucleons together. Because energy is released when a nucleus forms from separate neutrons and protons, the mass of a nucleus is less than the mass of the separated nucleons.

The mass defect Δm of a nucleus is defined as the difference between the mass of the separated nucleons and the mass of the nucleus.

For a nucleus $^{A}_{Z}X$ of mass M_{NUC}, **its mass defect $\Delta m = Zm_p + (A - Z)m_n - M_{NUC}$**

where m_p and m_N represent the masses of the proton and the neutron respectively.

The binding energy of the nucleus is the work that must be done to separate a nucleus into its constituent neutrons and protons.

For a nucleus which has a mass defect Δm, **its binding energy $= \Delta mc^2$**

The binding energy curve

The **binding energy per nucleon** of a nucleus is the average work done per nucleon to remove all the **nucleons** (protons and neutrons) from a nucleus – it is therefore a measure of the stability of a nucleus. If the binding energies per nucleon of two different nuclides are compared, the nucleus with more binding energy per nucleon is the more stable of the two nuclei.

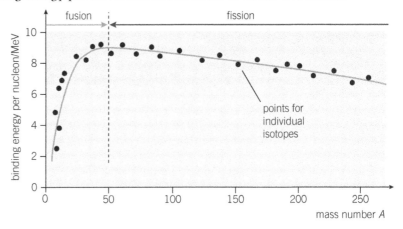

▲ **Figure 1** *Binding energy per nucleon for all known nuclides*

Figure 1 shows a graph of the binding energy per nucleon against mass number A for all the known nuclides. This graph is a curve which has a maximum value of 8.7 MeV per nucleon between $A = 50$ and $A = 60$. Nuclei with mass numbers in this range are the most stable nuclei.

When energy is released by a nucleus, its binding energy per nucleon increases. Energy is released in:

- **nuclear fission**, the process in which a large unstable nucleus splits into two fragments which are more stable than the original nucleus. The binding energy per nucleon increases in this process, as shown in Figure 1.

- **nuclear fusion**, the process of making small nuclei fuse together to form a larger nucleus. The product nucleus has more binding energy per nucleon than the smaller nuclei. So the binding energy per nucleon also increases in this process, provided the nucleon number of the product nucleus is no greater than about 50.

Worked example

The mass of a nucleus M_{NUC} of the iron isotope $^{56}_{26}$Fe is 55.920 66 u. Calculate the binding energy of this nucleus in MeV.

mass of a proton, m_p = 1.007 28 u

mass of a neutron, m_n = 1.008 67 u

1 u is equivalent to 931.5 MeV

Mass defect $\Delta m = 26m_p + (56 - 26)m_n - M_{NUC} = 0.528\,72$ u

Therefore, binding energy = 0.528 72 u × 931.5 MeV/u = 493 MeV

Binding energy per nucleon = $\dfrac{493\,\text{MeV}}{56}$ = 8.8 MeV

Summary questions

mass of a proton m_p = 1.007 28 u, mass of a neutron m_n = 1.008 67 u, 1 u is equivalent to 931.5 MeV

1 **a** Explain what is meant by the binding energy of a nucleus. (*2 marks*)
   b Describe and explain how the binding energy per nucleon of a heavy nucleus changes when it emits an alpha particle. (*2 marks*)

2 Calculate the binding energy per nucleon, in MeV per nucleon, of:
   a a $^{3}_{2}$He nucleus (*3 marks*)
   b a $^{238}_{92}$U nucleus. (*3 marks*)

   mass of an $^{3}_{2}$He nucleus = 3.0159 75 u

   mass of a $^{238}_{92}$U nucleus = 238.050 74 u

3 **a** Write down an equation to show the fusion reaction that occurs when a proton and a $^{2}_{1}$H nucleus fuse together to form a $^{3}_{2}$He nucleus. (*2 marks*)
   b Calculate the energy released in the above reaction. (*3 marks*)

   mass of $^{2}_{1}$H nucleus = 2.013 55 u   mass of $^{3}_{2}$He nucleus = 3.014 93 u

27.3 Fission and fusion

Specification reference: 3.8.1.7; 3.8.1.8

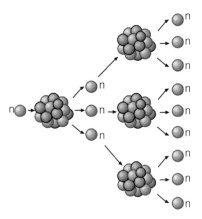

▲ **Figure 1** *A chain reaction in a nuclear reactor*

Revision tip

1 The energy released in a fission or fusion reaction may be calculated using $E = mc^2$ in the form $Q = \Delta mc^2$, where Δm is the difference between the total mass before and after the event.

2 In each type of reaction, the total number of neutrons and protons is unchanged but in a fusion reaction in which a β particle is emitted, a neutron changes into a proton or a proton changes into a neutron.

$$_1^1p + {}_1^1p \rightarrow {}_1^2H + {}_{+1}^0\beta + 17.6\,MeV$$

Induced fission

Fission of a nucleus occurs when a nucleus splits into two approximately equal fragments. This happens when the uranium isotope $_{92}^{235}U$ is bombarded with neutrons. This process is known as **induced fission**.

Each fission event releases energy and two or three neutrons referred to as **fission neutrons**. The neutrons are each capable of causing a further fission event as a result of a collision with another $_{92}^{235}U$ nucleus. A **chain reaction** is therefore possible in which fission neutrons produce further fission events which release fission neutrons and cause further fission events and so on. Complete fission of $_{92}^{235}U$ would release about a million times more than the energy released as a result of burning a similar mass of fossil fuel.

Many fission products are possible when a fission event occurs. The following equation shows a possible fission event in which a $_{92}^{235}U$ nucleus is split into a barium $_{56}^{144}Ba$ nucleus and a krypton $_{36}^{90}Kr$ nucleus and two neutrons are released.

$$_{92}^{235}U + {}_0^1n \rightarrow {}_{56}^{144}Ba + {}_{36}^{90}Kr + 2_0^1n + \text{energy released, } Q$$

Nuclear fusion

Fusion takes place when two nuclei combine to form a bigger nucleus. The binding energy per nucleon of the product nucleus is greater than of the initial nuclei. As a result, energy is released equal to the increase of binding energy.

Nuclear fusion can only take place if the two nuclei that are to be combined collide at high speed. This is necessary to overcome the electrostatic repulsion between the two nuclei so they can become close enough to interact through the strong nuclear force. For example, the reaction below represents the fusion of a nucleus of deuterium (the hydrogen isotope $_1^2H$) and a nucleus of tritium (the hydrogen isotope $_1^3H$).

$$_1^2H + {}_1^3H \rightarrow {}_2^4He + {}_0^1n + 17.6\,MeV$$

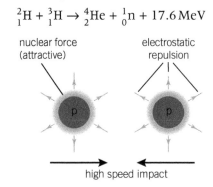

▲ **Figure 2** *Fusion of two protons*

Fusion power

The temperature of the material containing the nuclei to be fused must be of the order of $10^8\,K$ so that the nuclei are moving fast enough to fuse when they collide with each other. The material at this temperature is called a plasma as it consists of positive ions and unattached electrons.

A very strong magnetic field is used to confine the plasma gas so it doesn't make contact with the inside of its container and lose its energy. The plasma is heated by passing a very large current through it. At present, confinement can only be achieved for a relatively short duration during which time more power is released than is supplied to the plasma.

Summary questions

$1u \equiv 931\,MeV$

1 a Explain why the protons in a nucleus do not leave the nucleus even though a repulsive force acts between them due to their positive charge. *(2 marks)*

  b Explain why the mass of a nucleus is less than the mass of the separated protons and neutrons from which the nucleus is composed. *(3 marks)*

2 The equation below represents a reaction that takes place when a neutron collides with a nucleus of the uranium isotope $^{235}_{92}U$.

$$^{235}_{92}U + ^{1}_{0}n \rightarrow ^{140}_{54}Xe + ^{93}_{38}Sr + 3\,^{1}_{0}n + \text{energy released, } Q$$

Calculate the energy, in MeV, released in this fission reaction. *(4 marks)*

mass of $^{235}_{92}U$ nucleus $234.943\,42\,u$

mass of $^{140}_{54}Xe$ nucleus $139.891\,92\,u$

mass of $^{93}_{38}Sr$ nucleus $92.893\,13\,u$ mass of a neutron $1.008\,67\,u$

3 In a fusion reactor, a hydrogen $^{3}_{1}H$ nucleus is fused with a hydrogen $^{2}_{1}H$ nucleus to form a helium $^{4}_{2}He$ nucleus. In the process another particle X is released.

  a Write down an equation to represent this fusion reaction and to identify X. *(2 marks)*

  b Calculate the energy released in this reaction using the appropriate data below. You will need to use some or all of this data. *(4 marks)*

mass of a β particle $0.000\,55\,u$ mass of a proton $1.007\,28\,u$

mass of a neutron $1.008\,67\,u$

mass of $^{3}_{1}H$ nucleus $3.015\,50\,u$ mass of $^{2}_{1}H$ nucleus $2.013\,55\,u$

mass of $^{4}_{2}He$ nucleus $4.001\,50\,u$

27.4 The thermal nuclear reactor

Specification reference: 3.8.1.7; 3.8.1.8

Inside a nuclear reactor

A **thermal nuclear reactor** in a nuclear power station consists of a steel vessel known as the **reactor core** connected by steel pipes to a **heat exchanger**. The reactor core contains:

1 fuel rods spaced evenly in a **moderator** such as water under pressure or graphite

2 the moderator, which is necessary to slow the fission neutrons so they can cause further fission

3 **control rods** made of material such as boron which absorbs neutrons. The depth of the rods in the moderator is automatically adjusted so that the rate of fission events is constant

4 a **coolant** (e.g., water or carbon dioxide gas) that is pumped through steel pipes between the reactor core and the moderator so that it transfers energy from the reactor vessel to the heat exchanger. The coolant must flow easily and it must not be corrosive.

Fission facts

• The fuel rods contain enriched uranium mostly of the non-fissionable uranium isotope $^{238}_{92}U$, with about 2–3% of the uranium isotope $^{235}_{92}U$ which is fissionable. In comparison, natural uranium contains 99% U-238.

• The depth of the control rods in the core is automatically adjusted to keep the number of neutrons in the core constant so that exactly one fission neutron per fission event on average goes on to produce further fission. This condition keeps the rate of release of fission energy constant.

• The fission neutrons are slowed down by repeated collisions with the moderator atoms to increase the probability of induced fission. The process is most effective if the mass of the moderator atom is as close as possible to the mass of the neutron.

• For a chain reaction to occur, the mass of the fissile material (e.g., U-235) must be greater than a minimum mass, referred to as the **critical mass**. If the mass of fissile material is less than the critical mass needed, too many of the fission neutrons escape because the surface area to mass ratio of the material is too high.

▼ **Table 1** *Comparison of thermal reactors*

	Advanced gas-cooled reactor	Pressurised water reactor
Fuel	uranium oxide in stainless steel cans	uranium oxide in zirconium alloy cans
Moderator	graphite	water
Coolant	CO_2 gas	water
Coolant temperature/K	900	600
Typical power output/MW	1300	700

Safety features

1 The reactor core is a thick steel vessel that absorbs β radiation and some of the γ radiation and neutrons from the core.

2 The core is surrounded by very thick concrete walls which absorb the neutrons and γ radiation that escape from the reactor vessel.

3 Every reactor has an emergency shut-down system designed to insert the control rods fully into the core to stop fission completely.

4 The sealed fuel rods are inserted and removed from the reactor by means of remote handling devices. The rods are much more radioactive after removal than before because the used fuel rods emit β and γ radiation due to the many neutron-rich fission products that form.

Radioactive waste

High-level radioactive waste such as spent fuel rods contains many different radioactive isotopes, including fission fragments as well as unused uranium-235, uranium-238, and plutonium-239.

The spent fuel rods are removed by remote control and stored underwater in cooling ponds for up to a year because they continue to release heat due to radioactive decay.

The rods are then transferred in large steel casks to a reprocessing plant where the unused uranium and plutonium is then removed and stored in sealed containers for further possible use.

The rest of the material is radioactive waste and is stored for many years (as it contains long-lived radioactive isotopes) in sealed containers in deep trenches or in geologically-stable underground caverns. In some countries, high-level radioactive waste is vitrified by mixing it with molten glass and then stored as glass blocks in underground caverns.

Intermediate-level waste such as radioactive materials with low activity and containers of radioactive materials are sealed in drums that are encased in concrete and stored in specially constructed buildings with walls of reinforced concrete.

Low-level waste such as laboratory equipment and protective clothing is sealed in metal drums and buried in large trenches.

Summary questions

1 a What is the function of the moderator in a nuclear reactor?

(2 marks)

b Describe how the moderator works. *(5 marks)*

2 Describe and explain how the control rods of a nuclear reactor are used when a nuclear reactor is:

a releasing energy at a constant rate *(3 marks)*

b shut down. *(4 marks)*

3 Explain why the spent fuel rods from a nuclear reactor are more radioactive after removal from the reactor than they were before they were used in the reactor. *(5 marks)*

$1 u \equiv 931.5 \, \text{MeV}$

1 A nucleus $_Z^A X$ has a mass M. The mass of a proton is m_p and the mass of a neutron is m_n. The mass defect of the nucleus can be written as $rA + sZ + tM$. Which one of the following alternatives, **A–D**, gives the correct terms for r, s, and t? *(1 mark)*

	r	s	t
A	m_p	$m_p + m_n$	1
B	m_p	$m_p + m_n$	−1
C	m_p	$m_p - m_n$	1
D	m_p	$m_p - m_n$	−1

2 A nuclear power station has an electrical power of 1200 MW and an efficiency of 40%. It uses uranium fuel that contains the fissile isotope $_{92}^{235}U$. Each fission event releases approximately 200 MeV. Which one of the following alternatives, **A–D**, gives the nearest estimate of the number of fission events per second?

A 10^{12} **B** 10^{14} **C** 10^{16} **D** 10^{20} *(1 mark)*

3 A $_{92}^{235}U$ nucleus undergoes induced fission to form two nuclei X and Y with the emission of 2 fission neutrons. The mass number of X is P and its atomic number is Q. Which one of the following alternatives, **A–D**, gives the mass number and atomic number of Y? *(1 mark)*

	mass number	atomic number
A	233 − P	91 − Q
B	234 − P	91 − Q
C	233 − P	92 − Q
D	234 − P	92 − Q

4 The nuclide $_{29}^{64}Cu$ can decay by beta plus (β^+) emission to form a nickel (Ni) nucleus.

a Copy and complete the equation below to represent this decay.

$$_{29}^{64}Cu \rightarrow \, _{\_}^{0}\beta + \, _{\_}^{\_}Ni + \, \_ + Q \qquad \textit{(2 marks)}$$

b The energy, Q, released in the above decay is 0.66 MeV. Use this information and the data below to calculate the mass of the nickel nucleus in atomic mass units. *(3 marks)*

mass of $_{29}^{64}Cu$ nucleus = 63.913 84 u   mass of an electron = 0.000 55 u

c Figure 1 represents the above change in terms of quarks. Identify the particles a, b, c, and d and copy and complete the table below.

	Name	Symbol
a		
b		
c		
d		

(2 marks)

d State one similarity and one difference between a β^+ particle and the particle in Figure 1 labelled:

   i   d

   ii   c.   *(4 marks)*

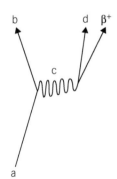

▲ **Figure 1**

5 a i Sketch a graph to show how the binding energy per nucleon depends on the nucleon number A for all known nuclides. Show appropriate values on the axes of your graph. *(4 marks)*

ii Use your graph to explain why energy is released when two small nuclei are fused together in a fusion reaction. *(4 marks)*

b i The mass of a helium $^4_2$He nucleus is 4.001 50 u. Use this information and the data below to calculate the binding energy per nucleon of a helium nucleus.

mass of a proton = 1.007 28 u; mass of a neutron = 1.008 67 u

(3 marks)

ii Mark and label the position of a $^4_2$He nucleus on your graph.

(1 mark)

iii The binding energy of a $^2_1$H nucleus is about a sixth of the binding energy of an alpha particle. Explain why the emission of an alpha particle from a large unstable nucleus is much more likely than the emission of a $^2_1$H nucleus. *(3 marks)*

6 a Explain what is meant by induced fission. *(2 marks)*

b i Describe what is meant by a chain reaction of fission events. *(5 marks)*

ii Explain why there needs to be a minimum mass of fissile material for a chain reaction to occur. *(4 marks)*

c In addition to fuel rods, a thermal nuclear reactor contains control rods and a coolant.

i State the function of the control rods, describe how they are used, and give an example of a suitable material from which they are manufactured. *(5 marks)*

ii The moderator is used to reduce the kinetic energy of the neutrons produced by the fission reactions. State the reason why this is necessary and explain why the process of reducing the kinetic energy of a fission neutron is more effective if the atomic mass of the moderator atoms is as close as possible to the mass of a neutron. *(5 marks)*

iii State the essential properties of the coolant and describe how it is used to transfer energy from the nuclear reactor. *(6 marks)*

7 a The polonium nucleus $^{210}_{84}$Po emits an alpha particle and forms a lead (Pb) nucleus.

i Copy and complete the equation below to represent this reaction.

$$^{210}_{84}\text{Po} \rightarrow \underline{}\text{Pb} + \underline{}\alpha + Q$$ *(2 marks)*

ii Use the data below to calculate the energy Q released in MeV in this reaction.

mass of $^{210}_{84}$Po nucleus = 209.936 75 u

mass of lead nucleus formed = 205.929 45 u

mass of alpha particle = 4.001 50 u *(3 marks)*

b i The energy released is shared as kinetic energy between the lead nucleus and the alpha particle. Explain why the alpha particle does not take all the energy released in the reaction. *(2 marks)*

ii Calculate the kinetic energy of the alpha particle immediately after it was emitted by the nucleus. *(6 marks)*

The notes below highlight key points in your option. Use these revision notes alongside the notes for your option in your AQA A Level Physics Student Book and the detailed notes on the option in the corresponding Kerboodle resources. Use the questions provided here and in the Kerboodle resources to check your knowledge and skills.

CHAPTER 1 TELESCOPES

1.1 Lenses

The **lens formula** $\frac{1}{u} + \frac{1}{v} = \frac{1}{f}$ where u is the distance from the lens to the object, v is the distance from the lens to the image, and f is the focal length of the lens.

Linear magnification $= \dfrac{\text{height of object}}{\text{height of image}} = \dfrac{v}{u}$

> **Revision tip**
> Make sure you can draw a ray diagram to scale to locate an image formed by a converging lens.

Summary questions

1 An object of height 18.0 mm is placed perpendicular to the axis of a converging lens of focal length 0.120 m at a distance of 0.080 m from the lens.
   a Calculate the image distance and the height of the image formed by the lens. *(2 marks)*
   b Describe the image. *(1 mark)*

1.2–1.5 Telescope design

The refracting telescope

In normal adjustment:

- the image of a distant object is at infinity
- the distance between the two lenses $= f_0 + f_e$
- the angular magnification $M = \dfrac{f_0}{f_e}$.

The Cassegrain telescope has a parabolic concave mirror as the objective and a small 'secondary' convex mirror to reflect light into the eyepiece.

> **Revision tip**
> Make sure you can draw a ray diagram for a refracting telescope in normal adjustment and for a Cassegrain telescope.

> **Revision tip**
> 1 To resolve two stars that are almost in the same direction, their angular separation $\geq \dfrac{\lambda}{D}$ where D is the telescope diameter.
> 2 Advantages of a CCD:
> - Quantum efficiency of a CCD pixel $\approx 70\%$ compared with $\approx 4\%$ for photographic film.
> - Wavelength sensitivity (≈ 100 nm to 1100 nm) > the human eye (≈ 350 nm to 650 nm).
> - Images recorded continuously, stored, and processed electronically.

Summary questions

1 a A telescope consists of two converging lenses of focal lengths 60 mm and 480 mm. It is used to view a distant object that subtends an angle of 0.150° to the telescope. Calculate:
   i the angular magnification of the telescope *(1 mark)*
   ii the angle subtended by the virtual image seen by the viewer. *(1 mark)*
   b Explain why a Cassegrain reflector with a parabolic mirror does not cause spherical or chromatic aberration whereas a refractor telescope does. *(2 marks)*
   c Two nearby stars are observed using a telescope with an objective of diameter 80 mm. Assuming light from them has an average wavelength of 500 nm, calculate their angular separation in degrees. *(2 marks)*

CHAPTER 2 SURVEYING THE STARS
2.1 Star magnitudes
2.2 Classifying stars
Star distances and magnitudes

The **parallax angle** θ of a star is the angle subtended by the star to the line between the Sun and the Earth when the line is perpendicular to the line of sight to the star.

The distance, in parsecs, from a star to the Sun $= \dfrac{1}{\theta(\text{in arcseconds})}$

Note: 1 arc second $= \dfrac{1}{3600}$ degree. Also 1 parsec = 3.26 light years.

Stellar spectral classes

The spectrum of light from a star is used to classify it. Make sure you know the main star classes (OBAFGKM), their approximate temperatures and colour. Hydrogen absorption lines (only in O, B, and A stars) are due to excitations from the $n = 2$ energy level (The Balmer lines).

Summary questions

1 a State what is meant by the absolute magnitude of a star. (*1 mark*)
   b A star has an apparent magnitude of + 4.8 and its distance from Earth is 34 pc. Calculate its absolute magnitude. (*2 marks*)

2.3 The Hertzsprung-Russell (HR) diagram

Figure 1 shows the HR diagram. Make sure you can draw and the label the HR diagram and that you know the main characteristics of each type of star shown on the diagram.

Key terms

Formation of a star: Dust and gas clouds → protostar due to gravity; core heats up due to nuclear fusion of hydrogen into helium; outer layers of the protostar (the photosphere) heat up and emit light.

Main sequence: Inward gravitational attraction balanced by the outward radiation pressure. The star has a constant luminosity; hydrogen in the core → helium.

Red giants: Core collapses as hydrogen all converted to helium; star becomes a giant or a supergiant as the outer layers of the star expand and cool. A star of:

- mass less than 8 solar masses → a white dwarf
- higher mass → **a supergiant** which explodes as a supernova.

White dwarfs: Cools and its core contracts, throwing off its outer layers. If its mass is:

- less than 1.4 solar masses, it becomes stable, as **a white dwarf** which cools and eventually fades out
- greater than 1.4 solar masses, it explodes as a **supernova**.

2.4 Supernovas, neutron stars, and black holes

In a supernova event, the core of a star collapses and becomes denser and denser until it is rigid. The matter collapsing onto the core rebounds off the core, sending matter outwards into space.

Type 1a supernovas each have a peak luminosity about $M = -18$ and a strong silicon absorption line. They are used to find the distance to their host galaxy.

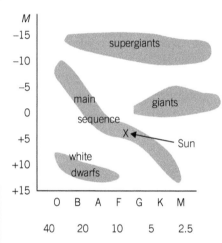

▲ **Figure 1** *The Hertzsprung-Russell diagram*

Summary questions

Two stars P and Q have surface temperatures of 3000 K and 10 000 K respectively and absolute magnitude −4, and +4 respectively.

1 a State which star has the greater power output and state the further evolutionary stages of each star. (*4 marks*)
   b Calculate the ratio of the diameter of P relative to Q. (*4 marks*)

Revision tip

The tell-tale sign of a supernova is a sudden change of about 20 in its absolute magnitude over about 24 hours and then a gradual decrease over several years.

Synoptic link

Elements heavier than iron are formed by nuclear fusion in a supernova explosion. See Topic 27.3, Fission and fusion.

2.4 Summary questions

1 a Describe the astronomical observations that indicate that a supernova event has occurred. (*3 marks*)

   b Explain the preceding changes in a star that causes a supernova event. (*3 marks*)

3.1 Summary questions

$c = 3.0 \times 10^8\,\text{ms}^{-1}$

1 A spectral line of a certain star in a binary system changes from its laboratory wavelength of 620 nm by ± 0.072 nm with a time period of 2.5 years. Calculate the orbital speed of the star and the radius of its orbit. (*4 marks*)

Neutron stars and black holes

A **neutron** star is the core of a supernova after all the surrounding matter has been thrown off into space.

A **black hole** is an object so dense that not even light can escape from it. The main physical property of a black hole is its mass.

The event horizon of a black hole is a sphere surrounding the black hole from which nothing can ever emerge. The radius of its event horizon (its Schwarzschild radius) $R_s = \dfrac{2GM}{c^2}$.

Supermassive black holes of enormous mass are thought to exist at the centres of many galaxies where they pull in gas and stars from their 'host' galaxy.

CHAPTER 3 COSMOLOGY

3.1 The Doppler effect

For a source of light moving at speed v relative to an observer,

$$\text{the change of wavelength of the light } \Delta\lambda = \pm\frac{v}{c}\lambda$$

The formula above holds provided $v \ll$ the speed of light c.

The **Doppler shift**, z, in frequency (or wavelength) is the fractional change $\dfrac{\Delta f}{f}$ (or $\dfrac{\Delta\lambda}{\lambda}$).

For a star or galaxy that is moving:

- away from the Earth, a **red shift** is observed (i.e., an increase of wavelength)
- towards the Earth, a **blue shift** is observed (i.e., a decrease of wavelength).

3.2 Hubble's Law and beyond

3.3 Quasars

Galaxies

Hubble's Law states that the distant galaxies are receding from us and their speed of recession v is directly proportional to the distance, d, to it.

$$v = Hd$$

where H, the constant of proportionality, is referred to as the Hubble constant.

Notes:

1 The red shift of a distant galaxy is used to calculate its speed of recession.

2 Its distance from Earth is determined by observing the period of individual Cepheid variable stars in the galaxy and using the known relationship between the period of a Cepheid variable and its mean absolute magnitude.

The Big Bang theory

The Big Bang theory states that the Universe was created in a massive 'primordial' explosion. The main evidence for the Big Bang theory is:

- **cosmic microwave background radiation** (CMBR), which is due to high-energy radiation created in the Big Bang that has been travelling through the Universe ever since the Universe became transparent

- **the relative abundance of hydrogen** and helium: stars and galaxies contain a ratio of about 7 to 1. This ratio is because the rest energy of a neutron is slightly greater than that of the proton.

The age of the Universe $= T \approx \dfrac{1}{H}$ (as the distance travelled by light in time T $(= cT) \approx$ maximum expansion distance $\dfrac{c}{H}$ given by Hubble's Law).

Dark energy

Distance measurements to type 1a supernovae by the red-shift method and the luminosity method give different results. This means that type 1a supernovae are much much further away than expected so they must be accelerating. Scientists think an unknown type of force is at work, releasing hidden energy referred to as **dark energy**.

At very large distances, dark energy is more prominent than gravity. Current theories suggest dark energy makes up about 70% of the total energy of the Universe.

Quasars

A quasar has a large red shift, indicating its distance is between 5000 and 10 000 light years away. Its light output is much greater than the Milky Way galaxy but its size is of the same order as the size of the Solar System.

Quasars are often found in or near distorted galaxies. These 'active' galaxies may have a **supermassive black hole** at their centres. Overheating of hot glowing gas as it is attracted towards the black hole results in the gas being thrown back out along its axis of rotation into space.

Summary questions

$c = 3.0 \times 10^8 \text{ ms}^{-1}, H = 65 \text{ kms}^{-1}\text{Mpc}^{-1}$

The absolute magnitude of any type 1a supernova is -18. A certain type 1a supernova has an apparent magnitude of $+22$.

1 a Calculate the distance to the supernova in parsecs. *(3 marks)*

  b Outline why measurements on type 1a supernovas have led to the conclusion that the expansion of the Universe is accelerating. *(2 marks)*

3.4 Exoplanets

Planets outside the Solar System are known as **exoplanets**.

- Many exoplanets are gas giants like Jupiter.
- Some exoplanets orbit within the habitable zone near a star (the zone in which liquid water may exist on a planet).

Methods of detection of exoplanets

1 **The radial velocity method:** a periodic Doppler shift is observed in the light from the star due to a planet and the star orbiting each other.

2 **The transit method:** the intensity of the light from the star regularly dips when an exoplanet passing 'in front' of the star blocks out some of the light from the star.

Summary questions

1 An exoplanet causes an intensity dip of 10% in the light intensity of a star. Estimate the radius of the exoplanet in terms of the radius of the parent star. *(3 marks)*

Medical physics

CHAPTER 1 PHYSICS OF THE EYE
1.1 Physics of vision

The ciliary muscle – concentric round the rim of the eye lens. To view a near object, the fibres contract to make the eye lens thicker. To view a distant object, the opposite happens. Adjustment is automatic and is called **accommodation.**

The iris – its concentric fibres contract and its radial fibres relax to make the eye pupil narrower so less light passes through it. In dim light, the opposite happens so more light passes through it.

The retina consists of rods and cones. Cones are colour sensitive and predominate at the fovea. Rods are sensitive to low intensity and are not colour sensitive.

To resolve two nearby point objects, their diffracted images on the retina need to be separated by at least two retinal cells.

The quantum efficiency of the eye is about 1 to 2% compared with $\approx 70\%$ for a CCD.

1.2 Lenses
1.3 Defects of vision
Lenses

The power of a lens in dioptres is defined as $\dfrac{1}{\text{its focal length in metres}}$.

The lens formula $\dfrac{1}{u} + \dfrac{1}{v} = \dfrac{1}{f}$ where u is the distance from the lens to the object and v is the distance from the image to the lens.

Linear magnification = height of image / height of object = $\dfrac{v}{u}$.

Sight defects

The normal eye has a near point of 25 cm and a far point at infinity.

* **Myopia or short-sight** – cannot focus on distant objects.
* **Hypermetropia or long-sight** – cannot focus on nearby objects.
* **Astigmatism** – images sharper in focus in one direction than in other directions.

Summary questions

1   **a**   With the aid of a diagram, describe how myopia is corrected.
                                                            (4 marks)
   b   A short-sighted eye has a far point 3.00 m away. Calculate the power of the correcting lens to give a corrected far point at infinity.
                                                            (2 marks)
   c   A long-sighted eye has a near point at 0.50 m away. Calculate the power of the correcting lens to give a corrected near point at 25 cm from the eye.   *(2 marks)*

CHAPTER 2 PHYSICS OF THE EAR
2.1 The structure of the ear

The normal human ear has a frequency range from about 20 Hz to 18 000 Hz.

The middle ear – the three bones: the hammer, the anvil, and the stirrup, transmit the vibrations of the tympanic membrane to the oval window of the inner ear. They:

- increase the force of the vibrations on the oval window
- filter out unwanted noise
- protect the ear from high-sound levels.

The inner ear – receives sound vibrations from the middle ear at its **oval window,** causing vibrations in the **cochlea** fluid which make the **basilar membrane** vibrate. Tiny sensitive 'hair cells' on the membrane then send nerve impulses to the brain.

2.2 Sound measurements
2.3 Frequency response

The decibel scale

The intensity of a sound wave (and of any type of wave) is the energy per second per unit area incident normally on a surface. The unit of intensity is $Js^{-1}m^{-2}$ or Wm^{-2}.

The threshold intensity of hearing I_0 is the least intensity the human ear can hear at a frequency of 1000 Hz (= 1 picowatt / m²).

Intensity level in decibels dB = $10\log\dfrac{I}{I_0}$

The dBA scale matches the normal ear's response at different frequencies.

Hearing loss in dB at any given frequency is the difference between the threshold intensity level for normal hearing and for the person being tested.

CHAPTER 3 BIOLOGICAL MEASUREMENT
3.1 ECG signals

Figure 1 shows a typical ECG trace (or electrocardiogram) for a healthy person.

The main features of the trace are labelled as P, Q, R, S, and T.

- **P** is when the atria contract. Between P and Q, the ventricles fill.
- **QRS** is when the ventricles contract and the atria relax.
- **T** is when the ventricles relax.

An ECG amplifier needs an even voltage gain of ≈ 1000 up to about 20 Hz, a high signal-to-noise ratio, and a high input impedance.

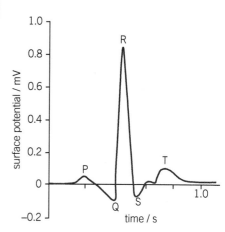

▲ **Figure 1** *An ECG trace of a healthy person*

1 a i Draw a labelled diagram of an ECG trace obtained from a person with a healthy heart, showing how the pd varies with time during one heartbeat. *(2 marks)*

  ii Label the main features on your trace and describe how they relate to the changes in the heart that take place during one heartbeat. *(5 marks)*

 b State two essential characteristics of an ECG amplifier other than high input impedance, giving a reason why each characteristic is essential. *(4 marks)*

CHAPTER 4 NON-IONISING IMAGING

4.1 Ultrasonic imaging

In an **A-scan system**, an oscilloscope is used to measure the time taken t by the reflected pulse to travel from the probe to the reflecting boundary and back. The distance, x, from the body surface to the reflecting boundary $= \frac{1}{2}ct$, where c is the speed of ultrasonic waves in the body.

In an **B-scan system**, the probe contains several transducers which transmit pulses simultaneously. As the probe is moved across the patient, the B-scan system therefore gives a two-dimensional image.

Reflection of ultrasonic waves

The **acoustic impedance** $Z = \rho c$ where ρ is the density of the substance and c is the speed of ultrasonic waves in it. The unit of acoustic impedance is $kg\,m^{-2}s^{-1}$.

- The intensity of waves reflected at a boundary, $I_R = \dfrac{(Z_2 - Z_1)^2}{(Z_2 + Z_1)^2} \times I_0$ where I_0 is the incident intensity and where Z_1 and Z_2 are the acoustic impedances of the incident and reflected intensities respectively.

- A gel is applied to the skin where the probe is applied to prevent reflection of ultrasonic waves at the body surface.

4.2 Endoscopy

The principle of the endoscope

The **incoherent** bundle of an endoscope transmits light into the body. A lens over the end of the **coherent bundle** is used to form an image of the body cavity on the end of the fibre bundle. The fibre ends are in the same relative positions at each end so the image formed by the lens inside the body is seen as a 'coherent' image on the end of the bundle outside the body.

The effect of different refractive index values

The angle of the viewing cone $= 2i_{max}$ where i_{max} is given by $n_o \sin i_{max} = n_1 \sin (90 - i_c)$ where n_o and $n_1 =$ refractive indices of air and the core respectively.

For no light loss from a fibre of diameter d, the maximum radius of curvature $R = \dfrac{d}{\dfrac{n_1}{n_2} - 1}$

where $n_2 =$ refractive index of the cladding.

1 a Explain the functions of the backing block in an ultrasonic probe. *(2 marks)*

 b Explain why a gel must be applied between an ultrasonic probe and the skin when the probe is used. *(2 marks)*

 c An organ has a density of $1070\,kg\,m^{-3}$ and the speed of sound through it is $1580\,ms^{-1}$. The corresponding data for surrounding soft tissue is $1050\,kg\,m^{-3}$ and $1550\,ms^{-1}$. Calculate the reflection coefficient of the boundary between the organ and the surrounding soft tissue. *(2 marks)*

Synoptic link

See Topic 5.3, Total internal reflection, to remind yourself about the law of refraction, total internal reflection, and optical fibres.

1 a State the function of a coherent bundle and of an incoherent bundle in an endoscope. *(3 marks)*
 b Explain why the optical fibres in the coherent bundle of an endoscope need to be very thin and why they should not be bent too much. *(2 marks)*
 c Calculate the angle of the viewing cone for a fibre which has a core of refractive index 1.60 and a refractive index 1.52. *(4 marks)*

4.3 The MR scanner

A hydrogen nucleus precesses in a magnetic field at two different energy levels. The frequency of precession depends on the magnetic flux density. When a pulse of radio waves is applied, at any point where the precession frequency is equal to the radio wave frequency:

- nuclear magnetic resonance occurs and nuclei in the lower energy state flip into the higher energy level
- excited nuclei flip back to the lower energy state, each emitting a radio wave photon.

In the MR scanner

In a scan, the magnetic field is designed to change the location of the excited nuclei systematically The signal detected after each pulse is from that location. The detector is connected to a computer which creates a visual image of a cross-section of the patient.

Tissue discrimination is possible because the rate of decay of the signal, the relaxation time, depends on the type of molecules surrounding the hydrogen molecules.

MRI scans are non-ionising radiation unlike X-ray scans.

Summary questions

1 a Explain how an MR scanner is used to detect the location of hydrogen nuclei in the body. *(5 marks)*
 b What is meant by relaxation time in the context of an MR scan and why is it an important consideration in an MR scan? *(3 marks)*

CHAPTER 5 X-RAY IMAGING

5.1 The physics of diagnostic X-rays

The anode pd controls the maximum energy of the X-ray photons from an X-ray tube. The minimum wavelength $\lambda_{MIN} = \frac{hc}{eV}$.

The electron beam current controls the intensity of the X-ray beam.

The electrical power supplied to an X-ray tube = beam current $I \times$ anode pd V.

Revision tip
Make sure you know how an X-ray tube works and that you can draw a graph to show how the relative intensity of the X-rays varies with wavelength for a certain tube voltage V.

Summary questions

$e = 1.6 \times 10^{-19}$ C, $h = 6.63 \times 10^{-34}$ Js, $c = 3.0 \times 10^{-8}$ ms^{-1}
1 a Sketch a graph to show how the intensity of the X-rays from an X-ray tube varies with wavelength. *(3 marks)*
 b An X-ray tube operates at a pd of 60 kV. Calculate the minimum wavelength of the X-rays emitted. *(2 marks)*

5.2 X-rays and matter

When a beam of X-rays passes through an 'absorber' of thickness x, the transmitted intensity I is given by

$$I = I_0 e^{-\mu x}$$

where I_o is the incident intensity of the beam and μ is the **attenuation coefficient** (sometimes called the absorption coefficient) of the absorber substance. The unit of μ is metre^{-1}.

The half-value thickness, $X_{\frac{1}{2}}$, of an absorber of X-rays is defined as the thickness required to reduce the intensity by 50%. Note that $X_{\frac{1}{2}} = \dfrac{\ln 2}{\mu}$

The mass attenuation coefficient of a substance $\mu_m = \dfrac{\mu}{\rho}$. The unit of μ_m is m^2 kg^{-1}.

Summary questions

1 Bone has a density of 1900 kgm^{-3} and a half-value thickness for 50 keV X-ray photons of 15 mm.
 a Calculate the mass attenuation coefficient of bone. *(3 marks)*
 b Calculate the % reduction in intensity when 50 keV X-rays pass through 4.0 mm of bone. *(3 marks)*

5.3 Image enhancement

Image quality

The image sharpness is improved by making the anode surface at an angle of about 70° to the electron beam so that the X-rays effectively originate from a much smaller area than the impact area of the beam.

The image contrast is improved by using:

- **a lead collimator grid** to prevent X-rays scattered in the patient from reaching the film
- **a contrast medium** (e.g., barium) as it is a good absorber of X-rays and can make the image of certain organs stand out.

Image intensification

The intensifying screen – a double-sided film between two sheets of fluorescent material that absorb X-rays photons and re-emit many more light photons.

The image intensifier allows the radiation dose to be reduced by a factor of 1000 and enables a 'real time' image to be observed and recorded.

A flat panel (FTP) detector used in place of a photographic film gives a digital image on a separate LCD screen.

The CT scanner

A CT (computed tomography) scanner consists of an X-ray tube and a ring of thousands of small solid-state detectors linked to a computer. The tube moves round the ring with the beam always directed at the patient at the centre of the ring. The detector signals are processed to display a 2D digital image of the cross-section of the patient.

An X-ray tube in a CT scanner operates at potentials between 100 and 150 kV and the X-rays are filtered so they are monochromatic (i.e., monoenergetic).

CHAPTER 6 RADIONUCLIDE IMAGING AND THERAPY

6.1 Radionuclide imaging

Properties of radioisotopes

The effective half-life T_{eff} of a radioactive isotope in the body is the time taken for the activity of the radioactive isotope in the body to reduce to half its initial activity.

$$\frac{1}{T_{eff}} = \frac{1}{T_b} + \frac{1}{T_p}$$

where T_p is **the physical half-life** of the radioactive isotope and T_b is the **biological half-life** of the substance in the body.

Most radioisotopes are produced by exposing stable isotopes to neutron radiation in a nuclear reactor or from a particle accelerator so the exposed isotope becomes unstable and emits γ radiation after β emission. The technetium isotope $^{99}_{43}Te$ is used for many diagnostic purposes as it forms in a 'metastable' state, usually referred to as Te-99m, which then decays with a short half-life of 6 hours by emitting β radiation with a much longer half-life.

A **gamma camera** is used to detect the γ photons from the radioisotopes. It uses **photomultipliers** in which each γ photon incident on the crystal of the gamma camera generates an electronic pulse.

The PET (positron emission tomography) scanner

In a PET scan:

- Each positron travels less than a millimetre because it meets an electron and they annihilate each other, producing two γ photons travelling in opposite directions.

- A ring of detectors connected to a computer registers a positron emission when opposite detectors detect a γ photon at the same time.

6.2 Radiation therapy

X-ray tubes for therapy do not operate above about 300 kV because their insulation breaks down at such high potential differences. Their X-rays are not penetrating enough to reach tumours deep inside the body.

Electron accelerators can generate X-ray beams with energies up to 25 MeV which are capable of destroying deep tumours. Electron accelerators can be switched on and off as required, enabling exposure times to be carefully controlled. In addition, they do not contain highly-radioactive sources and extensive shielding is not required round the source of radiation (i.e., the target).

Radioactive implants are used to destroy malignant cells in a tumour. Radioactive isotopes used include:

- isotopes that emit monoenergetic gamma photons so they have little effect beyond their intended range

- beta-emitting isotopes with half-lives of the order of days.

Summary questions

1 a A radioactive isotope has a physical half-life of 12.0 hours and a biological half-life of 24 hours. Calculate the effective half-life of this isotope in the body. *(2 marks)*

  b The technetium isotope $^{99}_{43}Te$ is metastable and has a half-life of 6.0 hours. State two reasons why this isotope is often used for diagnostic purposes. *(2 marks)*

Summary questions

1 a Give two reasons why X-rays from a 100 kV X-ray tube are less suitable for medical therapy compared with X-ray radiation from an electron accelerator. *(2 marks)*

  b The lutecium $^{177}_{71}Lu$ isotope, a β⁻ emitter with a half-life of 6.7 days, given intravenously, attaches itself to certain cancer cells and is used in the treatment of tumours that have spread into different parts of the body. The isotope emits 0.5 MeV β⁻ particles. Explain why 0.5 MeV X-ray photons from an electron accelerator could not be suitable for this purpose. *(2 marks)*

Engineering physics

CHAPTER 1 ROTATIONAL DYNAMICS

1.1 Angular acceleration

For constant angular acceleration of a rotating body. If the change of angular velocity is $\Delta\omega$ in a time interval Δt

$$\text{its angular acceleration, } \alpha = \frac{\Delta\omega}{\Delta t}$$

The unit of angular displacement is the radian per second (rad s^{-1}) where 2π radians = 360°. In a rotating body, for a point (i.e., small 'element') in the body at perpendicular distance r from the axis of rotation:

- its speed $v = \omega r$
- its tangential acceleration $a = a\,r$.

Equations for constant angular acceleration

For any body rotating with a constant angular acceleration α about a fixed axis with an initial angular speed ω_1, and an angular speed ω_2, and angular displacement θ time t later:

- $\omega_2 = \omega_1 + \alpha t$
- $\theta = \frac{1}{2}(\omega_2 + \omega_1)t$
- $\theta = \omega_1 t + \frac{1}{2}\alpha t^2$
- $\omega_2 = \omega_1^2 + 2\alpha\theta$

The average value of the angular speed $= \frac{1}{2}(\omega_2 + \omega_1)$

Angular motion graphs

Angular motion graph	Equivalent linear motion graph
1 Angular displacement θ against time t	**1 Displacement s against time t**
Gradient = angular velocity ω	Gradient = velocity v
2 Angular velocity ω against time t	**2 Velocity v against time t**
Gradient = angular acceleration α	Gradient = (linear) acceleration a
Area under the graph line = angular displacement	Area under the graph line = displacement s

Summary questions

1 A flywheel is accelerated by a constant torque for 38.2 s from rest. During this time it makes 24.6 turns. It then slows down to a standstill 92.0 s after the torque is removed, making 97.0 turns during this time.
   - **a** **i** Calculate its average angular acceleration when it was being accelerated.
     (3 marks)
     ii Calculate its average angular deceleration when it was slowing down.
     (2 marks)
   - **b** The moment of inertia of the flywheel was 290 kgm². Calculate the torque applied to it when it was being accelerated. *(3 marks)*

Summary questions

1 A flywheel accelerates from 2.0 to 17.0 revolutions per minute in 75.0 s. The radius of the flywheel is 0.036 m. Calculate:
   - **a** the angular acceleration of the flywheel *(2 marks)*
   - **b** the number of turns it makes. *(3 marks)*

1.2 Moment of inertia

Torque $T = F\,r$ where the force F that causes the torque acts at perpendicular distance r from the axis to the line of action of the force. The unit of torque is the newton metre (Nm).

The moment of inertia I of a body about a given axis is defined as $\Sigma m_i r_i^2$ for all the points in the body, where m_i is the mass of each point and r_i is its perpendicular distance from the axis. The unit of I is kgm².

When a body undergoes angular acceleration α, the resultant torque T acting on it is given by $T = I\alpha$. Two bodies of equal mass distributed in different ways will have different values of I.

In general, **the further the mass is distributed from the axis, the greater is the moment of inertia about that axis.**

1.3 Rotational kinetic energy

A body rotating at angular speed ω has kinetic energy $E_K = \frac{1}{2}I\omega^2$ where I is its moment of inertia about the axis of rotation. **The work done W** by a constant torque when the body is turned through angle $\theta = T\theta$. For a body rotating at constant angular speed ω, the **power P** delivered by a torque T acting on the body is given by

$$P = T\omega$$

Uses of flywheels

1 To smooth out the variations of the speed of a machine or an engine when the load varies.

2 To store kinetic energy when a vehicle's brakes are applied and it slows down.

1.4 Angular momentum

The angular momentum of a rotating object = $I\omega$ where I is the moment of inertia of the body about the axis of rotation and ω is its angular speed. The unit of angular momentum is $kg\,m^2\,rad\,s^{-1}$ or $N\,m\,s$.

Conservation of angular momentum

The total angular momentum of any system consisting of one or more rotating objects is constant, provided the resultant torque on the system is zero. **The angular impulse of a short-duration torque T is defined as $T\,\Delta t$.** If $\Delta\omega$ is the change of angular momentum due to this torque T, then **angular impulse $T\Delta t$ = change of angular momentum $I\Delta\omega$**

▼ **Table 1** *Comparison between linear and rotational motion*

Linear motion	Rotational motion
displacement s	angular displacement θ
speed and velocity v	angular speed ω
acceleration a	angular acceleration α
mass m	moment of inertia I
momentum mv	angular momentum $I\omega$
force F $F = ma$	torque $T = Fd$ $T = I\omega$
$F = \dfrac{d(mv)}{dt}$	$T = \dfrac{d(I\omega)}{dt}$
Impulse $F\Delta t = m\Delta v$	Angular impulse $T\Delta t = I\Delta\omega$
kinetic energy $= \frac{1}{2}mv^2$	kinetic energy $= \frac{1}{2}I\omega^2$
work done $= Fs$	work done $= T\theta$
power $= Fv$	power $= T\omega$

Summary questions

1 A metal disc X on the end of an axle rotates freely at 180 revolutions per minute. The moment of inertia of the disc is 0.034 kgm².

   a Calculate the angular momentum of the disc. *(2 marks)*

   b After a second disc Y that is initially stationary on the same axis is engaged by X, both discs rotate at 125 revolutions per second. Calculate the moment of inertia of Y. *(4 marks)*

Summary questions

1 A 1.85 kg object hanging from a string wrapped round the axle of a flywheel of diameter 12.0 mm is used to accelerate the flywheel which is on frictionless bearings from rest. The object falls through a vertical distance of 2.2 m in 9.4 s which is the time the string takes to unwrap from the axle. Calculate:

   a the potential energy lost by the object in descending 2.2 m
(1 mark)

   b the kinetic energy of the object 9.4 s after it was released from rest *Hint: Assume the acceleration of falling object is constant and calculate its speed from its initial speed, the distance it moved and the time it took.*
(2 marks)

   c i the kinetic energy gained by the flywheel
(1 mark)

     ii the moment of inertia of the flywheel. *(3 marks)*

Synoptic link

See Topic 11.1, Density, for internal energy.

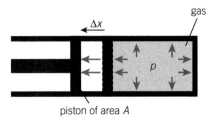

▲ **Figure 2** *Work done by a gas*

Revision tip

An energy transfer where no heat transfer occurs is called an **adiabatic change**.

An energy transfer where the temperature is constant is called an **isothermal change**.

Synoptic link

See Topic 20.2, The ideal gas law, to remind yourself about the properties of an ideal gas.

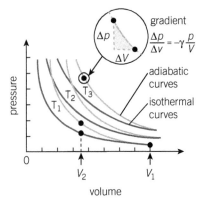

▲ **Figure 3** *Adiabatic curves*

CHAPTER 2 THERMODYNAMICS
2.1 The first law of thermodynamics
2.2 Thermodynamics of ideal gases

The first law of thermodynamics states that heat energy (energy transferred by heating) supplied to a system either increases the internal energy of the system or enables it to do work, or both. If the internal energy of the object increases by ΔU, and the work done **by** the system is W,

then the heat transfer into the system, $Q = \Delta U + W$

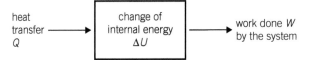

▲ **Figure 1** *Change of internal energy*

A negative value for ΔU means the internal energy has decreased. A negative value for Q means heat energy is transferred out of the system. A negative value for W means that work is done **on** the system.

Work done by an ideal gas

- Work is done on a gas when its volume is reduced (i.e., it is compressed).
- Work is done by a gas when it expands.
- Consider a tube of gas at pressure p trapped by a piston, as in Figure 2. When the gas expands,

work done by the gas, $W = p\Delta V$

Adiabatic changes of an ideal gas

For any adiabatic change of an ideal gas, its pressure p and volume V are given by the equation

$$pV^\gamma = \text{constant}$$

where γ is the adiabatic constant of the gas.

1 You will be given the value of γ in an examination question.

2 Adiabatic changes of an ideal gas can be plotted on a graph of pressure against volume. The curves are steeper than isotherms (which are constant temperature curves in accordance with $pV = \text{constant}$), as Figure 3 shows.

Pressure against volume curves

The area under any pV curve represents the work done on or by the gas.

Summary questions

1 a When a gas expands, it gains 100 J of internal energy and does 80 J of work. Calculate the heat transfer to or from the gas. *(1 mark)*

   b When a gas is compressed and 80 J of heat energy is transferred from it, it gains 100 J of internal energy. Calculate the work done on or by the gas. *(1 mark)*

2 a A cylinder contains 0.0420 m³ of air at a pressure of 153 kPa and a temperature of 320 K. The air in the cylinder expands adiabatically to a volume of 0.0542 m³. Calculate the pressure and the temperature of the air immediately after it has been compressed. $\gamma = 1.40$ for air *(5 marks)*

   b Estimate the work done by the gas. *(3 marks)*

2.3 Heat engines

A heat engine is designed to do useful work which is supplied to it as a result of heat transfer. For an engine to do work as a result of heat transfer from a high temperature source, we need a low temperature sink to draw the energy from the source. Assuming $\Delta U = 0$ for the engine,

$$Q_{\text{H}} = W + Q_{\text{C}}$$

where Q_{H} = heat transfer to the engine, W = work done by the engine, Q_{C} = heat losses to the surroundings.

The second law of thermodynamics states that it is impossible for heat transfer from a high temperature source to produce an equal amount of work.

The efficiency of an engine is defined as

$$\frac{\text{the work done by the engine}}{\text{energy supplied to the engine}}$$

Engine cycles
The petrol engine

The petrol engine uses air as the working substance, heating it by internal combustion. The indicator diagram in Figure 1 shows the pressure changes in one cylinder of a four-cylinder petrol engine in one complete cycle. The cylinders 'fire' sequentially to maintain the motion of each cylinder.

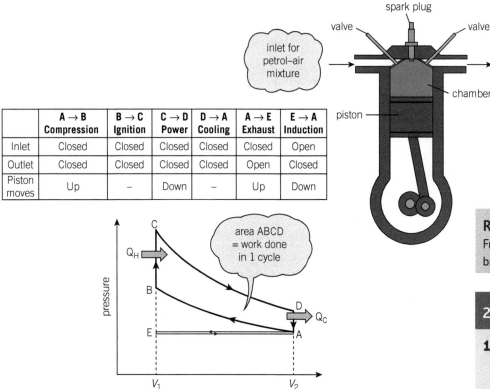

	A → B Compression	B → C Ignition	C → D Power	D → A Cooling	A → E Exhaust	E → A Induction
Inlet	Closed	Closed	Closed	Closed	Closed	Open
Outlet	Closed	Closed	Closed	Closed	Open	Closed
Piston moves	Up	–	Down	–	Up	Down

▲ **Figure 1** *The four-stroke petrol engine*

A → B Adiabatic compression No heat enters or leaves the cylinder. The work done on the air increases its internal energy.

B → C Ignition A spark ignites the mixture, heating the gas and raising its pressure and temperature in the cylinder very rapidly. Heat Q_{H} is gained by the gas during ignition.

C → D Power Work is done adiabatically by the expanding gas.

D → A Cooling The gas is cooled rapidly by contact with the internal walls of the cylinder block, reducing the pressure to the initial pressure. Heat Q_{C} is lost from the gas during cooling.

2.3 Summary questions

1 A vehicle engine with 4 cylinders uses fuel of calorific value 45 MJkg⁻¹ at a rate of 2.9×10^{-3} kgs⁻¹ when it operates at 37 cycles per second. In this situation, the work done by each cylinder in each engine cycle is 0.28 kW. Calculate:
   a the input power from the engine *(2 marks)*
   b the thermal efficiency of the engine. *(3 marks)*

2.4 Summary questions

1 A gas-powered engine operates between temperatures of 1100 K and 300 K. When the engine uses gas of calorific value 31 MJkg^{-1} supplied at a rate of 9.4 kgh^{-1}, the mechanical power output of the engine is 38 kW.

   a Calculate the theoretical efficiency of the engine. (2 marks)

   b i Show that the input power to the engine is 81 kW. (2 marks)

   ii Calculate the heat transferred by the heat engine to its low-temperature sink. (1 mark)

Summary questions

1 A heat pump fitted with a 120 W electric motor supplies 600 W of heat per second to a room from outdoors. Calculate:

   a its coefficient of performance (2 marks)

   b the heat gained from outdoors. (1 mark)

2 A refrigerator with a 5 W pump cools a plastic beaker containing 0.18 kg of water from 20°C to 0°C in 1200 s. Calculate:

   a the heat energy extracted from the water (2 marks)

   b the coefficient of performance of the refrigerator.

   specific heat capacity of water = 4200 Jkg^{-1}K^{-1} (2 marks)

The net work done W by one engine cylinder in one cycle = $Q_H - Q_C$ = the area of the indicator loop ABCD. The **indicated power** of the engine = the rate of net work done by the engine per second = area of indicator loop × number of cycles per second × number of cylinders.

The **input power** to the engine = the calorific value of the fuel (in Jkg^{-1}) × the fuel flow rate (in kgs^{-1}) where the **calorific value** of the fuel is the energy released per unit mass of fuel when the fuel is burnt.

The thermal efficiency of the engine = $\dfrac{\textbf{work done per second}}{\textbf{energy supplied per second}}$

$$= \dfrac{\textbf{indicated power}}{\textbf{input power from fuel combustion}}$$

- The mechanical efficiency of an engine
$$= \dfrac{\textbf{brake power } T\omega \textbf{ (i.e., output power)}}{\textbf{indicated power}}$$

- The overall efficiency = its mechanical efficiency × its thermal efficiency

2.4 Efficiency and thermodynamics

For any heat engine operating between a heat source and a heat sink:

- Its % efficiency of = $\dfrac{W}{Q_H} \times 100\% = \dfrac{Q_H - Q_C}{Q_H} \times 100\%$ (since $W = Q_H - Q_C$).
- The most efficient type of engine is a theoretical engine called **a reversible engine**.
- For a reversible engine, $\dfrac{T_c}{T_H} = \dfrac{Q_C}{Q_H}$.

The % efficiency of a reversible engine = $\dfrac{T_H - T_C}{T_H} \times 100\%$.

Efficiency in practice

Thermal efficiency is less than above because:

- the indicator loop area in practice does not have sharp corners as in the theoretical diagram so less work is done (due to the valves not opening and closing instantly)
- the maximum temperature is not reached as in perfect combustion
- the expansion and compression strokes are not perfectly adiabatic as assumed in calculating the theoretical thermal efficiency
- work is done by the engine to bring about the induction and exhaust strokes.

Mechanical efficiency is less than 100% because:

- friction between the moving parts in the engine cannot be eliminated
- oil used to lubricate the bearings in an engine are viscous which means they cause some resistance to the motion of the moving parts.

2.5 Heat pumps and refrigerators

A heat pump is a device that transfers energy from a cold space to a hot space.

Where work done W causes heat transfer Q_C from the cold space and heat transfer Q_H (= $Q_C + W$) into the hot space.

$$\textbf{its coefficient of performance} = \dfrac{Q_H}{W} = \dfrac{Q_H}{Q_H - Q_C}$$

If T_H and T_C are the temperatures of the hot 'space' and the cold 'space' respectively, its coefficient of performance $< \dfrac{T_H}{T_H - T_C}$

In a refrigerator, work is done in order to remove energy from food and drink in a cold space. The coefficient of performance of a refrigerator is defined as $\dfrac{Q_C}{W}$ (in comparison with $\dfrac{Q_H}{W}$ for a heat pump). Therefore,

$$\textbf{its coefficient of performance} = \dfrac{Q_C}{W} = \dfrac{Q_C}{Q_H - Q_C}$$

If T_H and T_C are the temperatures of the hot 'space' and the cold 'space' respectively, its coefficient of performance $< \dfrac{T_C}{T_H - T_C}$

Turning points in physics

CHAPTER 1 THE DISCOVERY OF THE ELECTRON

1.1 Thermionic emission of electrons

Cathode rays

When a sufficiently large pd is applied to a discharge tube, the gas conducts and emits light.

- Positive ions and electrons produced by ionisation near the cathode recombine and emit light photons.
- Electrons that do not recombine (called cathode rays) accelerate and move towards the anode. They excite gas atoms by collision further along the tube. These atoms de-excite and emit light photons.

The principle of thermionic emission

Electrons from a heated filament wire in a vacuum tube are attracted to the anode where some pass through a small hole to form a narrow beam. The speed v of each electron in the beam is given by $\frac{1}{2}mv^2 = eV$.

1.2 Deflection of an electron beam
1.3 Use of electric and magnetic fields to determine $\frac{e}{m}$

Using a uniform magnetic field

Electrons of speed v are deflected in a circular path. The centripetal force on each electron ($= \frac{mv^2}{r}$) is due to the magnetic force ($= Bev$) hence $r = \frac{mv}{Be}$.

Combining this equation with the anode pd equation $\frac{1}{2}mv^2 = eV_A$ gives $\frac{e}{m} = \frac{2V_A}{B^2r^2}$.

Using graphs

A graph of r against $\frac{1}{B}$ gives a straight line through the origin with a

gradient $k = \left(\frac{2mV_A}{e}\right)^{\frac{1}{2}}$.

The value of k is used to determine $\frac{e}{m}$.

Using a uniform electric field

The electric field is between two parallel plates of length L and spacing d at pd V_p, electrons of speed v are directed into the field perpendicular to the field and the deflection y of the beam is measured.

- The acceleration, a towards the positive plate $= \frac{2y}{t^2}$ (from $y = \frac{1}{2}at^2$) where $t = \frac{L}{v}$.

- The force on each electron, $F = \frac{eV_p}{d}$.

Therefore, the acceleration $a = \frac{F}{m} = \frac{eV_p}{md}$ so $\frac{e}{m} = \frac{ad}{V_p}$.

The significance of Thomson's determination of $\frac{e}{m}$

JJ Thomson showed that the electron's specific charge $\frac{e}{m}$ was $1860 \times$ larger than that of the hydrogen ion which had the largest known specific charge. But neither the mass nor the charge of the electron was known at that time.

1.4 The determination of the charge of the electron, e, by Millikan's method

For a droplet of charge Q and mass m between oppositely charged horizontal parallel metal plates, when the droplet is stationary, $\frac{QV_p}{d} = mg$. Hence $Q = \frac{mgd}{V_p}$.

Summary questions

1 a Explain why the positive ions in a discharge tube depend on the gas in the tube but the negative particles are identical. *(4 marks)*

b Electrons in a vacuum tube reach a speed of $3.6 \times 10^7\,\text{ms}^{-1}$ after being accelerated from rest through a potential difference V. Calculate the speed of the electrons if they are accelerated through a potential difference $\frac{1}{2}V$. *(2 marks)*

Synoptic link

The speed of the electrons can be determined using the perpendicular E and B fields as in the velocity selector in Topic 24.2, Moving charges in a magnetic field.

Summary questions

1 Electrons moving at a constant speed are directed horizontally into a uniform electric field due to two parallel plates spaced 40.0 mm apart which have a pd of 5000 V between them.

a When a uniform magnetic field of flux density 3.25 mT is applied at right angles to the beam and to the electric field, the beam is undeflected. Calculate the speed of the electrons. *(2 marks)*

b When the electric field is switched off, the beam is deflected to form a circle of diameter 68.0 mm. Calculate the specific charge, $\frac{e}{m}$, of the electron. *(2 marks)*

To measure the droplet's mass, the electric field is switched off and the droplet's terminal speed is measured. At this speed, the droplet weight $(\frac{4}{3}\pi r^3 \rho g)$ = the drag force on it $(6\pi\eta rv)$ where η is the viscosity of air. Therefore $r^2 = \frac{9\eta v}{2\rho g}$ is used to calculate the droplet radius r and its mass.

The significance of Millikan's results

Millikan discovered that the electric charge Q is **quantised** in whole number multiples of 1.6×10^{-19} C. He therefore concluded that the charge of the electron is -1.6×10^{-19} C.

Summary questions

$g = 9.81 \text{ ms}^{-2}$, $e = 1.6 \times 10^{-19}$ C

1 A charged oil droplet was held stationary between oppositely charged metal plates 5.50 mm apart when the pd between the plates was 696 V. When the pd was switched off, the droplet fell at a terminal speed of 1.55×10^{-4} ms^{-1}. Calculate:
  a the mass of the droplet *(3 marks)*
  b the charge of the droplet. *(2 marks)*
Density of the oil = 960 kg m^{-3}, viscosity of air = 1.80×10^{-5} Nsm^{-2}

CHAPTER 2 WAVE PARTICLE DUALITY
2.1 Early theories of light

Newton's corpuscular theory of light	• Light consists of 'corpuscles' (i.e., tiny particles).	• reflection explained by assuming corpuscles bounce off a mirror without loss of speed. • refraction explained by assuming corpuscles travel faster in a refractive substance than in air.
Huygens' wave theory of light	• A wavefront propagates by creating wavelets that move forwards and recreate the wavefront.	• reflection explained by assuming wavefronts reflect off a mirror without loss of speed and the speed of the wavelets is unchanged. • refraction explained by assuming the waves travel **slower** in a transparent substance than in air.

Wave theory was rejected in favour of corpuscular theory because:
- *there was no experimental evidence to show light travels faster or slower in a transparent substance than in air*
- *Newton had a much stronger scientific reputation than Huygens*
- *light waves considered to be longitudinal so could not explain polarisation of light.*

Young's double slits experiment	• Light passed through double slits produces an interference pattern.	• Corpuscular theory of light incorrectly predicted there would be only two bright fringes. Wave theory can explain interference.

The wave theory of light was not accepted for many years until it was shown that light in water travels slower than in air.

Revision tip
Make sure you can explain Newton's ideas on reflection and refraction in terms of velocity components.

Synoptic links
Look back at Topic 5.1, Refraction of light, and Topic 5.4, Double slit Interference.

Summary questions

1 **a** Explain how Newton's corpuscular theory of light explains the refraction of a light ray when it passes from glass into air. *(3 marks)*
  b Explain why Newton's theory of light cannot explain Young's fringes pattern whereas Huygens' wave theory can. *(4 marks)*

2.2 The discovery of electromagnetic waves

Maxwell's theory Electromagnetic waves are transverse electric and magnetic waves that are in phase and perpendicular to each other and the direction of propagation.	Maxwell derived $c = \dfrac{1}{\sqrt{\mu_0 \varepsilon_0}}$ for their speed in free space and showed it is the same as the speed of light.
Hertz's discovery • A spark gap transmitter was used to produce radio waves and a metal loop to detect them. The waves caused sparks to jump across the gap in a detector loop. • Hertz also showed that radio waves • can be polarised • are reflected by a metal sheet and can form stationary waves.	• Hertz measured the distance between adjacent nodes ($= 0.5\,\lambda$) of stationary radio waves of known frequency and used $v = f\lambda$ to find their speed. Note that $t = 1$, $(2\,f_0\,N)$ as there are $2\,N$ teeth and gaps round the cogwheel perimeter and the cogwheel's period of rotation is $1/f_0$.
Fizeau's measurement of the speed of light A rotating cog wheel was used to chop a narrow beam of light into pulses of light. The pulses were reflected back by a distant reflector. As the rotation frequency of the cog wheel was increased, the reflected light was blocked at a certain frequency f_o, corresponding to the time taken, t, for a gap to be replaced by the next tooth.	$c = \dfrac{2D}{t} = 4Df_0 N$ where D is the distance from the cog wheel to the reflector and N is the number of teeth on the cogwheel.

2.3 The development of the photon theory of light

The ultraviolet catastrophe

Figure 1 shows how the intensity of the radiation from a 'black body' (i.e., a perfect absorber) varies with wavelength for different temperatures of the object. Wave theory predicts incorrectly that the intensity becomes infinite at smaller and smaller wavelengths. Planck solved the problem by introducing the idea that the energy of an atom vibrating at frequency f can only be an integral multiple of hf.

The discovery of photoelectricity

Experiments showed that for any given metal:

- no photoelectrons are emitted if the frequency of the incident light is below a threshold frequency
- photoelectric emission occurs instantly.

Wave theory predicts incorrectly that light of any frequency causes photoelectric emission and the lower the frequency of the light, the longer photoelectric emission takes.

Einstein's photon theory of light states that light consists of photons, each carrying energy hf, where f is the frequency of the radiation.

For a conduction electron to escape, Einstein said it needs to:

1 absorb a single photon of energy hf

2 use energy equal to the work function ϕ of the metal to escape.

Hence, the **threshold frequency** of the incident radiation $f_0 = \dfrac{\phi}{h}$ as $hf \geq \phi$.

Summary questions

1 **a** Hertz discovered that the strength of the radio signal detected from a radio wave transmitter aerial varies according to the orientation of the detector. Explain this effect and state the conclusion drawn by Hertz about the radio waves from the transmitter. *(3 marks)*

b In an experiment to measure the speed of light by Fizeau's method, the reflected light could not be observed when the cog wheel with 720 teeth was rotating at 12.0 Hz and the distance travelled by the light was 8.5 km. Use this data to calculate the speed of light in air. *(2 marks)*

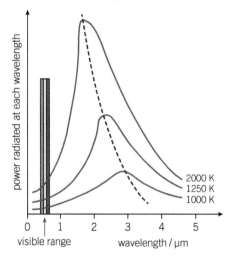

▲ **Figure 1** *Black-body radiation curves*

Using graphs

Photoelectric emission from a metal is stopped if its potential is increased to its **stopping potential**, V_s. At this potential,

$$eV_s = hf - \phi$$

Therefore, a graph of V_s (on the y-axis) against f is a straight-line graph with:

- $\dfrac{h}{e}$ for its gradient m
- $f_0 = \dfrac{\phi}{h}$ for its x-intercept.

Summary questions

$e = 1.6 \times 10^{-19}$ C, $h = 6.63 \times 10^{-34}$ Js, $c = 3.00 \times 10^8$ ms^{-1}

1 Light of wavelength 460 nm is directed at a metal surface that has a work function of 2.1 eV. Calculate:

   a   the energy of a photon of wavelength 460 nm   (2 marks)

   b   the maximum kinetic energy of the emitted photoelectrons.

                                                                    (3 marks)

2.4 Matter waves

De Broglie's hypothesis states that all matter particles have a wave-like nature and that the particle momentum mv is linked to its de Broglie wavelength λ by the equation

$$m v \times \lambda = h \text{ where } h \text{ is the Planck constant}$$

Combining the de Broglie equation with $\frac{1}{2}mv^2 = eV_A$ gives $\lambda = \dfrac{h}{\sqrt{(2meV_A)}}$.

The hypothesis was verified by the discovery that electrons in a beam are diffracted when they pass through a very thin metal foil.

The transmission electron microscope (TEM)

Electrons in a beam are scattered when they pass through the sample. Electromagnetic coils acting as 'magnetic lenses' focus the scattered electrons onto a fluorescent screen to form a magnified image of the sample structure.

Diffraction at the lens apertures reduces the amount of detail in the image. Increasing the anode pd gives better resolution because it reduces the de Broglie wavelength of the electrons. The amount of detail is also affected by sample thickness and lens aberrations.

The scanning tunnelling microscope (STM)

In the STM, electrons 'tunnel' across the small gap (≈ 1 nm) between a fine metal tip and the surface it scans.

- The tip is at a potential of about -1 V relative to the surface so that electrons only tunnel in one direction.
- Piezoelectric transducers are used to control the position of the tip.

In constant height mode, the tunnelling current varies as the tip scans across the surface and the gap width changes. The tunnelling current is recorded and used to map the height of the surface on a computer screen.

CHAPTER 3 SPECIAL RELATIVITY

3.1 About motion

Electromagnetic waves were thought to be vibrations in an invisible substance called **ether**. Detection of the ether was thought possible by comparing the time taken by light to travel the same distance in perpendicular directions. Michelson and Morley designed the interferometer in Figure 1 to measure such small differences.

The observer sees a pattern of interference fringes because of the difference in the path lengths of the two beams. Turning the apparatus through 90° in a horizontal plane swaps the beam directions relative to the Earth's motion through the ether. Ether theory predicted the difference in the travel times of the two beams would reverse, causing a measurable shift in the fringes. However, no fringe shift was observed. This 'null' result disproved the 'ether' theory.

3.2 Einstein's theory of special relativity

Einstein based his theory of special relativity on two postulates:

- **Physical laws have the same form in all inertial frames of reference** (frames of reference that move at constant velocity relative to each other).

- **The speed of light in free space, c, is invariant** (i.e., is always the same and is independent of the motion of the light source and the observer).

Relativistic effects

Time dilation: the time t between two events measured by a moving observer is stretched out ('dilated') compared with the proper time t_o (the time interval between the two events measured by an observer at rest relative to the events).	$t = \dfrac{t_o}{\sqrt{\left(1 - \dfrac{v^2}{c^2}\right)}}$
Length contraction: the length L of a rod moving in the same direction as its length is shorter than its proper length L_o (its length measured at rest).	$L = L_o \sqrt{\left(1 - \dfrac{v^2}{c^2}\right)}$
Relativistic mass: the mass m of a moving object is greater than its rest mass m_o and increases with speed. No material object can ever reach the speed of light as its mass would become infinite.	$m = \dfrac{m_o}{\sqrt{\left(1 - \dfrac{v^2}{c^2}\right)}}$
Mass and energy: Energy E and mass m are equivalent (i.e., interchangeable) on a scale given by $E = mc^2$	kinetic energy $E_k = mc^2 - m_o c^2$ $\left(\dfrac{1}{2}mv^2 \text{ at speeds } v \ll c\right)$

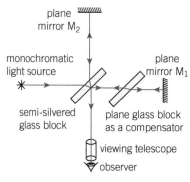

▲ **Figure 1** *The Michelson-Morley interferometer*

Revision tip

As speed $v \to c$, m increases gradually to about $2m_o$ at $v \approx 0.9\,c$ and then increases sharply and $\to$ **infinity** as v approaches c.

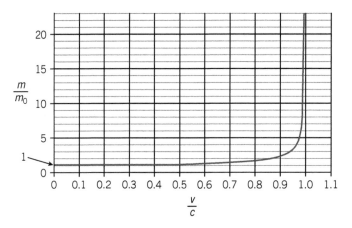

▲ **Figure 2** *Relativistic mass against speed*

Bertozzi's experiment

A linear accelerator was used to accelerate pulses of electrons from rest through a measured pd, V. He measured their speed v by timing each pulse between two electrodes a measured distance apart. A plot of $y = $ kinetic energy E_K against $x = \dfrac{v}{c}$ gives a curve that matches Figure 2 (whereas E_K against $\dfrac{1}{2}mv^2$ gives a much lower curve) that goes past $v/c = 1$.

Summary questions

$e = 1.6 \times 10^{-19}$ C, $c = 3.0 \times 10^8$ ms^{-1}

1 A beam of particles of rest energy 140 MeV travelling at a speed of 0.99 c travels in a straight line between two stationary detectors which are 200 m apart. Calculate:

   a the proper time between a particle passing between two detectors.

   (2 marks)

   b the kinetic energy of each particle in MeV at this speed. *(3 marks)*

Electronics

CHAPTER 1 DISCRETE SEMICONDUCTOR DEVICES

1.1 MOFSETs

An **n-channel enhancement MOSFET** (metal-oxide semiconducting field-effect transistor) is shown in the circuit in Figure 1. Figure 2 shows how the drain–source current, I_D, depends on the pd V_{DS} between the drain and the source for different values of the gate–source pd, V_{GS}.

- When $V_{GS} \leq$ the **threshold voltage** V_{Th}, $I_D = 0$ and the transistor is OFF.

- When $V_{GS} > V_{Th}$, increasing V_{DS} from zero increases I_D linearly then the rate of increase decreases and I_D becomes constant (= saturated current I_{DSS}).

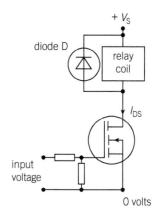

▲ **Figure 1** *A MOSFET in use*

Summary questions

1 a Explain why the relay switch in Figure 1 closes when the input voltage is made sufficiently positive. *(3 marks)*

b Sketch a graph to show how the drain–source current, I_D, for a MOSFET depends on the pd between the drain and the source, V_{DS}, when $V_{GS} > V_{Th}$. On your graph, mark the point P at which saturation occurs at minimum V_{DS}. *(3 marks)*

1.2 Zener diodes

Zener diode characteristics

A reverse-biased zener diode breaks down and conducts at a specific pd, its **breakdown voltage** V_Z. See Figure 1. A zener diode in series with a resistor R and a dc supply V_S is a **constant voltage source** as the output pd across the diode does not change when the current supplied to the output changes. The resistance of R is chosen between two limits:

- $R_{min} = \dfrac{(V_S - V_Z)}{I_{max}}$ where $I_{max} = \dfrac{\text{maximum power rating } P}{V_Z}$

- $R_{max} = \dfrac{(V_S - V_Z)}{I_{out}}$ where I_{min} = least current needed to operate the device at the output.

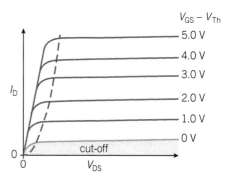

▲ **Figure 2** *Graph of I_D against V_{DS}*

Summary questions

1 A 6.0 V, 300 mW zener diode is reverse-biased and in series with a 60 kΩ resistor and a 9.0 V battery.

A 6.0 V, 30 mA lamp is connected across the zener diode. Calculate the current in:

a the resistor *(2 marks)*

b the zener diode. *(1 mark)*

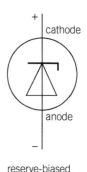

▲ **Figure 1** *The zener diode*

1.3 Photodiodes

Photodiode characteristics

A photodiode conducts when it is reverse-biased and light is incident on it. Figure 1 shows how the current varies with the pd across the photodiode for different light intensities. Note that the dark current (i.e., at zero light intensity) is negligible.

1 A reverse-biased photodiode is connected in series with a 9.0 V battery, a 4.7 kΩ resistor, and a milliammeter.

  a When light at a certain intensity is incident on the photodiode, the milliammeter reads 1.2 mA. Calculate the pd across the photodiode. (*2 marks*)

  b Calculate the photodiode pd when the light intensity is reduced by 50%. (*2 marks*)

1 a State the physical property that is measured by a Hall effect sensor. (*1 mark*)

  b When a Hall effect attitude device on the ground is orientated so its z-axis is vertical, it gives x, y, z readings of 3 μT, 4 μT, and 5 μT respectively.

    i Calculate the magnitude of the magnetic flux density B. (*2 mark*)

    ii Calculate the angle between the z-axis and the resultant magnetic field. (*1 mark*)

Uses of a photodiode

- to monitor light intensity changes
- to convert light pulses into electronic pulses
- to detect atomic particles when used with a scintillator.

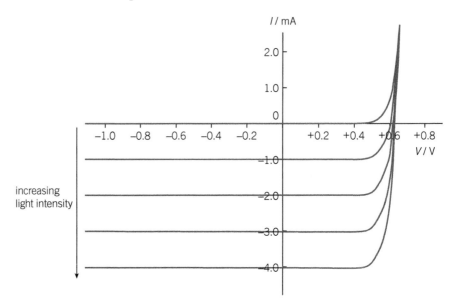

▲ **Figure 1** *A load line on the I against V graph for a photodiode*

1.4 Hall effect-sensors

The sensor is designed to detect, monitor, and measure magnetic fields. Charge carriers moving through the sample are forced by a magnetic field to one edge of the sample. The pd across the sample, the Hall voltage, V_H, is proportional to:

- the current I through the sample
- the component of the magnetic flux density B perpendicular to the sample.

Applications

- as a **proximity sensor**
- to monitor **'attitude'** or orientation of objects in a magnetic field relative to the field direction.

CHAPTER 2 SIGNALS AND SENSORS
2.1 Analogue and digital signals

Advantages of digital transmission

1 Elimination of noise by using regenerator amplifiers.

2 Digital signals can be compressed by using pulses of shorter duration so more bits per second can be sent.

Pulse code modulation is used to convert an analogue to a digital signal at a single output in a 3 stage process.

1 **Sampling** to produce a 'pulse amplitude modulated' (PAM) signal.

2 **Conversion to a data byte** (quantisation) on parallel outputs using an A/D converter.

3 **Parallel to serial conversion** to produce a sequence of bits at a single output.

Quality of conversion (back to an analogue signal)

1 **The frequency of the sampling process** (i.e., the sampling rate) must be at least twice the highest frequency of the analogue signal.

2 **The more bits in a sample** (i.e., the word-length), the smaller the spacing between adjacent levels and the greater the quality of the recovered analogue signal.

The number of bits transmitted per second = the sample frequency × the number of bits per sample.

2.2 Sensors
2.2.1 Sensor-operated systems

An electronic sensor contains a **sensing device** which supplies a signal to a **processing unit** which can operate an **output device** such as an indicator lamp or a relay.

Sensing devices in potential dividers

A potential divider that includes a thermistor or an LDR (light dependent resistor) can be used as a device to 'sense' temperature or light intensity changes.

> **Summary questions**
>
> 1 An LDR and a 220 kΩ resistor R are connected in series with each other and a 12V power supply, as shown in Figure 1. The LDR has a dark resistance of 900 kΩ and a resistance of 50 kΩ in a darkroom when the room lights are on. Calculate the output pd from this circuit in:
>   a darkness   (*2 marks*)
>   b in the dark room when the room lights are on.   (*2 marks*)

CHAPTER 3 ANALOGUE SIGNAL PROCESSING
3.1 LC resonance filters
The parallel *LC* circuit

Figure 2 shows how the rms pd across the parallel LC circuit in Figure 1 varies with frequency.

The pd peak occurs at **the resonant frequency** $f_{RES} = \dfrac{1}{2\pi\sqrt{(LC)^{\frac{1}{2}}}}$.

When a radio or TV receiver is 'tuned in' to a channel, the resonant frequency of the *LC* circuit is adjusted to the channel's radio frequency by changing the inductance *L* or the capacitance *C*.

- **The bandwidth f_B** of the circuit is defined as the difference between the frequencies (either side of the peak) at $\sqrt{2}$ of the maximum pd (i.e., the 50% energy points of the curve).
- **The Q factor** $= \dfrac{f_{RES}}{f_B}$.

> **Summary questions**
>
> 1 a Calculate the capacitance of the capacitor in a parallel *LC* circuit designed to detect radio signals of frequency 1.5 MHz if it has a 0.25 mH inductor.   (*2 marks*)
>   b Describe what is meant by the *Q* factor of a parallel *LC* circuit.
>   (*2 marks*)

> **Summary questions**
>
> 1 An audio signal with a maximum amplitude of 6.4 V is converted into a digital signal. Each sample of the audio signal is converted into an 8-bit byte.
>   a What is the least change of the audio voltage that would produce a different byte?   (*1 mark*)
>   b What byte is produced when the voltage of the audio signal is 4.1 V?
>   (*2 marks*)

> **Revision tip**
> Remember that ntc thermistors have a resistance that falls with increase of temperature.

> **Synoptic link**
> See Topic 13.5, The potential divider, for more about thermistors and LDRs in potential dividers.

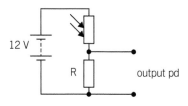

▲ **Figure 1** (2.2.1)

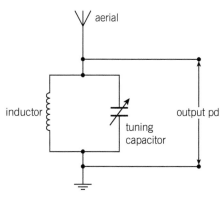

▲ **Figure 1** *A parallel LC circuit*

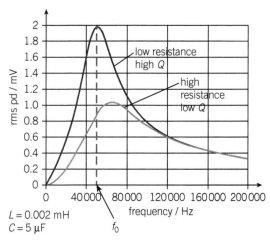

$L = 0.002$ mH
$C = 5\,\mu F$

▲ **Figure 2** *The frequency response of a parallel LC circuit*

Revision tip

Make sure you can explain how the comparator is used to switch a device (e.g., a buzzer or an LED on or off) in response to a change of pd from a sensor in a potential divider.

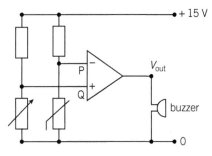

▲ **Figure 1** *A temperature-operated buzzer*

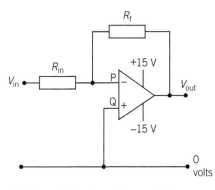

▲ **Figure 1** *The inverting amplifier*

Revision tip

The voltage gain of the inverting amplifier is independent of the open-loop voltage gain, A_0.

3.2 Ideal operation amplifiers

An operational amplifier has an inverting input, P, and a non-inverting input, Q. For pds V_- and V_+ applied to inputs P and Q respectively,

the output pd $\quad V_{OUT} = A_0(V_+ - V_-)$ **(the transfer function)**

where A_0 is the **open-loop gain** (typically of the order of 10^5).

1 The output pd cannot exceed the limits determined by the power supply. For a ±15 V supply, $(V_+ - V_-) < \pm 150\,\mu V$ if the output pd is not to be saturated (i.e., not reach ±15 V).

2 If either P or Q are earthed, the terminal that is not earthed is said to be at **virtual earth**, provided the output pd is not saturated.

3 An ideal op-amp has an infinite input resistance and zero output resistance.

The comparator

If the output pd V_{OUT} is:

- at positive saturation, then $V_+ > V_-$, (i.e., pd at Q > pd at P)
- at negative saturation, then $V_- > V_+$ (i.e., pd at P > pd at Q).

Summary questions

1 **a** Explain, in relation to an operational amplifier, what is meant by virtual earth. *(2 marks)*

 b i When the temperature of the thermistor in Figure 1 decreases, explain why the buzzer switches on at a certain temperature. *(3 marks)*

 ii Describe the adjustment you must make to the variable resistor so the buzzer switches on at a higher temperature. *(1 mark)*

3.3 Operational amplifiers circuits

Feedback is the process of taking some of the output signal from an amplifier and adding it to the input signal. Positive feedback increases the output and negative feedback reduces it. Resistors are used to supply negative feedback to reduce the gain of an op-amp.

The inverting amplifier

$$\text{Voltage gain} \quad \frac{V_{out}}{V_{in}} = -\frac{R_F}{R_{in}}$$

where R_F is the feedback resistor and R_{in} is the input resistor.

The summing amplifier

The circuit is like Figure 1 but with three input terminals and three input resistors.

$$\text{The output pd } V_{out} = -IR_F = -\left(\frac{V_1}{R_1} + \frac{V_2}{R_2} + \frac{V_3}{R_3}\right)R_F$$

The non-inverting amplifier

$$\text{Voltage gain} \quad \frac{V_{out}}{V_{in}} = 1 + \frac{R_F}{R_{in}}$$

where R_F is the feedback resistor and R_{in} is the input resistor.

The frequency response of an operational amplifier

For alternating input voltages, the voltage gain is constant up to a certain frequency and decreases at higher frequencies.

The bandwidth is the frequency range over which the voltage gain is constant.

The bandwidth × the voltage gain = constant.

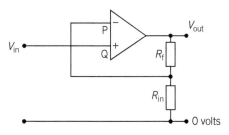

▲ **Figure 2** *The non-inverting amplifier*

Summary questions

1 a A sinusoidal alternating voltage with a peak voltage of 5.0 V is applied to the input of a non-inverting ±15 V operational amplifier which has a voltage gain of ×4. Draw graphs on the same axes to show how the input pd and the output pd vary with time. *(3 marks)*

b A summing amplifier has a 100 kΩ feedback resistor and four input terminals A to D with input resistances of 100 kΩ, 200 kΩ, 400 kΩ, and 800 kΩ respectively. A 4-bit byte 1011 is applied to inputs DCBA respectively where 1 represents a 4 V input. Calculate the output voltage. *(2 marks)*

Revision tip

You need to know how to derive the voltage gain equation for the inverting amplifier only. See the notes on the Electronics option in the Kerboodle resources.

CHAPTER 4 DIGITAL SIGNAL PROCESSING

4.1 Logic gates

Make sure you know the symbols and truth tables of NOT, OR, AND, NOR, EOR, and NAND gates. **In Boolean algebra**, + represents OR and a dot represents AND. Also, a bar above an input or output symbol represents NOT. For example, Q = (A + B).C means output Q = 1 if input A OR input B = 1 AND input C = 1.

4.2 Sequential logic

Summary questions

1 a Draw a truth table for the Boolean expression $C = \overline{\overline{A}.\overline{B}}$ *(4 marks)*

b Hence show that $\overline{\overline{A}.\overline{B}} = A + B$ *(1 mark)*

| Binary counter

See Figure 1 and Table 1	A series of identical circuits called T-type flip flops	Counts up repeatedly from 0000 to 1111 as the falling edge (1→0) at each Q output causes the next Q to switch.	A binary counter with n flip-flops goes through 2^n output states (the counter's modulus) before it resets itself.
BCD counter	Resets to 0 when the binary count reaches 1010 (=decimal 10)	The inputs of a 2-input AND gate to Q_4 and Q_2 and the AND output to the reset terminal.	Can be modified to count up to any number <10 by connecting the appropriate Q outputs to a 2 or 3 input AND gate which has its output gate are connected to the reset pin.
Johnson counter

See Figure 2 and Table 2 | A series of D-type flip flops with the last $\overline{Q}$ output connected to the first D input. | Each Q output is shifted to the next output when a clock pulse is applied | The clock pulses are applied, after the Q outputs have been reset to 0 . The sequence of changes shown in Table 2 takes place repeatedly. |

(a) circuit (reset terminals not shown)

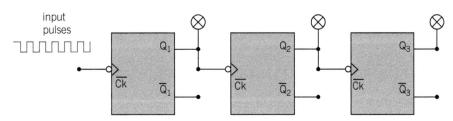

(b) input and output states

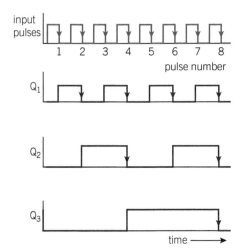

pulse number	Q_3	Q_2	Q_1
0	0	0	0
1	0	0	1
2	0	1	0
3	0	1	1
4	1	0	0
5	1	0	1
6	1	1	0
7	1	1	1
8	0	0	0

▲ **Figure 1** *A 3-bit binary counter (reset terminals not shown)*

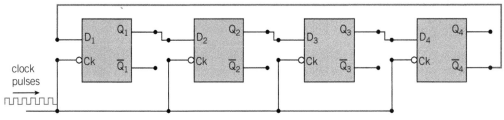

clock
pulses

▼ **Table 2** *A Johnson counter sequence*

▲ **Figure 2** *A Johnson counter (reset terminals not shown)*

Input pulse number	Q_1	Q_2	Q_3	Q_4
0	0	0	0	0
1	1	0	0	0
2	1	1	0	0
3	1	1	1	0
4	1	1	1	1
5	0	1	1	1
6	0	0	1	1
7	0	0	0	1
8	0	0	0	0

The astable oscillator

Clock pulses are provided by an **astable** oscillator. This is a digital circuit that repeatedly switches its output voltage between two voltage levels, as shown in Figure 3.

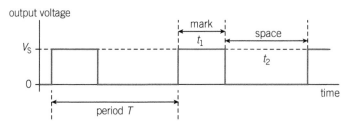

▲ **Figure 3** *The astable multivibrator output voltage*

The **mark-to-space ratio** (MSR) of the waveform $= \dfrac{t_1}{t_2}$

The **duty cycle** $= \dfrac{t_1}{T} \times 100\% \left(= \dfrac{\text{MSR}}{(\text{MSR} + 1)} \times 100\right)$

The frequency and the mark-to-space ratio are determined by the capacitance and resistance values of the capacitor and resistors in the oscillator.

CHAPTER 5 DATA COMMUNICATION SYSTEMS

5.1 Systems and links

The **bandwidth** of a transmission path is the range of frequencies that can be transmitted. The bandwidth is divided into frequency **channels** with a different signal allocated to each channel. For digital signals, each channel can carry a number of digital signals using **time-division multiplexing** (TDM). In this process, the duration of each bit is reduced and the time between one bit of a signal and the next is divided into a number of equal time slots which are used to carry bits from other signals.

Summary questions

1 **a** Explain how a binary counter can be adapted so it counts from 0 to 5 repeatedly.
(*3 marks*)

 b Use Figure 2 to explain how a Johnson counter with 3 flip-flops works.
(*4 marks*)

Synoptic link

See Topic 5.3, Total internal reflection for more about optical fibres.

Summary of the advantages and disadvantages of each type of communication link

Type of link	Twisted wire pair	TV coaxial cable	Radio/microwave link	Optic fibre
Typical attenuation	$\approx 5\times$ weaker at 10 MHz per 100 m	$\approx 10\times$ weaker at 100 MHz per 100 m	$\approx 10^8\times$ weaker at 1 GHz per km	$\approx 0.95\times$ weaker for infrared per km
Bandwidth	500 kHz	50 MHz	FM 200 kHz Microwaves 1 GHz	$\approx$ GHz for 1 km (bandwidth decreases with increasing distance)
Noise	electrical interference, cross-over from other wires	electrical interference from electric motors	electrical storms, cosmic sources	no electrical interference, noise removed by regenerators
Bit rate /10^6 bits per second	10	100	Digital TV 20 Microwave beam 70	$>10^6$
Repeater/regenerator spacing	short distances	up to 2–3 km	line of sight for microwaves	up to 100 km or more
Security	low, unless coded digital signals used			high

Note: Radio waves of frequencies < 30 MHz reflect from the ionosphere in the upper atmosphere back to the ground (and back to the ionosphere etc.).

5.1 Summary questions

1 **a** Which type of communication link is suitable for:
   i ship to shore communications *(1 mark)*
   ii signals to and from a satellite? *(1 mark)*
 b A digital signal consists of 16-bit words transmitted every 0.80 ms. Each bit is a pulse lasting 0.50 μs. How many such signals could be transmitted using time-division multiplexing in a single frequency channel and what would be the bit rate in the channel? *(3 marks)*

5.2 Modulation

Analogue signals are transmitted by **amplitude modulation (AM) or frequency modulation (FM)** where the carrier amplitude or frequency is modulated by the signal amplitude.

For a carrier wave of frequency f_c modulated by waves of frequency f_m, the **bandwidth** of the signal:

- $= 2 f_m$ **for amplitude modulation**
- $= 2 (\Delta f + f_M)$ **for frequency modulation** (where Δf is the maximum deviation of the sidebands from the carrier frequency).

The relative advantages of AM and FM transmission

Quality: FM less affected by noise than AM

Range: AM range greater than FM range for the same power

Relative advantages of Pulse Code Modulation (PCM)

- no noise and higher quality
- more than one signal can be transmitted using TDM on a single frequency channel
- a PCM signal can be security-coded.

The main disadvantages of PCM is its greater bandwidth.

5.2 Summary questions

1 **a** A radio transmitter broadcasts a frequency-modulated signal of bandwidth 500 kHz at a carrier frequency of 98.0 MHz. Explain what is meant by the bandwidth of a signal and calculate the upper and lower limits of the allocated frequency channel. Assume only the lower frequency band is used. *(2 marks)*
 b An FM radio station transmits its signal in a frequency channel of bandwidth 200 kHz at a carrier frequency of 102.25 MHz. The upper limit of the modulation frequency is 15 kHz. Calculate:
   i the maximum deviation of the signal frequency from the carrier frequency *(1 mark)*
   ii the carrier frequency of the next higher frequency channel. *(1 mark)*

Answers to practice questions

Chapter 1

1 a i 11 **ii** 12 **iii** 11 [1] *(1 mark)*

b Specific charge of the ion $= \dfrac{e}{23m_{nuc}}$ [1]

$= \dfrac{1.60 \times 10^{-19}\,\text{C}}{23 \times 1.67 \times 10^{-27}\,\text{kg}}$ [1]

$= 4.17 \times 10^6\,\text{C kg}^{-1}$ (where m_{nuc} is the mass of a nucleon) [1] *(3 marks)*

2 a Let N be the number of protons in the nucleus. Therefore, the specific charge $= \dfrac{Ne}{38m_{nuc}}$ [1] $= 4.30 \times 10^7\,\text{C kg}^{-1}$.

Hence, $N = \dfrac{38 \times 1.67 \times 10^{-27} \times 4.3 \times 10^7}{1.60 \times 10^{-19}}$ [1]

$= 17$ [1] *(3 marks)*

b $^{38}_{17}\text{Cl} \rightarrow {}^{38}_{18}\text{Ar} + {}^{0}_{-1}\beta$ [3] (1 mark for each correct term) *(3 marks)*

c Specific charge after emission $= \dfrac{18e}{38m_{nuc}}$

$= \dfrac{18 \times 1.60 \times 10^{-19}}{38 \times 1.67 \times 10^{-27}}$ [1] $= 4.54 \times 10^7\,\text{C kg}^{-1}$ [1] *(2 marks)*

3 a a = 206, b = 81, c = 4, d = 2 [1] for any 2 correct [1] for remaining 2 correct *(2 marks)*

b X = the strong nuclear force which acts between the nucleons in the nucleus [1]

c Y = the electromagnetic force (of repulsion) between the protons in the nucleus [1] *(2 marks)*

4 a $E = \dfrac{hc}{\lambda} = \dfrac{6.63 \times 10^{-34} \times 3.00 \times 10^8}{630 \times 10^{-9}} = 3.16 \times 10^{-19}\,\text{J}$ [1]

$= \dfrac{3.16 \times 10^{-19}}{1.60 \times 10^{-19}}\,\text{eV} = 1.97\,\text{eV}$ [1] *(2 marks)*

b Let N be the number of photons per second emitted. Therefore, $N\,E = 4.0 \times 10^{-3}\,\text{W}$

Hence, $N = \dfrac{4.0 \times 10^{-3}\,\text{W}}{3.16 \times 10^{-19}\,\text{J}}$ [1] $= 1.3 \times 10^{16}\,\text{s}^{-1}$ [1] *(2 marks)*

5 a It has an infinite range [1], it acts between charged particles and is attractive between oppositely charged particles and repulsive between particles with the same type of charge. [1] *(2 marks)*

b The force is due to the exchange of virtual photons between the two charged objects. [1] The photons are said to be virtual because they cannot be detected [1]. *(2 marks)*

6 a A positron is the antiparticle of the electron. [1] *(1 mark)*

b i A proton changes into a neutron and emits a positron in the process. So the number of protons decreases by one and the number of neutrons increases by 1. [1] *(1 mark)*

ii The weak nuclear force is responsible. The exchange particle is the W⁺ boson. [1] *(1 mark)*

iii a = proton, b = neutron [1], c = W⁺ boson [1], d and e = positron (β⁺ or e⁺) and a neutrino (n) in either order. [1] *(3 marks)*

7 a A proton in the nucleus interacts with an inner shell electron by emitting a W⁺ boson which is absorbed by the inner shell electron [1]. As a result, the proton changes into a neutron and the electron changes into a neutrino. [1] *(2 marks)*

b See Figure 1: any 2 correct [1] remaining two correct [1] *(2 marks)*

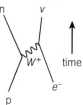

▲ **Figure 1**

8 a Pair production occurs when a photon of sufficient energy passing through matter near the nucleus of an atom [1] creates a particle and its corresponding antiparticle and ceases to exist. [1] An example is when a high-energy gamma photon creates an electron and a positron (or other correct example). [1] *(3 marks)*

b $2E_{ph} = 2 \times 0.51\,\text{MeV}$ [1] Therefore, $E_{ph} = 0.51\,\text{MeV}$ [1] *(2 marks)*

c Although annihilation is a process that turns matter into radiation and pair production turns radiation into matter, they are not reverse processes [1] because pair production involves a single photon passing near a nucleus whereas annihilation occurs whenever and wherever a particle and its corresponding antiparticle meet. [1] In both processes, energy and momentum are conserved but in the case of pair production, the nucleus is necessary to enable momentum to be conserved whereas for annihilation two photons are always produced in order to conserve momentum. [1] *(3 marks)*

Chapter 2

1 a $\bar{p}$ **b** e^+, v_e, μ^+ [1] **c** $\bar{p}$ **d** K^- [1] *(2 marks)*

2 a Hadrons can interact through the strong interaction. Leptons do not interact through the strong interaction. [1] *(1 mark)*

b i The weak interaction. [1]

ii The lepton numbers are −1 for the antimuon, the positron, and the muon antineutrino. The electron antineutrino has a lepton number of 1. The muon lepton number before the change is −1 and −1 after the change. Inserting these values into the table below shows that electron lepton number is conserved and muon lepton number is conserved.

	μ⁺	→	e⁺	v_e	$\overline{v_\mu}$	
Electron lepton number	0	=	−1	+1	0	[1]
Muon lepton number	−1	=	0	0	−1	[1]

iii Difference: the muon decays, the electron does not. Similarity: both are negatively charged. *(3 marks)*

3 a udd [1], the charge of an up quark is $+\frac{2}{3}$ and the charge of a down quark is $-\frac{1}{3}$ so the total charge of the udd combination is 0. [1] *(2 marks)*

b There are 2 non-strange charged mesons. [1] Their quark-antiquark compositions in brackets are π^- ($\overline{u}d$) and π^+ ($u\overline{d}$). The antiparticle of each of these mesons is therefore the other meson in the pair. [1] *(2 marks)*

4 a i π^+ [1]

ii The charge of an up quark is $+\frac{2}{3}$ and the charge of a down antiquark is $+\frac{1}{3}$ so the total charge of the ($u\overline{d}$) combination is +1. [1] *(2 marks)*

b i X is composed of 3 quarks because it is a baryon. [1] Two of the quarks are strange quarks because the strangeness of X is −2 and a strange quark has a strangeness of −1. [1]

The charge of a strange quark is $-\frac{1}{3}$ so the two strange quarks provide a charge of $-\frac{2}{3}$. The remaining charge is $-\frac{1}{3}$ because X has a charge of −1. [1] So the remaining charge is provided by a down quark. Therefore, X's quark composition is dss. [1] *(4 marks)*

ii All baryons eventually decay into a proton so X will decay into a proton. *(1 mark)*

5 a A proton in the nucleus changes into a neutron by emitting a W⁺ boson [1] which decays into a positron and an electron neutrino. [1] See Topic 2.4, Fig 2b. [1] *(3 marks)*

b Inserting values for baryon number, lepton number, and charge into the table below shows that electron lepton number is not conserved and muon lepton number is not conserved. So the decay will not happen. [1]

	μ+	→	e⁺	ν_e	$\overline{\nu}_\mu$
Charge	+1	=	+1	0	0
Baryon number	0	=	0	0	0
Electron lepton number	0	≠	−1	−1	0
Muon lepton number	−1	≠	0	0	1

Q and B rows correct [1] lepton rows correct [1] *(3 marks)*

6 a Strangeness is conserved as the reaction is a strong interaction. The total initial strangeness is −1 so the total final strangeness must also be −1. Therefore, the strangeness of X is −1 as the π^- has zero strangeness. [1] The charge of X is +1 as charge is conserved, the total initial charge is zero, and the π^- has a charge of −1. [1] The baryon number of the π^- meson and of the K^- meson is zero. Because baryon number is conserved and

the baryon number of the proton is +1, the baryon number of X must also be +1. [1] *(3 marks)*

b As X includes a strange quark in its composition, then another particle containing a strange antiquark must be produced in order to conserve strangeness (as the collision is a strong interaction). [1] So the other particle could not be a π meson which is a non-strange particle. [1] *(2 marks)*

7 a Similarity: both positively charged. Difference: an antimuon is an antiparticle, proton is a particle (or an antimuon is a lepton and a proton is a hadron). [1]

b Similarity: both are uncharged (or both are mesons/hadrons). Difference: a K^0 meson can decay into π mesons whereas a π^0 meson does not decay into K mesons. [1]

c Similarity: both are hadrons. Difference: the π^+ meson is charged whereas the neutron is uncharged (or the π^+ meson is a meson whereas the neutron is a baryon). [1] *(3 marks)*

8 a Q = 0, S = −2 [1] b Q = 0, S = −1 [1]
c Q = +1, S = +1 [1] d Q = +1, S = +1 [1] *(4 marks)*

Chapter 3

1 a The photoelectric effect is the emission of electrons from a metal surface when electromagnetic radiation is directed at the metal surface. [1] *(1 mark)*

b Electromagnetic radiation of a certain frequency f is composed of photons each of energy $E = hf$. [1] When photons of frequency f are directed at the surface, conduction electrons in the metal near the surface may each absorb a photon. [1] An electron that absorbs a photon therefore gains energy hf when it absorbs a photon. [1] An electron cannot escape if its energy is less than the work function ϕ of the metal. Therefore, electrons cannot escape if the frequency of the incident radiation is less than $\frac{\phi}{h}$. [1] *(4 marks)*

2 a Threshold frequency is the minimum frequency of incident electromagnetic radiation on a particular metal surface that will cause emission of electrons from the surface. [1] *(1 mark)*

b The work function of the metal $\phi = hf_{min} = 6.63 \times 10^{-34} \times 2.9 \times 10^{14} = 1.92 \times 10^{-19}$ J [1]

Photon energy $= hf = \frac{hc}{\lambda}$
$= \frac{6.63 \times 10^{-34} \times 3.00 \times 10^8}{560 \times 10^{-9}} = 3.54 \times 10^{-19}$ J [1]

Maximum kinetic energy of a photoelectron
$= hf - \phi = 3.54 \times 10^{-19} - 1.92 \times 10^{-19} = 1.6 \times 10^{-19}$ J
(to 2 significant figures). [1] *(3 marks)*

Note: The final answer is given to 2 significant figures (sf) because the work function is only given to 2 sf. All the working up to the final answer has been done to 3 sf to avoid intermediate rounding-off causing an arithmetical error in the final answer.

c Graph is a straight line with a positive gradient. [1] Line intercepts the x-axis at f_{min}. [1] *(2 marks)*

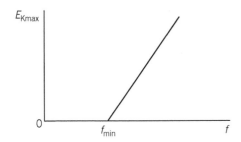

▲ **Figure 1**

3 a The threshold frequency $f_{min} = \dfrac{\phi}{h} = \dfrac{0.85\,eV}{h}$

$= \dfrac{0.85 \times 1.60 \times 10^{-19}\,J}{6.63 \times 10^{-34}}\,Js$ [1]

$= 2.05 \times 10^{14}\,Hz = 2.1 \times 10^{14}\,Hz$ (to 2 significant figures). [1]

(2 marks)

b Photon energy $= hf = \dfrac{hc}{\lambda}$

$= \dfrac{6.63 \times 10^{-34} \times 3.00 \times 10^{8}}{420 \times 10^{-9}} = 4.74 \times 10^{-19}\,J$ [1]

Work function $\phi = 0.85\,eV = 0.85 \times 1.6 \times 10^{-19}\,J$
$= 1.36 \times 10^{-19}\,J$ [1]

Maximum kinetic energy of a photoelectron
$= hf - \phi = 4.74 \times 10^{-19} - 1.36 \times 10^{-19}$
$= 3.38 \times 10^{-19}\,J$ [1]

(3 marks)

4 The frequency of blue light is greater than the frequency of red light. So a blue light photon has more energy than a red light photon. [1] In this example, the energy of a blue light photon is greater than the work function of the metal whereas the energy of a red light photon is less than the work function [1]. Therefore, when blue light is used, a blue light photon can give a conduction electron enough energy to overcome the work function of the metal and escape from its surface whereas a red light photon does not have enough energy to overcome the work function and escape. [1]

(3 marks)

5 a See Figure 1. Four energy levels shown and labelled correctly [1] gaps approx 1, 3, 5 [1]

(2 marks)

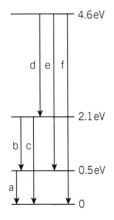

▲ **Figure 1**

b 6 photon energies *a–f* as follows, shown by the downward arrows on Figure 1 are possible:

a 0.5 eV, [1] *b* 1.6 eV (= 2.1 − 0.5 eV), *c* 2.1 eV, [1]
d 2.5 eV (= 4.6 − 2.1 eV), *e* 4.1 eV (= 4.6 − 0.5 eV),
f = 4.6 eV. [1]

(3 marks)

6 The wavelengths of the lines of a line spectrum of an element are characteristic of the atoms of that element because the energy levels of the atoms of an element are unique to that element. [1] When an electron in an atom moves from an energy level to a lower energy level, it releases a photon of energy equal to the energy difference between the two energy levels. [1] The photons that produce each line all have the same energy, which is different from the energy of the photons that produce any other line. The photon energies are therefore characteristic of the atom. [1] The wavelength of each photon depends on its energy so the photon wavelengths are characteristic of the atoms of the element. Therefore, the line spectrum is characteristic of the element. [1]

(4 marks)

7 a i particle nature of light

ii the wave nature of matter [1]

(1 mark)

b i momentum $p = \dfrac{h}{\lambda} = \dfrac{6.63 \times 10^{-34}}{500 \times 10^{-9}}$

$= 1.33 \times 10^{-27}\,kg\,m\,s^{-1}$ [1]

velocity $= \dfrac{momentum}{mass} = \dfrac{1.33 \times 10^{-27}}{9.11 \times 10^{-31}}$

$= 1460\,m\,s^{-1}$ [1]

ii momentum for the same de Broglie wavelength is the same, i.e.,
$1.33 \times 10^{-27}\,kg\,m\,s^{-1}$ [1]

velocity $= \dfrac{momentum}{mass} = \dfrac{1.33 \times 10^{-27}}{1.67 \times 10^{-27}}$

$= 0.80\,m\,s^{-1}$ [1]

(4 marks)

Chapter 4

1 a The vibrations of a polarised wave are perpendicular to the direction of travel of the wave along a line in one direction only. A transverse wave can therefore be polarised because its vibrations are perpendicular to the direction of travel of the wave. [1] A longitudinal wave cannot be polarised because its vibrations are along the direction of travel of the wave. [1]

(2 marks)

b The intensity of the light transmitted through both filters decreases as the filter is turned and becomes zero when the filter has been turned through 90°. [1] When the filter is turned through a further angle of 90°, the intensity increases from zero and then reaches a maximum again after being turned through a total angle of 180° from its initial position. [1] When the filter is turned through a further angle of 180°, the intensity varies in exactly the same way as it did when it was turned through 180° from its initial position. [1]

(3 marks)

2 a The microwaves reflected by the metal plate pass through the microwaves from the transmitter and so a stationary wave pattern is formed. [1] The maxima and minima are respectively the positions of the antinodes and nodes of the stationary wave pattern. [1]

Each node is where the two sets of waves always have equal and opposite displacements so the resultant amplitude is always zero. [1] Each antinode is a position where the wave peaks always arrive at the same time so the resultant amplitude is a maximum. [1] *(4 marks)*

b There are 4 nodes between the 1st and last node so the distance between the 1st and last node is equal to 5 node-to-node spacings. [1] The distance between adjacent nodes is therefore $\frac{82}{5}$ mm which is 16.4 mm. [1] Adjacent nodes are half a wavelength apart therefore the wavelength is equal to 32.8 mm = 2 × 16.4 mm. [1] *(3 marks)*

3 a i See Figure 1. Four equally-spaced loops [1] Antinodes (A) and nodes (N) correctly labelled [1] *(2 marks)*

▲ **Figure 1**

ii X and Y have the same amplitude [1], and their phase difference is π radians (or 180°) [1] *(2 marks)*

b i λ = 400 mm [1] as node-node spacing = 0.5λ = 200 mm [1] *(2 marks)*

ii The frequency of the first harmonic $f_1 = \frac{1}{4}$ × 200 Hz [1] (because the stationary wave pattern is the 4th harmonic).

Rearranging $f_1 = \frac{1}{2L}\sqrt{\frac{T}{\mu}}$ gives $T = (2Lf_1)^2\mu$ [1]

= $(2 × 0.800 × 50)^2 × 7.4 × 10^{-4}$ [1] = 4.7 N [1] *(4 marks)*

4 a $\lambda = \frac{c}{f} = \frac{1500}{1.6 × 10^6} = 9.4 × 10^{-4}$ m [1] *(1 mark)*

b i Period T = $\frac{1}{f} = \frac{1}{1.6 × 10^6} = 6.25 × 10^{-7}$ s = 0.625 μs [1]

Therefore, the number of waves in 10 μs = $\frac{10 \text{ μs}}{0.625 \text{ μs}}$ = 16 [1] *(2 marks)*

ii Any two of the following reasons: [2]

1 The waves spread out as they travel away from the probe and therefore their intensity decreases so their amplitude decreases.

2 The waves are only partially reflected as some of the wave energy is transmitted at each boundary. So the reflected waves are less intense and therefore have a smaller amplitude.

3 Some of the wave energy is absorbed by the body tissue as the waves travel through it. So the amplitude of the waves gradually decreases as they travel through the tissue. *(2 marks)*

5 a Coherent sources emit waves of the same frequency with a constant phase difference. [1] *(1 mark)*

b At certain points along XY, the waves from the two sources arrive out of phase by 180°. [1] Therefore, the waves partly or totally cancel each other out. So the amplitude is a minimum at these positions. [1] *(2 marks)*

6 a i The pipe length $L = \frac{1}{4}\lambda$ so λ = 4L = 4 × 380 mm = 1520 mm. [1] The speed of sound in the pipe = $f\lambda$ = 225 Hz × 1.52 m = 342 m s^{-1} [1]

ii The amplitude decreases gradually from the open end to the closed end where it is zero. [1] *(3 marks)*

b At 675 Hz, the frequency is 3 times the first harmonic frequency of 225 Hz. The 3rd harmonic stationary wave pattern is set up when the pipe length = $\frac{3}{4}\lambda$ [1] and there is an extra node and antinode in the pipe compared with the 1st harmonic, as shown in Figure 2. [1] *(2 marks)*

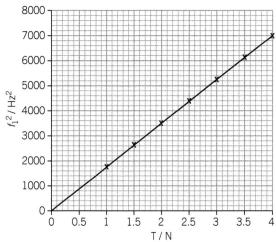

▲ **Figure 2**

7 a Suspend a mass of known weight W from the end of the string supported by the pulley. The tension T in the string is equal to W. [1] Increase the signal generator frequency from zero until the string vibrates with a node at either end and a single antinode in the middle. [1] This is the 1st harmonic mode of vibration. [1] Measure the signal generator frequency and the tension and record the measurements in a table. [1] Repeat the procedure for a range of different weights. [1] (max 4)

b i Square both sides of $f_1 = \frac{1}{2L}\sqrt{\frac{T}{\mu}}$ to give

$f_1^2 = \frac{1}{4L^2} × \frac{T}{\mu} = \frac{1}{4\mu L^2}T$ [1]. Therefore, a graph of $y = f_1^2$ against $x = T$ should be a straight line through the origin of the form $y = mx$ where the gradient $m = \frac{1}{4\mu L^2}$ [1] *(2 marks)*

ii Graph: suitable scales on both axes [1], points plotted correctly [1], best fit line [1]

Large gradient triangle and correct use [1] to give gradient in the range 1700 to 1850 (Hz^2 N^{-1}) [1]

Calculation of μ from $m = \frac{1}{4\mu L^2}$ to give

$\mu = \frac{1}{4\mu L^2} = \frac{1}{4 × 1770 × (0.64)^2} = 3.4$ to $3.5 × 10^{-4}$ kg m^{-1} [1] *(6 marks)*

Graph: f_1^2 / Hz^2 (y-axis) against T / N (x-axis)

▲ **Figure 1**

Chapter 5

1 a i See Figure 1 [1]   (*1 mark*)

ii $n_1 \sin\theta_1 = n_2 \sin\theta_2$ gives $1 \sin 35 = 1.55 \sin r$. [1]

Therefore, $\sin r = \dfrac{1\sin 35}{1.55} = 0.370$ so $r = 21.7°$ [1]
                    (*2 marks*)

b i Consider angles a and b in Figure 1:

$a = 90 - 21.7 = 68.3°$; $a + b + 60° = 180°$
therefore $b = 120 - a = 51.7°$

So the angle of incidence where the ray leaves the prism = $90 - 51.7 = 38.3°$ [1]

ii Applying $n_1 \sin\theta_1 = n_2 \sin\theta_2$ at the point where the ray leaves the prism gives

$1.55 \sin 38.3 = 1\sin r$. [1] Therefore, $\sin r = 1.55 \times \sin 38.3 = 0.961$ so $r = 73.9°$ [1]   (*3 marks*)

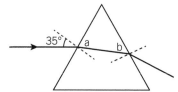

▲ **Figure 1**

2 a The core of a communications fibre needs to be narrow in order to ensure the light stays as near as possible to the axis of the fibres. [1] This is because non-axial rays take longer than axial rays to travel along the fibres and if this difference is too great, the pulses of light lengthen and merge before they reach the detector. [1]   (*2 marks*)

b i $n_1 \sin\theta_1 = n_2 \sin\theta_2$ at the air–core boundary
$1.55 \sin c = 1.45 \sin 90.0$. [1]

Therefore, $\sin c = 1.45 \div 1.55 = 0.935$
so $c = 69.3°$ [1]   (*2 marks*)

ii If the cladding was absent, light would pass between fibres where they are in contact instead of undergoing total internal reflection. [1] The cladding ensures total internal reflection occurs at the core–cladding boundary which means the light signals in each fibre are secure as they cannot pass between fibres that are in contact. [1]   (*2 marks*)

3 a i Applying $n_1 \sin\theta_1 = n_2 \sin\theta_2$ to the red component gives $1.52 \sin 40.0 = 1 \sin r$. [1] Therefore, $\sin r = 1.52 \times \sin 40.0 = 0.9770$ so $r = 77.7°$ [1]

ii Applying $n_1 \sin\theta_1 = n_2 \sin\theta_2$ to the blue component gives $1.55 \sin 40.0 = 1\sin r$. [1]

Therefore, $\sin r = 1.55 \times \sin 40.0 = 0.9963$ so $r = 85.1°$ [1]   (*4 marks*)

b The angle between the red and blue components = $85.1 - 77.7 = 7.4°$ [1]   (*1 mark*)

4 a The two slits emit light waves of the same frequency and with a constant phase difference. [1] Therefore, at any position on the screen where the waves overlap, the waves from the two slits have a constant phase difference that depends on the difference in the distance from that position to each slit (i.e., the path difference). [1]

At positions where the path difference is a whole number of wavelengths, the waves arrive in phase so they reinforce and a bright fringe is seen at that position. [1] Midway between the bright fringes, the path difference is a whole number of wavelengths plus one-half wavelength. [1] The waves therefore arrive out of phase by 180° so they cancel and a dark fringe is seen midway between adjacent bright fringes. [1]   (*4 marks*)

b 6 fringes mean that there are 5 fringe spacings in 65 mm so the fringe spacing between adjacent fringes $W = 65\,\text{mm} \div 5 = 13\,\text{mm}$ [1]

Using the equation $s = \dfrac{\lambda D}{W}$ gives

$\lambda = \dfrac{630 \times 10^{-9} \times 2.50}{13 \times 10^{-3}}$ [1]$= 1.2 \times 10^{-4}$ m [1] (*3 marks*)

5 a See Topic 5.6, Figure 1. Your graph should show each outer bright fringe is half the width of the central bright fringe [1] which should be at least 3 times higher than the nearest bright fringe on each side of the central fringe. [1] The graph should be symmetrical about a line through the centre of the central bright fringe. [1]   (*3 marks*)

b If the slit is made wider, the bright fringes become brighter and closer together. [1] The central bright fringe is still twice the width of the other bright fringes and is still about 3 times higher than the nearest fringe on each side. [1]   (*2 marks*)

6 a The maximum order number $= \dfrac{d}{\lambda}$ rounded down where $d = \dfrac{1}{600}\,\text{mm} = 1.67 \times 10^{-6}\,\text{m}$. [1] Use the largest wavelength value of 445 nm to give the smallest value of $\dfrac{d}{\lambda} = \dfrac{1.67 \times 10^{-6}}{445 \times 10^{-9}} = 3.75\,\text{m}$. [1]

Rounding this down gives 3 for the maximum order number. [1] Smaller wavelengths are diffracted less by the grating so the 3rd order will have the full range of wavelengths from 430 to 445 nm. [1]   (*4 marks*)

b Using the 3rd order equation $3\lambda = d \sin\theta$ for 430 nm and 445 nm in turn gives the angle of diffraction for these wavelengths.

For $\lambda = 430$ nm, $\sin\theta = \dfrac{3\lambda}{d} = \dfrac{3 \times 430 \times 10^{-9}}{1.67 \times 10^{-6}}$
$= 0.772$ [1] which gives $\theta = 50.57°$

For $\lambda = 445$ nm, $\sin\theta = \dfrac{3\lambda}{d} = \dfrac{3 \times 445 \times 10^{-9}}{1.67 \times 10^{-6}}$
$= 0.799$ [1] which gives $\theta = 53.07°$

The angular width of the 3rd order beam is therefore $53.07 - 50.57 = 2.50°$ [1]   (*3 marks*)

7 a i If slit S was not narrow enough, the bright fringes would be wider and the dark fringes narrower. [1] This is because adjacent strips of a wide slit would each produce diffraction patterns slightly displaced from each other. [1] The resultant pattern would therefore have reduced darkness between the bright fringes. [1] So S needs to be narrow enough to give bright and dark fringes of equal width. [1]   (*4 marks*)

ii S₁ and S₂ are coherent sources because each slit emits a wavefront every time a wavefront from S reaches S₁ and S₂. [1] So S₁ and S₂ emit waves with a constant phase difference and are therefore coherent sources. [1] *(2 marks)*

b The intensity of the fainter fringes has been suppressed by single slit diffraction at S₁ and S₂ [1] because these fringes are at a single slit diffraction minimum. [1] Adjacent fringes further from the centre are brighter because they are not at a single slit diffraction minimum. [1] *(3 marks)*

8 a $d = \dfrac{1}{600}$ mm $= 1.67 \times 10^{-6}$ m. [1] Using the 2nd order equation $2\lambda = d\sin\theta$ gives

$$\lambda = \frac{d\sin\theta}{2} = \frac{1.67 \times 10^{-6} \times \sin 40.0}{2} = 5.37 \times 10^{-7}\,\text{m}.\ [1]$$

(3 marks)

b The wavelength of the light in glass

$$= \frac{\text{wavelength in air}}{\text{refractive index of glass}} = \frac{5.37 \times 10^{-7}}{1.50} = 3.58 \times 10^{-7}\,\text{m}.$$

[1] Using the 2nd order equation gives

$$\sin\theta = \frac{2\lambda}{d} = \frac{2 \times 3.58 \times 10^{-7}}{1.67 \times 10^{-6}} = 0.429\ [1]\ \text{which gives}$$

$\theta = 25.4°$ [1] *(4 marks)*

Chapter 6

1 a See Figure 1 [1] $a = 56°$ [1] $b = 30°$ [1]. Note the construction arcs drawn using the same scale as the 6 N vector. *(3 marks)*

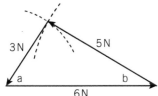

▲ **Figure 1**

b 5 N in the opposite direction to the 5 N vector in Figure 1. [1] *(1 mark)*

2 The tension in each section of the wire is the same because their horizontal components are equal and opposite to each other [1] and the angle between each wire and the horizontal is the same (i.e., $T_1\cos 15 = T_2\cos 15$) [1].

The sum of their vertical components is equal and opposite to the weight.

(i.e., $T_1\sin 15 + T_2\sin 15 = W$). [1] Hence, $2\,T_1\sin 15 = W$ so $T_1 = T_2 = \dfrac{W}{2\sin 15} = \dfrac{6.3}{2\sin 15} = 12.2\,\text{N}$ [1] *(4 marks)*

3 a See the parallelogram of forces in Figure 2 [1]. The angle between the resultant and the 7.2 kN force is 24° [1] and is 21° between the 8.0 kN force and the resultant. [1] *(3 marks)*

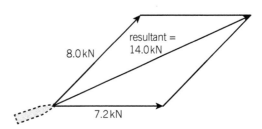

▲ **Figure 2**

b i The resultant is 1.75 times longer than the 8.0 kN force [1] so its magnitude is 14 kN [1] ($= 1.75 \times 8.0$ kN).

ii The drag force is 14.0 kN in the opposite direction to the resultant force. [1] *(3 marks)*

4 a See Figure 3: resolve T into a horizontal component $T\cos 60$ and a vertical component $T\sin 60$. Taking moments about P gives $T\sin 60 \times (0.950 - 0.350)$ [1] $= W \times (0.500 - 0.350)$ [1]

Therefore, $T = \dfrac{1.20 \times 0.150}{0.600\sin 60} = 0.346 = 0.35\,\text{N}$ [1] to 2 significant figures. *(3 marks)*

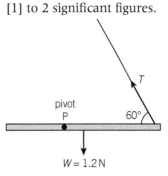

▲ **Figure 3**

b The horizontal component of the support force $S_x = T\cos 60 = 0.346 \times \cos 60 = 0.173$ N [1]

The vertical component of the support force $S_y = W - T\sin 60 = 1.2 - (0.346 \times \sin 60) = 0.900$ N [1]

The magnitude of the support force $S = (S_x^2 + S_y^2)^{1/2} = 0.92$ N [1] to 2 significant figures.

The angle of the line of action of S to the horizontal $= \arctan (S_y \div S_x) = 79°$ [1]

Note the line of action of S should pass through the point where the lines of action of W and T intersect which will be the position on the string vertically above the centre of mass. *(4 marks)*

5 a See Figure 4 where C is the centre of mass of the plank. [1] *(1 mark)*

b Distances XP $= 1.50 - 0.20 = 1.30$ m, PC $= 0.50$ m, and CY $= 2.00 - 0.20 = 1.80$ m

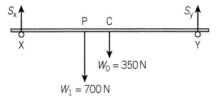

▲ **Figure 4**

Taking moments about Y to find the support force S_x at X gives:

Anticlockwise moment = (700 × PY) + (350 × CY)
= (700 × 2.30) + (350 × 1.80) = 2240 N m [1]

Clockwise moment = S_x × XY = 3.60 S_x [1]

Therefore, 3.60 S_x = 2240 hence $S_x = \dfrac{2240}{3.60} = 622$ N [1]

Since $S_x + S_y$ = the total weight = 700 + 350 = 1050 N, then S_y = 1050 − 622 = 428 N [1]

(4 marks)

6 a The total weight supported by the curtain brackets is 108 N (= 22 N + 86 N). [1] The brackets are equidistant from the centre of mass of the pole which is at the centre of the pole. The curtains are symmetrical about the centre of mass. [1] Therefore, each bracket supports half of the total weight which means the support force of each bracket on the pole is 54 N vertically upwards. [1] *(3 marks)*

b The support force S_x at X increases as the centre of mass of the curtain moves towards X. [1] When the curtain is at X, the bracket at X supports the weight of the curtain that has been pulled back and also the force of the pole on X due to the other curtain and the weight of the pole. [1] *(2 marks)*

c See Figure 5: M is the centre of mass of the plank. W_1 and W_2 represent the weight of each curtain (= 43 N each).

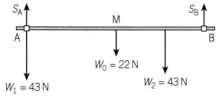

▲ **Figure 5**

The line of action of W_1 passes through bracket A and the line of action of W_2 passes through the point on the curtain pole midway between M and B.

To determine the support force S_A at A, take moments about B:

Distance AB = 2.80 − (2 × 0.05) = 2.70 m,
AM = MB = $\dfrac{2.80}{2}$ − 0.05 = 1.35 m.

The perpendicular distance from B to the line of action of $W_2 = \dfrac{1}{2}$MB = 0.675 m.

Taking moments about B to find the support force S_A at A gives:

Clockwise moment = S_A × AB = 2.70 S_A [1]

Anticlockwise moment = (43 × AB) + (22 × MB) + (43 × $\dfrac{1}{2}$ MB)
= (43 × 2.70) + (22 × 2.30) + (43 × 0.675) [1]
= 196 N m [1]

Therefore, 2.70 S_A = 196 [1] hence $S_A = \dfrac{196}{2.70} = 73$ N [1] (to 2 significant figures)

(5 marks)

7 Figure 6 shows the forces acting at P.

Resolving the forces vertically and horizontally gives:

a Horizontally: $T_1 \cos 20 = T_2 \cos 15$ [1] hence $T_1 = \dfrac{\cos 15}{\cos 20} T_2 = 1.03 T_2$ [1] *(2 marks)*

b Vertically: $T_1 \sin 20 + T_2 \sin 15 = W$ hence $(1.03 \sin 20 + \sin 15) T_2 = 30$ [1]

Therefore, $0.611 T_2 = 30$ [1] so $T_2 = \dfrac{30}{0.611} = 49.1$ N [1] to 2 significant figures.

Since $T_1 = 1.03 T_2$ then $T_2 = 1.03 × 49.1 = 51$ N [1] to 2 significant figures. *(4 marks)*

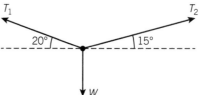

▲ **Figure 6**

8 a Applying the principle of moments about the wheel axis when a horizontal force F is applied to the handle gives $F × 0.85$ m $= W × 0.16$ m [1] hence $F = \dfrac{0.16}{0.85} × 220$ N = 41 N [1] *(2 marks)*

b The suitcase is unstable in this position because its weight will have a non-zero moment about the wheel axis if it is displaced slightly from this position [1]. As a result, it will topple forwards or backwards if it is displaced slightly from this position. [1] *(2 marks)*

c An upward force is necessary to provide a moment in the opposite direction to the moment of the weight about the wheel axis [1] which acts in a clockwise direction. The force on the handle must therefore be upwards to provide an anticlockwise moment about the wheel axis. [1]

(2 marks)

Chapter 7

1 $u = 98$ km h^{-1} = 27.2 m s^{-1}, $v = 0$, $s = 73$ m

a To find a, rearrange $v^2 = u^2 + 2as$ to give $a = \dfrac{v^2 - u^2}{2s} = \dfrac{0 - 27.2^2}{2 × 73}$ [1] $= -5.07$ m s^{-2} = 5.1 m s^{-2} [1] to 2 significant figures *(2 marks)*

b To find t, rearrange $v = u + at$ to give $t = \dfrac{v - u}{a} = \dfrac{0 - 27.2}{-5.07}$ [1] = 5.4 s [1] *(2 marks)*

2 a The gradient and therefore the acceleration is greatest in magnitude at $t = 30$ s. [1]

Drawing a tangent at this point gives the maximum acceleration = gradient of the tangent. [1]
$= \dfrac{0 - 80}{55 - 4} = -1.54$ m s^{-2} = −3.25 m s^{-2} [1] *(3 marks)*

b Distance travelled = area under the line (= 89 + 76 + 59 + 37 + 17 + 4) = 282 small squares where 25 small squares = 1 large square = 200 m. [1] Therefore, the distance travelled = (282 ÷ 25) × 200 m = 175 N. [1] *(2 marks)*

3 a i For the first 1.2 s, $u = 0$, $t = 1.20\,\text{s}$, $v = 9.60\,\text{m s}^{-1}$

Rearranging $v = u + at$ gives $a = \dfrac{v - u}{t} = \dfrac{9.60 - 0}{1.20}$

$= 8.00\,\text{m s}^{-2}$ [1]　　　　　*(1 mark)*

ii Distance travelled in first 1.20 s, $s = ut + \dfrac{1}{2}at^2$

$= 0 + \dfrac{1}{2} \times 8.00 \times 1.20^2 = 5.76\,\text{m}$ [1]

Therefore, the distance travelled at $9.6\,\text{m s}^{-1}$

$= 100\,\text{m} - 5.76\,\text{m} = 94.24\,\text{m}$

Time taken to run this distance $= \dfrac{\text{distance}}{\text{speed}}$

$= \dfrac{94.24}{9.6} = 9.82\,\text{s}$ [1]

Therefore, the total time taken $= 1.20 + 9.82$

$= 11.02\,\text{s}$ [1]　　　　　*(3 marks)*

b Let V represent Y's maximum speed.

The time taken at constant speed $V = 11.02 - 1.30$
$= 9.72\,\text{s}$

So the distance moved at constant speed
$V = \text{speed} \times \text{time} = 9.72V$ [1]

The distance moved by Y in 1.3 s is given by

$s = \dfrac{u + v}{2}t = \dfrac{0 - V}{2} \times 1.30 = 0.65V$ [1]

Therefore, the total distance moved by
$Y = 9.72V + 0.65V = 10.37V = 100\,\text{m}$

Hence, $V = \dfrac{100}{10.37} = 9.64\,\text{m s}^{-1}$ [1]　　*(3 marks)*

4 a $u = 1.80\,\text{m s}^{-1}$, $v = 0$, $t = 12.0\,\text{s}$

Its rate of change of velocity $= a = \dfrac{v - u}{t} = \dfrac{0 - 1.80}{12.0}$

[1] $= -0.150\,\text{m s}^{-2}$ [1]　　　　*(2 marks)*

b $u = 1.80\,\text{m s}^{-1}$, $a = -0.150\,\text{m s}^{-2}$, $t = 16.0\,\text{s}$

i To calculate v, using $v = u + at$ gives

$v = 1.80 + (-0.150 \times 16.0)$ [1] $= -0.60\,\text{m s}^{-1}$

The negative sign means the wagon is moving
down the incline. [1]　　　　　*(2 marks)*

ii To calculate s, using $s = ut + \dfrac{1}{2}at^2$ gives

$s = (1.80 \times 16.0) + (0.5 \times -0.150 \times 16.0^2)$
$= 9.60\,\text{m}$ [1]

The position of the wagon is therefore 9.60 m
from the bottom of the incline. [1]　*(2 marks)*

5 a i $u = 0$, $t = 20\,\text{s}$, $v = 8.0\,\text{m s}^{-1}$

Therefore, $s = \dfrac{u + v}{2}t = \dfrac{0 + 8.0}{2} \times 20 = 80\,\text{m}$ [1]
(1 mark)

ii $u = 8.0\,\text{m s}^{-1}$, $s = 20.0\,\text{m}$, $v = 0$

Rearrange $s = \dfrac{u + v}{2}t$ to give $t = \dfrac{2s}{u + v} = \dfrac{2 \times 20.0}{0 + 8.0}$

$t = 5.0\,\text{s}$ [1]　　　　　　*(1 mark)*

b See Figure 1: PQR correct [1] RS correct [1] *(2 marks)*

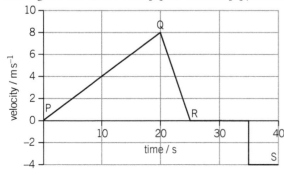
▲ **Figure 1**

c i $u = 0$, $v = 8.0\,\text{m s}^{-1}$, $t = 5.0\,\text{s}$

acceleration $a = \dfrac{v - u}{t} = \dfrac{8.0 - 0}{5.0} = 1.6\,\text{m s}^{-2}$ [1]
(1 mark)

ii Displacement = area between v–t line and axis.

Displacement PR $= \dfrac{1}{2} \times 8.0\,\text{m s}^{-1} \times 25.0\,\text{s}$
$= 100\,\text{m}$ [1]

Displacement RS $= -4.0\,\text{m s}^{-1} \times 5.0\,\text{s} = -20\,\text{m}$

Therefore, displacement PS $= 100 - 20 = 80\,\text{m}$ [1]
(2 marks)

6 a Vertical motion $u = 0$, $a = -9.81\,\text{m s}^{-2}$, $t = 2.0\,\text{s}$

Vertical displacement $y = \dfrac{1}{2}a_y t^2$ gives

$y = \dfrac{1}{2} \times -9.81 \times 2.0^2$ [1] $= -19.6\,\text{m}$ [1]

Horizontal motion $u = 14\,\text{m s}^{-1}$, $a_x = 0$, $t = 2.0\,\text{s}$

Horizontal displacement $x = u_x t = 14 \times 2.0 = 28\,\text{m}$ [1]

Distance $s = \sqrt{x^2 + y^2} = \sqrt{28^2 + (-19.6)^2} = 34\,\text{m}$ [1]
to 2 significant figures　　　　*(4 marks)*

b The stone would have experienced a drag force
due to air resistance. [1] The drag force increases
with speed. [1] The effect of drag on the vertical
component of velocity would be to slow its descent
and increase the time taken to fall to the ground. [1]

Compared with the motion without air resistance,
the horizontal component of velocity is reduced
by the drag force more than the time to fall
through a given vertical distance is increased.
[1] So the effect of air resistance is to reduce the
horizontal component of displacement to the
point where the stone hit the water. [1]　*(5 marks)*

7 a Horizontal component $u_x = 2.60\cos 30.0$
$= 2.25\,\text{m s}^{-1}$ [1]

Vertical component $u_y = 2.60\sin 30.0 = 1.30\,\text{m s}^{-1}$ [1]
(2 marks)

b Consider the vertical motion: $u_y = 1.30\,\text{m s}^{-1}$,
$t = 3.90\,\text{s}$, $a_y = -9.81\,\text{m s}^{-2}$ (– for downwards)

i To calculate the vertical component of

displacement, y, using $y = u_y t + \dfrac{1}{2}a_y t^2$ gives

$y = (1.30 \times 3.90) + (\dfrac{1}{2} \times -9.81 \times 3.90^2)$ [1]
$= -69.5\,\text{m}$.

Therefore, the vertical distance
fallen = 69.5 m. [1]　　　　　*(2 marks)*

ii Consider the horizontal motion:
$u_x = 2.25\,\text{m s}^{-1}$, $a_x = 0$, $t = 3.90\,\text{s}$

Therefore, the horizontal displacement $x = u_x t$
$= 2.25 \times 3.90 = 8.78\,\text{m}$ [1]　　*(1 mark)*

8 a i Horizontal motion: $a_x = 0$, $x = 85.5\,\text{m}$,
$t = 4.24\,\text{s}$, $u_x = ?$

Horizontal displacement $x = u_x t$ therefore the
horizontal component of the initial velocity

$u_x = \dfrac{x}{t} = \dfrac{85.5}{4.24} = 20.2\,\text{m s}^{-1}$ [1]　*(1 mark)*

ii Vertical motion: $t = 4.24\,\text{s}$, $y = -1.80\,\text{m}$,
$a_y = -9.81\,\text{m s}^{-2}$ (– for downwards), $u_y = ?$

To calculate the vertical component of the

initial velocity, rearrange $y = u_y t + \dfrac{1}{2}a_y t^2$ to give

$u_y t = y - \frac{1}{2}a_y t^2 = -1.80 - (\frac{1}{2} \times -9.81 \times 4.24^2)$ [1]

$= 86.4\,\text{m}$ [1]

Therefore, $u_y t = 86.4\,\text{m}$ hence $u_y = \dfrac{86.4}{t} = \dfrac{86.4}{4.24}$

$= 20.4\,\text{m s}^{-1}$ [1]

(3 marks)

b i Initial speed of javelin $= \sqrt{u_x^2 + u_y^2} = \sqrt{20.2^2 + 20.4^2}$

$= 28.7\,\text{m s}^{-1}$ [1]

(1 mark)

ii $\tan\theta = \dfrac{u_y}{u_x} = \dfrac{20.4}{20.2} = 1.01$ therefore $\theta = 45.3°$[1]

(1 mark)

Chapter 8

1 a i $u = 0, v = 8.8\,\text{m s}^{-1}, t = 40\,\text{s}, m = 1800\,\text{kg}$

acceleration $a = \dfrac{v - u}{t} = \dfrac{8.8 - 0}{40} = 0.22\,\text{m s}^{-2}$ [1]

force $F = ma = 1800 \times 0.22 = 400\,\text{N}$ [1]

ii $a = 0.22\,\text{m s}^{-2}, m = 300\,\text{kg}$

tension $T = ma = 300 \times 0.22 = 66\,\text{N}$ [1] *(3 marks)*

b force = 396 N

acceleration $= \dfrac{\text{engine force}}{\text{mass}} = \dfrac{396}{1500} = 0.26\,\text{m s}^{-2}$ [1]

$u = 0, v = 8.8\,\text{m s}^{-1}, a = 0.26\,\text{m s}^{-2}$

To calculate t, rearrange $v = u + at$ to give $t = \dfrac{v - u}{a}$

$= \dfrac{8.8 - 0}{0.26} = 33.8\,\text{s} = 34\,\text{s}$ to 2 significant figures. [1]

(2 marks)

2 a $u = 0, v = 8.3\,\text{m s}^{-1}, s = 150\,\text{m}, m = 61\,\text{kg}$

To calculate s, rearrange $v^2 = u^2 + 2as$ to give

$a = \dfrac{v^2 - u^2}{2s} = \dfrac{8.3^2 - 0}{2 \times 74}$ [1] $= 0.465\,\text{m s}^{-2}$ [1]

resultant force = mass × acceleration

$= 61 \times 0.465 = 28\,\text{N}$ [1] *(3 marks)*

b Component of weight acting down the slope

$= mg\sin 5°$ [1] $= 61 \times 9.81 \times \sin 5.0 = 52.2\,\text{N}$ [1]

resultant force $= 52.2 + 28.3 = 80.5\,\text{N}$[1]

acceleration $= \dfrac{\text{resultant force}}{\text{mass}} = \dfrac{80.5}{61}$

$= 1.3\,\text{m s}^{-2}$ [1] *(4 marks)*

3 a maximum acceleration = initial gradient

(= gradient of tangent at $t = 0$) [1]

$= \dfrac{26 - 0}{5.0} = 5.2\,\text{m s}^{-2}$ [1] *(2 marks)*

b i Resultant force = driving force – drag force and the drag force is zero at $t = 0$. [1]

So at $t = 0$, the driving force = resultant force = mass × acceleration $= 1200 \times 5.2 = 6200\,\text{N}$ [1]

(4 marks)

ii At constant velocity, the resultant force is zero as the acceleration is zero. [1] Therefore, the drag force at this velocity is equal to the driving force. [1] So the drag force is 650 N in the opposite direction to the direction of motion of the car. [1] *(3 marks)*

4 a i $u = 18\,\text{m s}^{-1}, v = 0.5\,\text{m s}^{-1}, s = 28\,\text{m}, m = 750\,\text{kg}$

To calculate a, rearrange $v^2 = u^2 + 2as$ to give

$a = \dfrac{v^2 - u^2}{2s} = \dfrac{0.5^2 - 18^2}{2 \times 28}$ [1] $= -5.8\,\text{m s}^{-2}$ (2 s.f.) [1]

ii resultant force = mass × acceleration

$= 750 \times -5.78 = -4300\,\text{N}$

The resultant force is 4300 N acting down the slope. [1] *(3 marks)*

b i Component of weight acting down the slope $= mg\sin 32°$ [1] $= 750 \times 9.81 \times \sin 32$ $= 3900\,\text{N}$[1]

ii Resultant force down slope = component of weight down the slope + average resistive force. Average resistive force = resultant force – component of weight [1] $= 4330 - 3900$ $= 430\,\text{N}$. [1] *(4 marks)*

5 a The acceleration changes rapidly from zero to a large negative value when the parachute opens [1] and a large upwards force acts on the parachute due to the increase of air resistance when it opens [1]. The acceleration decreases as the parachute speed decreases and the resistive force on it decreases and therefore the upwards force on it gradually decreases [1] until the resistive force is equal to the weight. [1] The resultant force and the acceleration are then zero and the velocity hence the speed is constant. [1] (max 4)

b i See Figure 1: steeper deceleration [1] lower final speed [1]

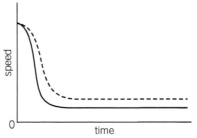

▲ **Figure 1**

ii The deceleration would be greater because the mass would be less and the initial upward force due to the parachute would be the same. [1] The constant speed at which the upward force is equal and opposite to the weight would be less so the flat section of the graph would be lower. [1] *(4 marks)*

6 a i The overall resultant force is the difference in the weights which is 40 N (= 290 N – 250 N). [1]

Mass of the load $m_1 = \dfrac{250}{g} = 25.5\,\text{kg}$: mass of the counterweight $m_2 = \dfrac{290}{g} = 29.6\,\text{kg}$ [1]

Therefore, the acceleration $a = \dfrac{\text{resultant force}}{\text{total mass}}$

$= \dfrac{40}{29.6 + 25.5}$ [1] $= 0.73\,\text{m s}^{-2}$ [1] *(4 marks)*

ii Considering the forces acting on the load, the resultant force = load weight – cable tension T [1]

Therefore, $T = m_1 g - m_1 a$ [1] $= m_1(g - a)$ $= 29.6 \times (9.81 - 0.73) = 270\,\text{N}$ [1]

(or using the counterweight equation above $T = m_2 a + m_2 g = m_2(a + g)$ [1] $= 25.5 \times (9.81 + 0.73) = 270\,\text{N}$ (2 s.f.) [1])

(2 marks)

b Support force on the pulley $= 2T + W_0$ [1] where W_0 = the weight of the pulley and the cable.

Therefore, the support force = (2 × 269) + 55
= 590 N [1]   (2 marks)

7 a i $u = 30\,\text{m s}^{-1}, v = 0, s = 0.80\,\text{m}$,

impact time $t = \dfrac{2s}{u+v} = \dfrac{2 \times 0.80}{30+0} = 0.053\,\text{s}$ [1]

ii Acceleration $a = \dfrac{v-u}{t} = \dfrac{0-30}{0.0533} = -563\,\text{m s}^{-2}$ [1]

impact force $F = ma = 1800 \times -563$
$= -1.0 \times 10^{6}\,\text{N}$ (2 s.f.) [1] (3 marks)

b $u = 30\,\text{m s}^{-1}, v = 0, s = 0.80 + 0.40 = 1.20\,\text{m}, m = 71\,\text{kg}$

To calculate s, rearrange $v^2 = u^2 + 2as$ to give

$a = \dfrac{v^2 - u^2}{2s} = \dfrac{0 - 30^2}{2 \times 1.20}$ [1] $= -375\,\text{m s}^{-2}$ [1]

impact force $F = ma = 71 \times -375 = -2.67 \times 10^{4}\,\text{N}$ [1]
                                             (3 marks)

8 a i $u = 0, t = 5.6\,\text{s}, s = 15\,\text{m}$

To calculate acceleration a, rearrange

$s = ut + \dfrac{1}{2}at^2$ with $u = 0$ to give

$a = \dfrac{2s}{t^2} = \dfrac{2 \times 15}{5.6^2}$ [1] $= 0.957\,\text{m s}^{-2} = 0.96\,\text{m s}^{-2}$ to
2 significant figures [1]

ii Resultant force $F = ma = 38 \times 0.957 = 36.4\,\text{N}$
$= 36\,\text{N}$ to 2 significant figures [1]   (3 marks)

b i Component of weight down the slope
$= mg\sin\theta = 38 \times 9.81 \times \sin 6.1$ [1]
$= 39.6\,\text{N} = 40\,\text{N}$ to 2 significant figures [1]
                                             (2 marks)

ii Resistive forces due to frictional forces at the bearings of the skateboard wheels and air resistance act in the opposite direction to the direction of motion. [1]

The resultant force = the component of the weight acting down the slope – the total resistive force, so the resultant force is less than the component of weight acting down the slope. [1] The difference is the resistive force which is therefore equal to 3.2 N
(= 39.6 N – 36.4 N). [1]   (2 marks)

Chapter 9

1 a For any system of interacting objects, their total momentum remains constant, provided no external resultant force acts on the system. [1]   (1 mark)

b i Let V = the velocity of the vehicles after the collision

Total initial momentum = (1200 × 21.0) +
(1600 × 3.0) = 30 000 kg m s⁻¹ [1]

Total final momentum = (1200 + 1600) × V
$= 2800\,V$ [1]

Therefore, $2800V = 30\,000$ which gives
$V = \dfrac{30\,000}{2800} = 11\,\text{m s}^{-1}$ (2 s.f.) [1]

ii Kinetic energy before collision =
$\left(\dfrac{1}{2} \times 1200 \times 21.0^2\right) + \left(\dfrac{1}{2} \times 1600 \times 3.0^2\right)$
$= 2.72 \times 10^{5}\,\text{J}$ [1]

Kinetic energy after collision $= \dfrac{1}{2} \times$
$(1200 + 1600) \times 10.7^2 = 1.61 \times 10^{5}\,\text{J}$ [1]

Loss of kinetic energy $= 2.72 \times 10^{5}\,\text{J} -$
$1.61 \times 10^{5}\,\text{J} = 1.1 \times 10^{5}\,\text{J}$ [1]   (6 marks)

2 a Component of momentum perpendicular to the wall = 0.44 × 24 × cos 30 = 9.1 kg m s⁻¹ [1] (1 mark)

Change of momentum = 2 × 9.1 kg m s⁻¹
= 18 kg m s⁻¹ (2 s.f.) [1]

b Force = $\dfrac{\text{change in momentum}}{\text{contact time}} = \dfrac{18.2}{0.095} = 190\,\text{N}$ [1]
                                             (2 marks)

3 a An elastic collision is one in which the total kinetic energy after the collision is equal to the total kinetic energy before the collision.   (1 mark)

b i Let V = the velocity of Y after the collision

Total initial momentum = (0.12 × 0.58) –
(0.10 × 0.50) = 1.96 × 10⁻² kg m s⁻¹ [1]

Total final momentum = (0.12 × 0.15)
+ (0.10 × V) = 0.018 + 0.10V [1]

Therefore, $0.10\,V + 0.018 = 1.96 \times 10^{-2}$ which

gives $V = \dfrac{1.96 \times 10^{-2} - 0.018}{0.10} = 0.016\,\text{m s}^{-1}$

The velocity of Y is therefore 0.016 m s⁻¹ in the same direction as X initially. [1]   (3 marks)

ii Kinetic energy before collision =
$\left(\dfrac{1}{2} \times 0.12 \times 0.58^2\right) + \left(\dfrac{1}{2} \times 0.10 \times -0.50^2\right)$
$= 3.27 \times 10^{-2}\,\text{J}$ [1]

Kinetic energy after collision $= \left(\dfrac{1}{2} \times 0.12 \times 0.15^2\right)$
$+ \left(\dfrac{1}{2} \times 0.10 \times 0.016^2\right) = 1.36 \times 10^{-3}\,\text{J}$ [1]

The collision is inelastic as the total final kinetic energy is less than the total initial kinetic energy. [1]   (3 marks)

4 a i Let V = velocity of A after A and B move apart

Total final momentum = (1.20 × V) +
(0.80 × 0.15) = 1.2V + 0.12

Total initial momentum = 0 [1]

Therefore, $1.20V + 0.12 = 0$ [1] which gives
$V = -\dfrac{0.12}{1.20} = -0.10\,\text{m s}^{-1}$

The velocity of A = 0.10 m s⁻¹ in the opposite direction to that of B. [1]   (3 marks)

ii Kinetic energy of A $= \dfrac{1}{2} \times 1.20 \times 0.10^2$
$= 6.0 \times 10^{-3}\,\text{J}$ [1]

Kinetic energy of B $= \dfrac{1}{2} \times 0.80 \times 0.15^2$
$= 9.0 \times 10^{-3}\,\text{J}$ [1]

Total kinetic energy $= 6.0 \times 10^{-3} + 9.0 \times 10^{-3}$
$= 0.015\,\text{J}$ [1]   (3 marks)

b Let V = the speed of B

Kinetic energy of B $= \dfrac{1}{2} \times 0.80 \times v^2$ [1] = 0.015 J

Therefore, $v^2 = \dfrac{2 \times 0.015}{0.80} = 0.0375$ so
$v = 0.19\,\text{m s}^{-1}$ [1]   (2 marks)

5 a [1] for both mass numbers correct, [1] for Th atomic number correct.   (2 marks)
$^{210}_{83}\text{Bi} \rightarrow {}^{4}_{2}\alpha + {}^{206}_{81}\text{Th}$

b i Let V = recoil velocity of the thallium nucleus

Total final momentum = $(4u \times 1.5 \times 10^7 \, \text{m s}^{-1})$ + $(206u \times V)$

Total initial momentum = 0 [1]

Therefore, $(4u \times 1.5 \times 10^7 \, \text{m s}^{-1}) + (206u \times V) = 0$ [1]

which gives $V = \dfrac{-(4u \times 1.5 \times 10^7) \, \text{m s}^{-1}}{206u}$

$= -2.9 \times 10^5 \, \text{m s}^{-1}$

The recoil velocity of the thallium nucleus $= 2.9 \times 10^5 \, \text{m s}^{-1}$ [1]   *(3 marks)*

ii Kinetic energy of α particle $= \dfrac{1}{2} \times 4 \times 6.7 \times 10^{-27} \times (1.5 \times 10^7)^2 = 3.0 \times 10^{-12} \, \text{J}$ [1]

Kinetic energy of thallium nucleus $= \dfrac{1}{2} \times (206 \times \dfrac{1}{4} \times 6.7 \times 10^{-27}) \times (2.9 \times 10^5)^2 = 1.5 \times 10^{-14} \, \text{J}$ [1]

Therefore, kinetic energy of the α particle as a % of the total kinetic energy

$= \dfrac{3.0 \times 10^{-12} \, \text{J}}{(3.0 \times 10^{-12} \, \text{J}) + (1.5 \times 10^{-14} \, \text{J})} \times 100\%$

$= 99.5\%$ [1]   *(3 marks)*

6 a The impulse (or the change of momentum) of the ball. [1]   *(1 mark)*

b i Change of momentum = area under curve $= 520 \times 10^{-3} \, \text{Ns}$. Answer in the range 520 ± 40. [1] Answer in the range 520 ± 20. [2]   *(2 marks)*

ii Initial momentum = 0 therefore final momentum = 0.52 N s [1]

Velocity after impact $= \dfrac{\text{change of momentum}}{\text{mass}}$

$= \dfrac{0.52 \, \text{N s}}{0.056 \, \text{kg}} = 9.3 \, \text{m s}^{-1}$ [1]   *(2 marks)*

iii Impact time ≈ 30 ms [1]

Average acceleration $= \dfrac{\text{change of velocity}}{\text{time taken}} \approx$

$\dfrac{9.3 \, \text{m s}^{-1}}{30 \times 10^{-3} \, \text{s}} = 310 \, \text{m s}^{-2}$ [1]   *(2 marks)*

7 a i $u = 0$, $s = -0.80 \, \text{m}$, $a = -9.81 \, \text{m s}^{-2}$, $v = ?$

Using $v^2 = u^2 + 2as$ gives $v^2 = 2 \times -9.81 \times -0.80$ $= 15.7 \, \text{m}^2 \, \text{s}^{-2}$ [1] hence $v = 4.0 \, \text{m s}^{-1}$ [1]

Momentum just before impact $= 4000 \times 4.0$ $= 1.6 \times 10^4 \, \text{kg m s}^{-1}$ [1]   *(3 marks)*

ii Total momentum just after impact = $(4000 + 2000) \, V = 6000V$ [1] where V is the velocity of the hammer and the pile just after impact.

Using the principle of conservation of momentum gives $6000V = 1.6 \times 10^4$ [1]

Therefore, $V = \dfrac{1.6 \times 10^4}{6000} = 2.7 \, \text{m s}^{-1}$ [1]   *(3 marks)*

b i $u = 2.7 \, \text{m s}^{-1}$, $s = -0.020 \, \text{m}$, $v = 0$, $a = ?$

Rearranging $v^2 = u^2 + 2as$ gives

$a = \dfrac{-u^2}{2s} = \dfrac{-2.7^2}{2 \times 0.020}$ [1] $= 180 \, \text{m s}^{-2}$ (2 s.f.) [1]

ii The total mass decelerated by the frictional force = 6000 kg

Frictional force $F = ma = 6000 \times 180$ $= 1.1 \times 10^6 \, \text{N}$ [1]   *(3 marks)*

8 a Mass per second transferred $= \dfrac{2.7}{60} = 0.045 \, \text{kg s}^{-1}$ [1]

Momentum transferred per second = mass transferred per second × velocity of the water jet $= 0.045 \times 400$ [1] $= 18 \, \text{N}$ [1]   *(3 marks)*

b The force of the surface on the jet = momentum loss per second = 18 N [1]

The force of the jet on the surface is equal and opposite to the force of the surface on the jet so the force of the jet on the surface is 18 N. [1]   *(2 marks)*

c Area of cross-section of the jet $= \dfrac{1}{4}\pi d^2$

$= 0.25 \times \pi \times (3.8 \times 10^{-4})^2 = 1.1 \times 10^{-7} \, \text{m}^2$ [1]

pressure $= \dfrac{\text{force}}{\text{area}} = \dfrac{18 \, \text{N}}{1.1 \times 10^{-7} \, \text{m}^2} = 1.6 \times 10^8 \, \text{Pa}$ (or N m^{-2}) [1]   *(2 marks)*

Chapter 10

1 a i Loss of potential energy $= mg\Delta h = 52.0 \times 9.81 \times 2.50 = 1280 \, \text{J}$ [1]

ii Gain of kinetic energy (KE) $= \dfrac{1}{2}mv^2 =$ $0.5 \times 52.0 \times 2.12^2 = 117 \, \text{J}$ [1]   *(2 marks)*

b Work done to overcome friction $= mg\Delta h - \dfrac{1}{2}mv^2$ [1] $= 1280 - 117 = 1163 \, \text{J}$ [1]

Using work done = force × distance gives:

frictional force $= \dfrac{\text{work done to overcome friction}}{\text{distance moved}} = \dfrac{1163 \, \text{J}}{10.2 \, \text{m}} = 114 \, \text{N}$ [1]   *(3 marks)*

2 a When she accelerates, she loses gravitational potential energy (GPE) and gains kinetic energy. [1] When she decelerates, energy is transferred to the trampoline as she loses all her kinetic energy as she slows down to a standstill [1] and she loses gravitational potential energy as she descends to her lowest position. [1]   *(3 marks)*

b i gain of kinetic energy $\dfrac{1}{2}mv^2$ = loss of potential energy $mg\Delta h$ [1]

Rearranging this equation gives $v = \sqrt{2g\Delta h} = \sqrt{(2 \times 9.81 \times (0.90 - 0.16)} = 3.8 \, \text{m s}^{-1}$ [1]

ii Maximum energy transferred = total loss of gravitational potential energy $= mg\Delta h$ [1]

$= 35 \times 9.81 \times 0.90 = 310 \, \text{J}$ [1] (to 2 significant figures)   *(4 marks)*

c Tension T at maximum extension $= k \, \Delta L = 3500 \times 0.042 = 147 \, \text{N}$ [1] (where extension $\Delta L = 0.042 \, \text{m}$)

Maximimum energy stored in each spring $= \dfrac{1}{2}T\Delta L$ $= 0.5 \times 147 \times 0.042 = 3.1 \, \text{J}$

Total energy stored = 90 × 3.1 J = 280 J to 2 significant figures [1]   *(2 marks)*

3 a Kinetic energy $= \dfrac{1}{2}mv^2 = 0.5 \times 12\,000 \times 18^2$ $= 1.9 \, \text{MJ}$ [1]   *(1 mark)*

b i Distance travelled per second = 18 m therefore height gain per second = 18 sin 5.0° [1]

Gain of potential energy each second $= mg\Delta h$ $= 12000 \times 9.81 \times 18 \sin 5.0 = 1.9 \times 10^5 \, \text{J}$ [1]   *(2 marks)*

ii Rearranging $P = Fv$ gives $F = \dfrac{P}{v} = \dfrac{228000}{18}$ [1]

$= 1.3 \times 10^4\,\text{N}$ [1] (*2 marks*)

iii Component of weight acting down the slope
$= mg \sin 5°$

$= 12\,000 \times 9.81 \times \sin 5 = 1.03 \times 10^4\,\text{N}$ [1]

Total resistive force = output force of engine –
component of weight acting down the slope [1]

$= 1.27 \times 10^4\,\text{N} - 1.03 \times 10^4\,\text{N} = 2.4 \times 10^3\,\text{N}$ [1]

(*3 marks*)

Alternative method: Power wasted due to
resistive force = output power – GPE gain
per second [1] $= 228\,\text{kW} - 185\,\text{kW} = 43\,\text{kW}$.
Therefore, total resistive force
$= \dfrac{\text{power wasted}}{\text{speed}} = \dfrac{43\ \text{kW}}{18\ \text{m s}^{-1}}$ [1] $= 2.4 \times 10^3\,\text{N}$ [1]

4 a i Volume of trapped water = area × depth
$= 150 \times 10^6 \times 5.0 = 7.5 \times 10^8\,\text{m}^3$ [1]

Mass of trapped water = volume × density [1]
$= 7.5 \times 10^8 \times 1050 = 7.9 \times 10^{11}\,\text{kg}$ [1] (*2 marks*)

ii Height change Δh of trapped water due to
release $= 0.5 \times 5.0\,\text{m} = 2.5\,\text{m}$ [1]

Loss of GPE $= mg\Delta h = 7.9 \times 10^{11} \times 9.81 \times 2.5$
$= 1.94 \times 10^{13}\,\text{J}$ [1]

Loss of GPE per second $= \dfrac{\text{loss of GPE}}{\text{time taken}}$

$= \dfrac{1.94 \times 10^{13}}{6.0 \times 3600} = 900\,\text{MW}$ [1] (*3 marks*)

b Area of panels needed $= \dfrac{900\ \text{MW}}{220\ \text{W m}^{-1}}$

$= 4.1 \times 10^6\,\text{m}^2$ [1] (*1 mark*)

5 a Gain of potential energy $= mg\Delta h = 1200 \times 9.81$
$\times 15 = 177\,\text{kJ}$ [1]

Minimum output power $= \dfrac{\text{gain of GPE}}{\text{time taken}} = \dfrac{177\ \text{kJ}}{20\ \text{s}}$

$= 8.89\,\text{kW}$ [1]

Input power $\approx \dfrac{\text{minimum output power}}{\text{efficiency}} = \dfrac{8.89\ \text{kW}}{0.58}$

$=15\,\text{kW}$ [1] (*3 marks*)

b Assuming the gain of kinetic energy = loss of
potential energy [1]

$\dfrac{1}{2}mv^2 - \dfrac{1}{2}mu^2 = mg\Delta h$ where $u = 1.2\,\text{m s}^{-1}$ and
$mg\Delta h = 177\,\text{kJ}$

Therefore, $0.5 \times 1200 \times v^2 - (0.5 \times 1200 \times 1.2^2)$
$= 177\,\text{kJ}$[1]

$600v^2 = 177\,\text{kJ} + (600 \times 1.2^2) = 178\,\text{kJ}$ [1] hence

$v = \sqrt{\dfrac{178000}{600}} = 17\,\text{m s}^{-1}$ [1] (*4 marks*)

c i Energy transferred to the water = loss of KE of
carriage in the water

$= 178\,\text{kJ} - (0.5 \times 1200 \times 11.5^2)$ [1] $= 178\,\text{kJ} - 79\,\text{kJ} = 99\,\text{kJ}$ [1]

ii Assume the work done to overcome resistive
forces = energy transferred to the water, then
using work done = force × distance gives

average resistive force $= \dfrac{\text{work done}}{\text{distance moved}}$

$= \dfrac{99\ \text{kJ}}{18.4\ \text{m}} = 5.4\,\text{kN}$ [2] (*4 marks*)

6 a i Kinetic energy $= \dfrac{1}{2}mv^2 = 0.5 \times 250\,000 \times 150^2$
$= 2.8 \times 10^9\,\text{J}$ [1]

ii Assume work done = gain of kinetic
energy. As work done = force × distance,

force $= \dfrac{\text{gain of kinetic energy}}{\text{distance}}$ [1] $= \dfrac{2.81 \times 10^9}{2400}$

$= 1.2\,\text{MN}$ [1] (*3 marks*)

b i Gain of GPE $= mg\Delta h = 250\,000 \times 9.81 \times 8500$
$= 2100\,\text{MJ}$[1]

ii Kinetic energy at $250\,\text{m s}^{-1} = \dfrac{1}{2}mv^2 = 0.5 \times$
$250\,000 \times 250^2 = 7800\,\text{MJ}$ [1] (*2 marks*)

c Minimum energy needed $= 2100 + 7100$
$= 9890\,\text{MJ}$ [1]

Useful energy from each kilogram of fuel
$= 0.55 \times 30\,\text{MJ} = 16.5\,\text{MJ}$ [1]

Mass of fuel used $\approx \dfrac{9900\ \text{MJ}}{16.5\ \text{MJ kg}^{-1}} = 600\,\text{kg}$ [1]

(*3 marks*)

7 a GPE per second transferred by the water
$= \dfrac{450\ \text{MW}}{0.80} = 563\,\text{MW}$ [1]

Therefore, $\dfrac{mg\Delta h}{t} = 563\,\text{MW}$ where $\Delta h = 390\,\text{m}$
and m is the mass of water that passes through
the turbine in time t. [1]

So the mass flow per second $= \dfrac{m}{t}$

$= \dfrac{\text{GPE per second transferred}}{g\Delta h}$ [1]

$= \dfrac{563 \times 10^6}{9.81 \times 390} = 1.47 \times 10^5\,\text{kg s}^{-1}$ [1]

Volume of water per second passing through the
turbines $= \dfrac{\text{mass per second}}{\text{density of water}}$ [1]

$= \dfrac{1.47 \times 10^5\ \text{kg s}^{-1}}{1000\ \text{kg m}^{-3}} = 150\,\text{m}^3\,\text{s}^{-1}$ [1] (*6 marks*)

b i To generate 100 MJ, the GPE transferred by the
water flowing downhill $= \dfrac{100}{0.8} = 125\,\text{MJ}$. [1]

To pump the water uphill, the electrical
energy needed $= \dfrac{125\ \text{MJ}}{0.8} = 156\,\text{MJ}$. [1]

Therefore, the % of the energy supplied that
is wasted $= \dfrac{56}{156} \times 100\% = 36\%$ [1] (*3 marks*)

ii The efficiency of generating electrical energy
is 80%. The efficiency of storing water in the
uphill reservoir must be less than 100% [1]
as energy will always be dissipated due to
friction in the bearings of the turbines and
the generators and due to resistance heating
of the electricity cables and wires. [1] As the
electrical energy generated from pumped
storage is less than 100% of 80%, the overall
efficiency of the process is less than 80%. [1]

(*3 marks*)

Chapter 11

1 **a** Density is defined as mass per unit volume of a substance [1] *(1 mark)*

b **i** Sphere radius $r = 7.10\,\text{mm}$

Volume V of sphere $= \frac{4}{3}\pi r^3 \times L = \frac{4}{3} \times \pi$

$\times (7.10 \times 10^{-3})^3 = 1.50 \times 10^{-6}\,\text{m}^3$ [1]

Density of metal $= \frac{\text{mass}}{\text{volume}} = \frac{11.80 \times 10^{-3}\,\text{kg}}{1.50 \times 10^{-6}\,\text{m}^3}$
[1] $= 7870\,\text{kg}\,\text{m}^{-3}$ [1] *(2 marks)*

ii % uncertainty in radius $= \pm \frac{0.01\,\text{mm}}{7.10\,\text{mm}} \times 100$
$= \pm 0.14\%$

Therefore, % uncertainty in volume
$= 3 \times \pm 0.14\% = \pm 0.42\%$ (as $V \propto r^3$) [1]

% uncertainty in mass $= \pm \frac{0.04\,\text{g}}{11.80\,\text{g}} \times 100$
$= \pm 0.34\%$ [1]

Total % uncertainty $= \pm (0.42 + 0.34)$
$= \pm 0.76\%$

Uncertainty in density $= \pm \frac{0.76}{100} \times 7870\,\text{kg}\,\text{m}^{-3}$
$= \pm 60\,\text{kg}\,\text{m}^{-3}$ [1] *(3 marks)*

2 **a** Use a top pan balance to measure the mass m of the wire. Measure the length L of a sample of the wire just less than 1 m long using a metre rule. [1]

Use a micrometer to measure the diameter of the wire in 5 different places and calculate the mean diameter d. [1]

Calculate the volume V of the sample using the formula $V = \frac{1}{4}\Delta d^2 L$ then calculate the density ρ of the material using the formula $\rho = \frac{m}{V}$. [1] *(3 marks)*

b **i** Mean thickness $= 2.11\,\text{mm}$, uncertainty
$= \frac{1}{2} \times \text{range} = +0.5 \times (2.16 - 2.06)$
$= +0.05\,\text{mm}$ [1] *(1 mark)*

ii Volume of plate $= 0.050\,\text{m} \times 0.050\,\text{m}$
$\times 2.11 \times 10^{-3}\,\text{m} = 5.3 \times 10^{-6}\,\text{m}^3$ [1]

iii Density of metal $= \frac{\text{mass}}{\text{volume}} = \frac{14.20 \cdot 10^{-3}\,\text{kg}}{5.28 \cdot 10^{-6}\,\text{m}^3}$.
[1] $= 2700\,\text{kg}\,\text{m}^{-3}$ [1] *(3 marks)*

c % uncertainty in thickness $t = \pm \frac{0.05}{2.11} \times 100$
$= 2.4\%$

% uncertainty in plate length $L = \pm \frac{0.5}{50} \times 100$
$= 1.0\%$

% uncertainty in mass $m = \pm \frac{0.10}{14.20} \times 100$
$= 0.7\%$ [1]

Volume $= L^2 t$ so density $\rho = \frac{m}{L^2 t}$

Therefore, % uncertainty in density $= (2 \times 1.0\%)$
[1] $+ 2.4\% + 0.7\% = 5.1\%$ [1]

(Note % uncertainty in L^2
$= 2 \times$ % uncertainty in L) *(3 marks)*

3 **a** **i** Spring constant $k = $ gradient of line [1]
$= \frac{5.00 - 0.25\,\text{N}}{490 - 300\,\text{mm}} = 25.0\,\text{N}\,\text{m}^{-1}$ [1] *(2 marks)*

ii Weight of hanger $= 0.25\,\text{N}$ therefore extension
due to hanger $= \frac{0.25\,\text{N}}{25.0\,\text{N}\,\text{m}^{-1}}$

$= 0.010\,\text{m} = 10\,\text{mm}$. [1] Unstretched length of spring $= 300\,\text{mm} - 10\,\text{mm} = 290\,\text{mm}$ [1]
(2 marks)

b Extension ΔL of spring at $450\,\text{mm} = 450 - 300$
$= 150\,\text{mm}$ [1]

Energy stored in spring $= \frac{1}{2}k\Delta L^2 = 0.5 \times 25.0 \times$
0.150^2 [1] $= 0.28\,\text{J}$ [1] *(3 marks)*

4 **a** **i** Rearranging the Young modulus equation
$E = \frac{TL}{A\Delta L}$ gives $\Delta L = \frac{TL}{AE}$ [1] $=$

$\frac{250 \times 0.320}{1.23 \times 10^{-6} \times 2.1 \times 10^{11}}$ [1] $= 3.1 \times 10^{-4}\,\text{m}$ [1]
(3 marks)

ii For tension $T = 250\,\text{N}$, work done $= \frac{1}{2}T\Delta L$
$= 0.5 \times 250 \times 3.1 \times 10^{-4} = 0.039\,\text{J}$ [1] *(1 mark)*

b **i** $E_K = \frac{1}{2}mv^2 = 0.5 \times 0.170 \times 1.40^2 = 0.167\,\text{J}$ [1]
(1 mark)

ii Rebound kinetic energy $= 0.5 \times 0.170$
$\times 1.15^2 = 0.112\,\text{J}$ [1]

% kinetic energy retained $= \frac{0.112}{0.167} \times 100$
$= 67\%$ [1] *(2 marks)*

iii The cord was stretched beyond its limit of proportionality [1] and some of the energy supplied to the cord by the hammer was not stored as elastic energy and was dissipated to the surroundings [1] [OR the tension against extension line for unloading was below the loading line [1] and the area between the two lines corresponds to energy dissipated [1].
(2 marks)

5 **a** **i** See Figure 1. Points plotted correctly; suitable scales; best fit line. [3]

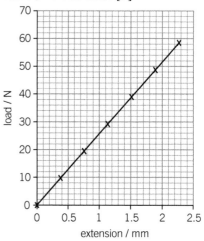

▲ **Figure 1**

ii Suitable gradient triangle shown and used correctly. [1] *(3 marks)*

Area of cross-section of wire $= \frac{1}{4}\pi d^2 = \frac{1}{4} \times \pi$
$\times (0.71 \times 10^{-3})^2 = 3.96 \times 10^{-7}\,\text{m}^2$ [1]

Gradient $= \frac{51.5 - 8.0\,\text{N}}{2.00 - 0.30\,\text{mm}} = 25.6\,\text{N}\,\text{mm}^{-1}$
$= 2.56 \times 10^4\,\text{N}\,\text{m}^{-1}$ [1]

$\left(\text{As the gradient} = \frac{T}{\Delta L}\right)$, Young modulus

equation $E = \frac{TL}{A\Delta L} = \text{gradient} \times \frac{L}{A}$

$= 2.56 \times 10^4 \times \dfrac{2.92}{3.96 \times 10^{-7}}$ [1] $= 1.9 \times 10^{11}$ Pa [1]

(5 marks)

b Tension = stress × area of cross–section so maximum load

$= 530 \times 10^6$ Pa $\times 3.96 \times 10^{-7}$ m^2 = 210 N [1]. (1 mark)

6 a P: The Young modulus is equal to the gradient of the initial straight section of each line [1] and P has the largest initial gradient [1]. (2 marks)

b i Q: A brittle material breaks without stretching significantly beyond its limit of proportionality [1] which is what happens to Q but not to P or R. [1] (2 marks)

ii R: A ductile material stretches considerably before it snaps without much extra force after its limit of proportionality is reached. [1] R is the most ductile because it stretches more than P or Q [1]. (2 marks)

c P: The energy stored per unit volume is given by the area under the line. [1] This area up to the limit of proportionality is greatest for P. [1] (2 marks)

7 a Tension T in each cable $= \dfrac{16300\,\text{N}}{6} = 2720\,\text{N}$ [1]

Area of cross-section A of each cable $= \dfrac{1}{4}\pi d^2 = \dfrac{1}{4} \times$

$\pi \times (6.0 \times 10^{-3})^2 = 2.83 \times 10^{-5}$ m^2 [1]

Rearranging the Young modulus equation

$E = \dfrac{TL}{A\Delta L}$ gives $\Delta L = \dfrac{TL}{AE}$ [1]

$= \dfrac{2720 \times 35.0}{2.83 \times 10^{-5} \times 210 \times 10^9}$ [1] $= 1.60 \times 10^{-2}$ m [1]

(5 marks)

b i Acceleration $= \dfrac{\text{change of velocity}}{\text{time taken}} = \dfrac{0.72 - 0}{8.0}$

$= 0.090$ m s^{-2} [1] (1 mark)

ii Total mass accelerated $= \dfrac{16300\,\text{N}}{g} = 1660$ kg

Resultant force due to extra tension

$= ma = 1660 \times 0.090 = 149$ N [1]

Extra tension in each cable $= \dfrac{149\,\text{N}}{6} = 25.0$ N

The stiffness constant k of the cable

$= \dfrac{AE}{L} = \dfrac{T}{\Delta L} = \dfrac{2720\,\text{N}}{1.60 \times 10^{-2}\,\text{m}} = 1.70 \times 10^5$ N m^{-1} [1]

Therefore, the extra extension of each cable [1]

$= \dfrac{\text{extra tension}}{k} = \dfrac{25.0\,\text{N}}{1.70 \times 10^5\,\text{N m}^{-1}}$

$= 1.5 \times 10^{-4}$ m (to 2 significant figures) [1]

(4 marks)

Alternative method: The cable tension when accelerating $= mg + ma$. [1] Therefore, the increase of

$\dfrac{\text{cable tension}}{\text{cable tension at constant velocity}} = \dfrac{ma}{mg} = \dfrac{a}{g}$ [1]

The extension is proportional to the cable tension [1] therefore the increase of the extension =

$\dfrac{a}{g} \times$ extension $= \dfrac{0.09}{9.81} \times 1.60 \times 10^{-1}$ m

$= 1.5 \times 10^{-4}$ m.] [1]

Chapter 12

1 a Resistance of wire $R = \dfrac{V}{I} = \dfrac{6.0\,\text{V}}{2.6\,\text{A}} = 2.31\,\Omega$ [1]

Area of cross-section $= \dfrac{1}{4}\pi d^2 = 0.25\pi \times$

$(0.38 \times 10^{-3})^2 = 1.13 \times 10^{-7}$ m^2 [1]

Rearranging $\rho = \dfrac{RA}{L}$ gives $L = \dfrac{RA}{\rho}$

$= \dfrac{2.31 \times 1.13 \times 10^{-7}\,\text{m}^2}{4.8 \times 10^{-7}\,\text{m}}$ [1] $= 0.54$ m [1]

(4 marks)

b i $\Delta Q = I\Delta t = 2.6$ A $\times 600$ s $= 1560$ C [1]

ii Energy transferred $= IV\Delta t = 2.6$ A $\times 6.0$ V $\times 600$ s $= 9400$ J (2 s.f.) [1] (2 marks)

c The conduction electrons collide with positive ions in the resistor and transfer kinetic energy to them so the ions vibrate more. [1] The internal energy of the resistor increases so its temperature increases [1] and it loses energy by heating the surroundings. [1] (3 marks)

2 a i Resistance at 6.00 V $= \dfrac{V}{I} = \dfrac{6.00\,\text{V}}{1.20\,\text{A}} = 5.0\,\Omega$ [1]

ii Resistance at 12.00 V $= \dfrac{V}{I} = \dfrac{12.00\,\text{V}}{1.72\,\text{A}} = 6.98\,\Omega$ [1]

(2 marks)

b i Energy transferred $= IV\Delta t = 1.72$ A $\times 12.00$ V $\times 300$ s $= 6190$ J [1]

ii $\Delta Q = I\Delta t = 1.72$ A $\times 300$ s $= 516$ C [1] (2 marks)

c $\dfrac{\text{Power at 12 V}}{\text{Power at 6 V}} = \dfrac{12.00 \times 1.72}{6.00 \times 1.20}$ [1] $= 2.9$ [1] (2 marks)

3 a See Figure 1: correctly plotted points [1] suitable scales [1] best fit line [1] (3 marks)

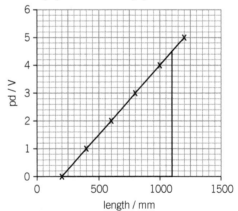

▲ **Figure 1**

b i Substituting $R = \dfrac{V}{I}$ into $\rho = \dfrac{RA}{L}$ gives $\rho = \dfrac{VA}{IL}$

Rearranging this equation gives $V = \dfrac{\rho IL}{A}$ [1]

(1 mark)

ii Equation above is the equation of a straight line through the origin $y = mx$

where $y = V$, $x = L$, and m = gradient [1]

Gradient of graph $m = \dfrac{V}{L} = \dfrac{\rho I}{A}$ so resistivity

ρ = gradient $\times \dfrac{A}{I}$ [1]

Gradient $m = \dfrac{4.50 - 0}{1100 - 200} = 5.0 \times 10^{-3}$ V mm^{-1}

$= 5.0$ V m^{-1} [1]

Area of cross-section $A = \frac{1}{4}\pi d^2 = 0.25\pi \times$
$(0.26 \times 10^{-3})^2 = 5.31 \times 10^{-8}\,\text{m}^2$ [1]

Resistivity $\rho = \text{gradient} \times \frac{A}{I} = 5.0\,\text{V m}^{-1} \times$
$\frac{5.31 \times 10^{-8}\,\text{m}^2}{0.41\,\text{A}} = 6.5 \times 10^{-7}\,\text{m}$ [1]   (5 marks)

c   A thinner wire has a smaller cross-sectional area A
and the resistivity and the current are the same. [1]
As the graph gradient $m = \frac{\rho I}{A}$, the gradient would be
greater as A is smaller. [1]   (2 marks)

4   a   See Figure 2 for circuit diagram: correct position
of voltmeter and variable resistor [1]

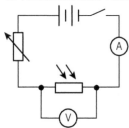

▲ Figure 2

Set up the circuit and close the switch with the
LDR in daylight. Use the variable resistor to adjust
the current to give a suitable reading on the
voltmeter without overloading the ammeter. [1]
Use the ammeter and the voltmeter to measure
the current and pd. [1] Open the switch and
cover the LDR completely. Close the switch again
and adjust the variable resistor to give the same
reading on the voltmeter as before. Measure the
current at this pd. [1] Since the pd is the same
for both sets of readings, the ratio of the dark
resistance to the daylight resistance is equal to the
daylight current /the dark current. [1]   (5 marks)

b   i   When the light intensity increases, the LDR
resistance decreases. [1] As the pd across
the LDR is constant and the LDR resistance
becomes smaller, the current in the circuit
therefore increases. [1]   (2 marks)

ii $R = \frac{V}{I} = \frac{4.5\,\text{V}}{1.2 \times 10^{-3}\,\text{A}} = 3800\,\Omega$ [1]   (1 mark)

iii   The number of charge carriers in the LDR
increases when the light intensity on the LDR
increases. [1] For any given pd across the LDR,
the current therefore becomes greater because
there are more charge carriers in the LDR. [1]
Since the resistance is equal to the pd /current,
the resistance therefore becomes smaller when
the light intensity is increased. [1]   (3 marks)

5   a   The forward voltage is the least pd across the
diode needed to make it conduct when it is in a
circuit in the forward direction. [1]   (1 mark)

b   i $\Delta Q = I\Delta t = 0.17\,\text{A} \times 10 \times 10^{-3}\,\text{s} = 1.7 \times 10^{-3}\,\text{C}$ [1]

ii $P = IV = 0.17\,\text{A} \times 2.2\,\text{V} = 0.37\,\text{W}$ [1]   (2 marks)

c   i   Maximum number of photons of energy E
emitted $= \frac{P}{E} = \frac{0.37\,\text{W}}{3.5 \times 10^{-19}} \approx 10^{18}$ [1]

ii   Some photons would be absorbed by the
LED material (or energy is dissipated by the
LED due to its resistance when current passes
through it). [1]   (2 marks)

6   a   i   The transition temperature is the temperature
at and below which the material has zero
resistivity. [1]   (1 mark)

ii   See Figure 3.

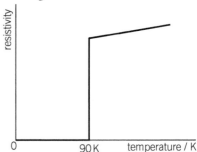

▲ Figure 3

Correct graph with vertical step at 90 K [1],
resistance below 90 K clearly zero. [1]   (2 marks)

b   i   Below the transition temperature, the resistance
R of the wire is zero so the pd across it is zero
as $V = IR$. The current is non-zero because the
circuit is a complete circuit. [1]   (1 mark)

ii   The ammeter reading decreases because
the wire has non-zero resistance and the
circuit resistance is therefore greater. [1] The
voltmeter reading is no longer zero because
the wire has non-zero resistance and a current
passes through it. [1]   (2 marks)

7   a   Resistance of wire $R = \frac{V}{I} = \frac{230\,\text{V}}{10.8\,\text{A}} = 21.3\,\Omega$ [1]

Area of cross-section $= \frac{1}{4}\pi d^2 = 0.25\pi$
$\times (0.30 \times 10^{-3})^2 = 7.07 \times 10^{-8}\,\text{m}^2$ [1]

Rearranging $\rho = \frac{RA}{L}$ gives $L = \frac{RA}{\rho}$
$= \frac{21.3\,\Omega \times 7.07 \times 10^{-8}\,\text{m}^2}{5.6 \times 10^{-8}\,\Omega\,\text{m}}$ [1] $= 27\,\text{m}$ (2 s.f.) [1]   (4 marks)

b   i   Since $R = \frac{\rho L}{A}$ and the resistance and resistivity
are unchanged, then $\frac{L}{A}$ is unchanged. [1] The
diameter of X is twice that of W so the area
of cross-section of X is 4 times that of W. For
$\frac{L}{A}$ to be unchanged, the length L of X would
therefore need to be 4 times that of W [1]
which would mean the length of X would
need to be 108 m. [1]   (3 marks)

ii   As X is thicker, there could be half as many
turns per centimetre on the tube. [1] Because
X is four times longer than W, four times as
many turns round the tube would be needed
for X compared with W. [1] So the tube
would need to at least twice as long and this
is unlikely to be realistic unless the heater is
redesigned [1]. Also, the element may not be as
hot because the energy supplied to it would be
dissipated over a greater length [1]   (4 marks)

Chapter 13

1 a i Total resistance $R = \left(\dfrac{1}{4.0} + \dfrac{1}{6.0}\right)^{-1} + 3.0 = 5.4\,\Omega$ [1]

Current $= \dfrac{\text{battery emf}}{\text{total resistance}} = \dfrac{6.0\,\text{V}}{5.4\,\Omega} = 1.1\,\text{A}$ [1]
(2 marks)

ii Pd across internal resistance $= 1.1\,\text{A} \times 3.0\,\Omega$
$= 3.3\,\text{V}$ [1]

Therefore, pd across parallel combination
$=$ battery emf $-$ pd across internal resistance
$= 6.0 - 3.3 = 2.7\,\text{V}$ [1] *(2 marks)*

b The circuit resistance increases so the current in the battery decreases. [1] Therefore, the pd across the internal resistance decreases. [1] Since the pd across the external resistor is equal to the battery pd $-$ the pd across the internal resistance, the pd across the battery increases. [1] *(3 marks)*

2 a i pd across silicon diode in the forward direction $= 0.6\,\text{V}$ [1]

pd across resistor $=$ battery pd $-$ pd across diode $= 9.0\,\text{V} - 0.6\,\text{V} = 8.4\,\text{V}$ [1] *(2 marks)*

ii Current in diode $=$ current in resistor
$= \dfrac{\text{resistor pd}}{\text{resistance of resistor}} = \dfrac{8.4\,\text{V}}{6.8\,\text{k}\Omega} = 1.2\,\text{A}$
(2 s.f.) [1] *(1 mark)*

b The pd across the diode is unchanged so the pd across the two parallel resistors is 8.4 V. [1] The current in each resistor is therefore 1.24 mA and therefore the current in the diode and the battery is 2.5 mA ($= 2 \times 1.24\,\text{mA}$). [1] *(2 marks)*

3 a Using the circuit shown in Figure 1, the variable resistor is used to change the current which is measured using the ammeter. The battery pd is measured using the voltmeter for at least six different currents. [1] (The lamp is included to limit the maximum value of the current.)

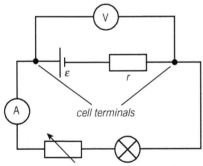

▲ Figure 1

The measurements are recorded and plotted on a graph of $y =$ battery pd V against $x =$ current I. The graph should be a straight line with a negative gradient in accordance with the equation $V = \varepsilon - Ir$ where ε is the battery emf and r is the internal resistance of the battery. [1] The gradient of the line is equal to $-r$ where r is the internal resistance of the battery. [1]

The internal resistance is determined by using a large section of the line to draw a gradient triangle and using the triangle to measure the change of pd ΔV and the corresponding change of

current ΔI. [1] The internal resistance r is equal to the magnitude of $\dfrac{\Delta V}{\Delta I}$. [1] *(5 marks)*

b i The external resistance R of the circuit
$= \left(\dfrac{1}{3.0} + \dfrac{1}{9.0}\right)^{-1} = 2.25\,\Omega$. [1] Since the pd across the external resistors is 5.4 V, the circuit current $= \dfrac{5.4\,\text{V}}{2.25\,\Omega} = 2.4\,\text{A}$. [1] *(2 marks)*

ii Using the equation $\varepsilon = V + Ir$ with $\varepsilon = 6.0\,\text{V}$, $V = 5.4\,\text{V}$, and $I = 2.4\,\text{A}$ gives
$6.0 = 5.4 + 2.4r$ [1] Therefore, $2.4r = 6.0 - 5.4$
$= 0.60$ hence $r = \dfrac{0.60}{2.4} = 0.25\,\Omega$ [1] *(2 marks)*

iii $\dfrac{\text{Power dissipated in the internal resistance}}{\text{Power supplied by the battery}}$
$= \dfrac{I^2 r}{I\varepsilon} = \dfrac{Ir}{\varepsilon} = \dfrac{2.4 \times 0.25}{6.0} = 0.10$ [1] *(1 mark)*

4 a i Net emf $= 5.0 - 2.0 = 3.0\,\text{V}$ [1] *(1 mark)*

ii Total circuit resistance $= \dfrac{3.0\,\text{V}}{0.50\,\text{A}} = 6.0\,\Omega$ [1]

Internal resistance of B $=$ total circuit resistance $-$ power supply internal resistance
$= 6.0\,\Omega - 0.9\,\Omega = 5.1\,\Omega$ [1] *(2 marks)*

b i The net emf of the two batteries decreases as the mobile phone battery emf increases. [1] The circuit resistance is unchanged so the current in the circuit decreases. [1] *(2 marks)*

ii Rearranging $\Delta Q = I\Delta t$ gives $I = \dfrac{\Delta Q}{\Delta t}$
$= \dfrac{1600\,\text{C}}{3 \times 3600\,\text{s}} = 0.15\,\text{A}$ [1] *(1 mark)*

5 a Energy supplied in 1 hour $= IV\Delta t = 3.0\,\text{A} \times 6.0\,\text{V} \times 3600\,\text{s} = 65\,000\,\text{J}$ [1] *(1 mark)*

b i Power supplied by panel $= 3.0\,\text{A} \times 6.0\,\text{V} = 18.0\,\text{W}$

Power produced by each cell $= 1.0\,\text{A} \times 1.2\,\text{V} = 1.2\,\text{W}$ [1]

Therefore, the number of cells $= \dfrac{18.0\,\text{W}}{1.2\,\text{W}} = 15$ [1]
Circuit diagram: the diagram should show 3 parallel rows of cells with 5 cells in each row acting in the same direction. [1] *(3 marks)*

ii The panel pd of 6.3 V when the appliance is disconnected is the net emf of the panel. [1]

When the panel is connected to the appliance, the lost voltage due to its internal resistance
$= 6.3\,\text{V} - 6.0\,\text{V} = 0.3\,\text{V}$. [1]

The current in the panel when it is connected to the appliance $= 3.0\,\text{A}$. Therefore, the internal resistance of the panel $= \dfrac{\text{lost pd}}{\text{current}} = \dfrac{0.3\,\text{V}}{3.0\,\text{A}} = 0.1\,\Omega$ [1] *(3 marks)*

6 a A potential divider is a circuit consisting of two resistors connected to a source of fixed pd. The potential difference of the source is divided or 'shared' between the two resistors in proportion to their resistances. [1] By changing the ratio of the resistances, the share of the potential difference across each resistor can be changed. [1] *(2 marks)*

b The pd across the LDR = battery pd – pd across R = 5.0 – 4.2 = 0.8 V [1]

Therefore, the ratio of the LDR resistance to the resistance of $R = \dfrac{0.8\,V}{4.2\,V} = 0.19$ [1]

Hence, the LDR resistance = 0.19 × 1.5 kΩ = 290 Ω [1] *(3 marks)*

c When the light intensity is reduced, the LDR resistance increases [1] so the pd across the LDR increases and the pd across R decreases. [1] *(2 marks)*

d Connect the LDR to the circuit using long connecting wires. [1]

Draw radial lines at 15° intervals on a large fixed sheet of white paper from a fixed point P at the centre of the sheet and place the lamp at P. [1]

With the LDR facing the lamp directly [1], move it along one of the lines until the voltmeter reading is exactly 2.5 V and mark the position of the LDR on the line. [1]

Repeat the procedure on an adjacent line and then on successive unmarked adjacent lines until all the radial lines are marked. [1]

Since the marks all correspond to the same voltmeter reading, their position along each line indicates where the light intensity is the same along each line. [1]

Joining the marks gives a 'constant intensity' line round the lamp which is non-circular if the light intensity varies with direction. [1]

The nearer a mark on a line is to the lamp, the lower the intensity of the light emitted by the lamp in that direction. [1] (max 6)

7 a When the temperature of T is increased, its resistance decreases so the share of the battery pd across it decreases. [1] Therefore, the potential difference across the variable resistor increases. [1] *(2 marks)*

b i The pd across T = 6.0 V – 2.0 V = 4.0 V [1]

$\dfrac{\text{resistance of the variable resistor}}{\text{resistance of T}}$

$= \dfrac{\text{pd across the variable resistor}}{\text{pd across T}} = \dfrac{2.0}{4.0} = 0.5$ [1]

So the resistance of the variable resistor = 0.5 × the resistance of T = 0.5 × 2.7 kΩ = 1.35 kΩ [1] *(3 marks)*

ii If the thermistor temperature increases above −5.0°C with the variable resistor set at 1.35 kΩ, the alarm switches on because the thermistor resistance decreases so the pd across the variable resistor increases above 2.0 V. [1]

If the variable resistance is decreased with the thermistor at −5.0°C, the voltage across the variable resistor would decrease below 2.0 V [1] so the alarm would be off at −5.0°C. [1] The thermistor would need to be warmer in order for the alarm to switch on [1] (or the alarm would switch on only when the thermistor temperature is above a higher temperature than −5.0°C). *(4 marks)*

Chapter 17

1 A: Speed of J = circumference/time for 1 orbit = $2\pi R_J / T_J$

Speed of E = circumference/time for 1 orbit = $2\pi R_E / T_E$

$\dfrac{\text{Speed of J}}{\text{Speed of E}} = \dfrac{2\pi R_J / T_J}{2\pi R_E / T_E} = \dfrac{R_J}{R_E} \times \dfrac{T_E}{T_J} \approx \dfrac{5 \times 1}{12} = 0.42$ [1]

2 C: $\dfrac{E_K}{a} = \dfrac{\frac{1}{2}mv^2}{v^2/r} = \dfrac{mr}{2}$ [1]

3 A: Support force $S = mg - ma = mg - \dfrac{mv^2}{2}$ (as mg – S = ma) [1]

4 a i Frequency of rotation = 1400/60 s = 23.3 Hz

Angular speed $\omega = 2\pi f = 2\pi \times 23.3$ Hz = 147 rad s⁻¹ = 150 rad s⁻¹ (2 s.f.) [1]

ii speed of point on wall = ωr = 147 rad s⁻¹ × 0.45 m = 66 m s⁻¹ [1]

iii centripetal acceleration = $\omega^2 r$ = (147 rad s⁻¹)² × (½ × 0.45 m) = 4900 m s⁻² [1]

b The clothing in the tub cannot leave the tub as it is forced to move round in circular motion by the wall and it cannot pass through the holes in the wall [1]

The water in the clothing passes through the holes in the wall because the clothing cannot provide enough force on the water to retain it. [1]

5 a The speed of a point on the tyre $v = \omega R = 2\pi f R$ where R = 0.60 m ÷ 2 = 0.30 m

So the frequency of rotation $f = v/2\pi r$ = 25 m s⁻¹/ (2π × 0.30 m) = 13.3 Hz (2 s.f.) [1]

b i Centripetal force $F = m\omega^2 r$ where r = 0.21 m and $\omega = 2\pi f = 2\pi \times$ (26.5 Hz) = 167 rad s⁻¹ [1]

Therefore F = 0.025 kg × (167 rad s⁻¹)² × 0.21 m = 146 N = 150 N (2 s.f.) [1]

ii Graph; (parabolic) curve through the origin with increasing gradient [1]

passes through F = 150 N, v = 25 m s⁻¹ [1] and through a further point consistent with $F \propto v^2$ [1]

6 a The speed of a point on the tyre $v = \omega r = 2\pi f r$ where r = 0.50 m ÷ 2 = 0.25 m

So the frequency of rotation $f = v/2\pi r$ = 8.2 m s⁻¹/ (2π × 0.25 m) = 5.2 Hz [1]

b Centripetal acceleration = v^2/r = (8.2 m s⁻¹)²/ 0.25 m = 270 m s⁻² [1]

7 a Use a stopwatch to time 20 complete turns of the turntable (by observing an object fixed to the turntable near its rim) [1]

Repeat the procedure 3 times and obtain the average timing; the time period is the average timing/20. [1]

b **i** Angular displacement $\Delta\theta$ in $180\,s$ = angular speed × time = $0.86\,rad\,s^{-1}$ × $180\,s$ = $155\,rad$. [1]
Number of turns = $\Delta\theta/2\pi$ = 24.7 turns = 25 turns (2 s.f.) [1]

ii Centripetal acceleration = $\omega^2 r$ = $(0.86\,rad\,s^{-1})^2$ × $(0.270\,m/2)$ = $0.10\,m\,s^{-2}$ [1]

8 **a** **i** Loss of PE = mgh = $2200\,kg$ × $9.81\,m\,s^{-2}$ × $61\,m$ = $1.3(2)\times10^6\,J$ [1]

ii Assuming gain of KE = loss of PE then $\frac{1}{2}mv_{max}^2$ = $1.32\times10^6\,J$ [1]
Therefore v_{max}^2 = $(2 \times 1.32\times10^6\,J)/2200\,kg$ = $1200\,m^2\,s^{-2}$
hence v_{max} = $34.6\,m\,s^{-1}$ = $35\,m\,s^{-1}$ (2 s.f.) [1]

b Centripetal force mv^2/r = $(2200 \times (34.6\,m\,s^{-1})^2/52\,m$ = $50\,600\,N$ [1]
Support force due to the track S = $mg + mv^2/r$ = $(2200 \times 9.81\,m\,s^{-2})$ + $50\,600\,N$ = $72\,000\,N$ [1]

Chapter 18

1 **D:** The maximum speed = $A\omega$ and ω = $\sqrt{(k/m)}$ [1]

2 **C:** A is incorrect because T does not change as the amplitude decreases
B is incorrect because any damped system gradually returns to equilibrium
D is incorrect because any system that is lightly damped can be made to resonate [1]

3 **B:** A is incorrect because E_K decreases
C is incorrect because E_P increases
D is incorrect because E_K increases and E^P decreases [1]

4 **a** T = $35\,s/20$ = $1.75\,s$ so ω = $2\pi/T$ = $2\pi/1.75\,s$ = $3.59\,rad\,s^{-1}$ [1]
For A = $22\,mm$, acceleration a = $-\omega^2 A$ = $-(3.59\,rad\,s^{-1})^2$ × $0.022\,m$ = $-0.28\,m\,s^{-2}$ [1]
magnitude of maximum acceleration = $0.28\,m\,s^{-2}$ [1]

5 **a** **i** $40\,mm$ [1]

ii Angular frequency ω = $5\,rad\,s^{-1}$ therefore T = $2\pi/\omega$ = $2\pi/5$ = $1.26\,s$ [1]

b At t = $0.90\,s$, displacement x = $40\,mm$ × $\cos 5t$ = $40\,mm$ × $\cos (5 \times 0.90\,s)$ = $-8.4\,mm$ [1]

6 Extension e = $(0.320\,m - 0.250\,m)$ = $0.070\,m$
The spring constant k = mg/e = $(0.15\,kg \times 9.81\,m\,s^{-2})/0.070\,m$ = $21\,N\,m^{-1}$ [1]
Therefore T = $2\pi\sqrt{\dfrac{0.15\,kg}{21\,N\,m^{-1}}}$ = $0.53\,s$ [1]

Alternative method $mg = ke$ therefore $m/k = e/g$ [1]
hence T = $2\pi\sqrt{\dfrac{e}{g}}$ [1]
= $2\pi\sqrt{\dfrac{0.070\,m}{9.81\,m\,s^{-2}}}$ = $0.53\,s$ [1]

7 **a** See Topic 18.5, Figure 2 Light damping curve: the time period should be unchanged as the amplitude decreases [1]
the amplitude should decrease exponentially [1]

b If the amplitude decreases by a factor f in 1 cycle, it decreases by a factor f^5 in 5 cycles therefore f^5 = 0.45 [1]
which gives f = 0.85 by trial and error
(or by calculation ;– $\ln f$ = $(\ln 0.45)/5$ = -0.160 which gives f = $e^{-0.160}$ = 0.85 [1])
Total energy $\propto A^2$ so total energy decreases by a factor f^2 = 0.72 or 72% [1]

8 **a** The periodic force due to the bumps has a time period of $4.0\,s$ (= $12\,m/3.0\,m\,s^{-1}$ and therefore the time period of the periodic force is $4.0\,s$ [1]

The carrier box oscillates at resonance due to the periodic force so the time period of the oscillations at the natural frequency of the trailer is $4.0\,s$ [1]

As there are two identical springs in parallel, the spring constant of the system = $2k$ where k is spring constant of each spring. Therefore the time period T = $2\pi\sqrt{m/2k}$ [1]
Rearranging T = $2\pi\sqrt{m/2k}$ gives k = $2\pi^2 m/T^2$ where T = $4.0\,s$ [1]
Therefore k = $2\pi^2 \times 210\,kg/(4.0\,s)^2$ [1] = $260\,N\,m^{-1}$ [1]

b No, the reduction in mass means the time period of natural oscillations will be less [1]
so the trailer will only resonate over these speed bumps at a higher speed [1]
as the time between successive bumps would need to be less. [1]

Chapter 19

1 **B:** Work is done on the air so D is incorrect. There is no change in the internal energy as the temperature at the end is the same as at the start so B must be correct. [1]

2 **B:** Energy is needed to melt a substance so A is incorrect. B is correct because the specific heat capacity is inversely proportional to the rate of increase of temperature and this is higher for the solid than the liquid (which means C is incorrect). The internal energy increases during melting so D is incorrect. [1]

3 **D:** Specific latent heat unit = $J\,kg^{-1}$ = $N\,m\,kg^{-1}$ = $(kg\,m\,s^{-2})\,m\,kg^{-1}$ = $m^2\,s^{-2}$ [1]

4 **a** Energy to heat mass m by temperature rise ΔT = $mc\Delta T$
Energy supplied = IVt
Assuming no energy loss to the surrounding, $mc\Delta T = IVt$ [1]
Therefore specific heat capacity c = $IVt/m\,\Delta T$ = $2.60\,A \times 12.2\,V \times 200\,s/(1.210 \times (33.5 - 18.0)°C)$ [1] = $338\,J\,kg^{-1}\,K^{-1}$ [1]

b i % uncertainty in current = $(0.05/2.60) \times 100\% = 1.9\%$
% uncertainty in pd = $(0.10/12.2) \times 100\% = 0.8\%$
% uncertainty in time = $(0.40/200) \times 100\% = 0.2\%$ [1 for all above correct]
Therefore % uncertainty in energy supplied = $(1.9 + 0.8 + 0.2)\% = 2.9\%$ [1]

ii % uncertainty in temperature change = $(1.0/15.5) \times 100\% = 6.5\%$ [1]

iii % uncertainty in mass = $(0.005/1.210) \times 100\% = 0.4\%$
% uncertainty in c = sum of above % measurements uncertainties = 9.8% [1]

c Measure the temperature at intervals after the heater is switched off [1]. If there is no energy loss, the temperature should be unchanged after the heater is switched off [1].

5 Let m = mass of ice melted, m_1 = mass of water in the calorimeter, m_2 = mass of the copper can, the specific heat capacity of water = c_1, and the specific heat capacity of copper = c_2.

Energy needed to melt the ice = ml = $3.40 \times 10^5 \,\text{J kg}^{-1} \times m$ [1]

Energy needed to raise the temperature of the melted ice to 12°C = $mc_1(T - 0)$ = $(4200 \,\text{J kg}^{-1}\text{K}^{-1} \times 12\,\text{K} \times m)$ = $5.04 \times 10^4 \,\text{J kg}^{-1} \times m$ [1]

Total energy needed
= $(3.40 \times 10^5 + 5.04 \times 10^4) \,\text{J kg}^{-1} m$
= $3.90 \times 10^5 \,\text{J kg}^{-1} m$

Energy released by the water and can = $m_1c_1(T_o - T)$ + $m_2c_2(T_o - T)$ [1] = $[(0.080\,\text{kg} \times 390 \,\text{J kg}^{-1}\text{K}^{-1} \times (15 - 12)\text{K}]$ + $[(0.120\,\text{kg} \times 4200 \,\text{J kg}^{-1}\text{K}^{-1} \times (15 - 12)\text{K}]$ = $94\,\text{J} + 1512\,\text{J} = 1606\,\text{J}$ [1]

Assuming energy needed = energy released,
$5.04 \times 10^4 \,\text{J kg}^{-1} \times m = 1606\,\text{J}$ [1]

Therefore $m = 1606\,\text{J}/3.90 \times 10^5 \,\text{J kg}^{-1} = 4.1 \times 10^{-3}\,\text{kg}$ [1]

6 a Energy transferred per second by the water = $(m/t)c(T_2 - T_1)$ = $0.042 \,\text{kg s}^{-1} \times 4200 \,\text{J kg}^{-1}\text{K}^{-1} \times (50°\text{C} - 44°\text{C})$ [1] = $1060 \,\text{J s}^{-1}$ = $1100 \,\text{J s}^{-1}$ (2 s.f.) [1]

b The average temperature of the radiator when it is hot = 47°C [1]

Energy E needed to heat the radiator from 15°C to 47°C = $m_{\text{rad}} c_{\text{rad}} \Delta T$ = $28\,\text{kg} \times 560 \,\text{J kg}^{-1}\text{K}^{-1} \times (47°\text{C} - 15°\text{C})$ = $5.0 \times 10^5 \,\text{J}$ [1]

In time t, energy transferred $E = Pt$ where P is the mean rate of energy transfer during the time the radiator heats up. Assume 50% of the energy supplied in this time is transferred to the surrounding and 50% is used to heat the radiator then $P = \frac{1}{2} \times 1060 \,\text{J s}^{-1} = 530 \,\text{J s}^{-1}$. [1]

Rearranging $E = Pt$ gives $t = E/P$ = $5.0 \times 10^5 \,\text{J}/530 \,\text{J s}^{-1} = 940\,\text{s}$. [1]

7 a The heater supplied energy at a constant rate. If the solid was fully-insulated (from the surroundings), its temperature would have increased at a constant rate. [1]

Because it was not fully insulated, energy was transferred from the solid to the surroundings by heating. The rate of transfer of energy would therefore have been less than if it had been fully insulated. [1]

As the solid became hotter, the rate of transfer of energy to the surroundings increased so its rate of increase of temperature decreases. [1]

b Initially the solid is at the same temperature as the surroundings so the rate of transfer to the surroundings is zero initially. [1]

Therefore, the energy per second supplied by the heater $P = mc(\Delta T/\Delta t)_o$ where $(\Delta T/\Delta t)_o$ = its initial rate of increase of temperature. [1]

From the graph, $(\Delta T/\Delta t)_o \approx 6\,\text{K}/60\,\text{s} = 0.10 \,\text{K s}^{-1}$ [1]

Therefore, the specific heat capacity of the solid $\approx P/m(\Delta T/\Delta t)_o = 50\,\text{W}/(0.58\,\text{kg} \times 0.10 \,\text{K s}^{-1})$ = $860 \,\text{kg s}^{-1}$. [1]

c The temperature would have increased for 300 s in the same way then it would have decreased [1] at a decreasing rate [1]

8 a Energy transferred to cool the water = $mc\Delta T$ = $0.070\,\text{kg} \times 4200 \,\text{J kg}^{-1}\text{K}^{-1} \times 10\,\text{K} = 2940\,\text{J}$ [1]

Rate of transfer of energy = $2940\,\text{J}/140\,\text{s} = 21 \,\text{J s}^{-1}$ [1]

b i Energy transferred to freeze the water = mL = $0.070\,\text{kg} \times 340 \,\text{kJ kg}^{-1} = 23\,800\,\text{J}$ [1]

Time taken to transfer energy = $23\,800\,\text{J}/21 \,\text{J s}^{-1} = 1130\,\text{s} = 1100\,\text{s}$ (2 s.f.) [1]

ii The rate of energy transfer is assumed to be the same when it freezes as when it cools [1]. This may not be correct because the rate of transfer during cooling changes. [1]

When ice forms in the beaker and water is still present, the rate of transfer of energy from the beaker is unchanged, transferring energy at the same rate as water [1] Ice does/may not transfer energy at the same rate as water does [1]

Chapter 20

1 D: Rearranging $pV = nRT$ gives $n_p = pV/RT$ for the gas in P.

For the gas in Q, $n_q = 0.60\,pV/0.75\,TR = 0.80\,pV/RT = 0.80\,n$ [1]

2 B: Mass of gas m = number of moles n × molar mass M
Volume $V = nRT/p$ (from $pV = nRT$)

Density $= \dfrac{m}{V} = \dfrac{nM}{(nRT/p)} = \dfrac{pM}{RT}$ [1]

3 A: T = constant so molar mass m × mean square speed = constant $\propto T$

rms speed for X = $(800 \,\text{m s}^{-1}/1600 \,\text{m s}^{-1}) = 0.5$ rms speed for Y

$M_Y/M_X = \dfrac{\text{mean square speed for X}}{\text{mean square speed for Y}} = 0.5^2 = 0.25$

$M_Y = 0.25\,M_X = 0.25\,M$ [1]

4 a Graph should be a straight line with a +gradient through origin [2]

b i as **a** with a steeper gradient [1]

  ii as **a** with a lesser gradient [1]

5 a i Rearrange $pV = nRT$ to give $n = pV/RT = 1.40 \times 10^5$ Pa $\times 0.020\,\text{m}^3/(8.31\,\text{J mol}^{-1}\,\text{K}^{-1} \times 285\,\text{K})$ = 1.18 mol = 1.2 mol (2 s.f) [2]

  ii Total KE $= \dfrac{3}{2} nRT = 1.5 \times 1.18 \times 8.31\,\text{J mol}^{-1}\,\text{K}^{-1} \times 285\,\text{K} = 4200\,\text{J}$ [1]

  iii Average KE of a molecule $= 3/2\,kT = 1.5 \times 1.38 \times 10^{-23}\,\text{J K}^{-1} \times 285\,\text{K} = 5.9 \times 10^{-21}\,\text{J}$ [1]

5 b Internal energy U = total KE of molecules [1] so $\Delta U = 3/2\,nR\,\Delta T = 1.5 \times 1.18 \times 8.31\,\text{J mol}^{-1}\,\text{K}^{-1} \times (320 - 285)\,\text{K} = 510\,\text{J}$ [1]

6 a The rms speed of the gas molecules = the square root of the mean value of the square of the molecule speeds [1]

  b i See SB p53 Figure 1; curve from 0 to max number then decreasing [1] and tending to zero [1]

  ii See SB p53 Figure 2; same general shape as above but with a lower max [1] further along speed axis [1]

  c Mass of a nitrogen molecule = $0.028\,\text{kg mol}^{-1}/6.02 \times 10^{23}\,\text{mol}^{-1} = 4.65 \times 10^{-26}\,\text{kg}$ [1]

Mean kinetic energy $\dfrac{1}{2}\,mc_{\text{rms}}^2 = \dfrac{3}{2}\,kT = 1.5 \times 1.38 \times 10^{-23}\,\text{J K}^{-1} \times 273\,\text{K} = 5.65 \times 10^{-21}\,\text{J}$ [1]

Therefore rms speed = $(2 \times \text{mean KE}/m)^{1/2} = (2 \times 5.65 \times 10^{-21}\,\text{J}/4.65 \times 10^{-26}\,\text{kg})^{1/2} = 490\,\text{m s}^{-1}$ [1]

7 a i The molecules move about unpredictably, colliding with each other unpredictably and changing directions when they collide [1]

  ii An elastic collision is one in which the total kinetic energy after the collision is the same as before the collision [1]

  b i Initial momentum = $5.3 \times 10^{-26}\,\text{kg} \times 1200\,\text{m s}^{-1} = 6.36 \times 10^{-23}\,\text{kg m s}^{-1}$ [1]

Final momentum = $-6.36 \times 10^{-23}\,\text{kg m s}^{-1}$ [1]

Change of momentum $\Delta p = -6.36 \times 10^{-23}\,\text{kg m s}^{-1} - 6.36 \times 10^{-23}\,\text{kg m s}^{-1} = (-)1.27 \times 10^{-23}\,\text{kg m s}^{-1} = (-)\,1.3 \times 10^{-23}\,\text{kg m s}^{-1}$ (2 s.f.) [1]

  ii Force F = pressure × area [1] = $120 \times 10^3\,\text{Pa} \times 1 \times 10^{-6}\,\text{m}^2 = 0.12\,\text{N}$ [1]

F = no. of impacts per sec × change of momentum [1]

No. of impacts per second = $F/\Delta p = 0.12\,\text{N}/1.27 \times 10^{-23}\,\text{kg m s}^{-1} = 9.4 \times 10^{21}\,\text{s}^{-1}$ [1]

Chapter 21

1 C: The distance r from the centre of the planet to a point at height R above the surface is $2R$. g is inversely proportional to r^2 so at $r = 2R$, $g = \frac{1}{4} \times g_s$ so A and B are incorrect. V is inversely proportional to r so at $r = 2R$, $V = \frac{1}{2} \times V_s = \frac{1}{2}\,g_s\,R$ so D is incorrect. [1]

2 A: The unit of gravitational potential gradient is the same as the unit of gravitational field strength which is the same as the unit of acceleration. So the base unit combination for gravitational potential gradient is m s^{-2} [1]

3 B: V is inversely proportional to r so at $r = 2R$, $V = \frac{1}{2} \times V_s$ so $V/V_s = \frac{1}{2}$. Curve B is therefore the correct curve as it passes through $V/V_s = \frac{1}{2}$ at $r = 2R$. [1]

4 a The mass of the planet $m = \dfrac{4}{3}\pi R^3 \times$ density ρ

$$g_s = \frac{GM}{R^2} = \frac{G \times \frac{4}{3}\pi R^3 \times \rho}{R^2} = \frac{4}{3}\pi G\rho R \text{ [1]}$$

Rearranging this equation gives $\rho = \dfrac{g_s}{\frac{4}{3}\pi GR}$

$$= \frac{1.62}{\frac{4}{3}\pi \times 6.67 \times 10^{-11} \times 1.74 \times 10^6} \text{ [1]} = 3330\,\text{kg m}^{-3}\text{ [1]}$$

b $\frac{1}{2}m v^2 \geq \dfrac{GMm}{r}$ gives escape speed $v_{\text{esc}} = \sqrt{\dfrac{2GM}{R}}$

$$= \sqrt{\frac{2g_s R^2}{R}} = \sqrt{2g_s R} \text{ [1]}$$

$$= \sqrt{2 \times 1.62 \times (1.74 \times 10^6)} = 2370\,\text{m s}^{-1} \text{ [1]}$$

5 a Rearranging $\dfrac{r^3}{T^2} = \dfrac{GM}{4\pi^2}$ gives $m = \dfrac{4\pi^2 r^3}{GT^2}$ where $T = 16.0 \times 24 \times 60 \times 60\,\text{s} = 1.38 \times 10^6\,\text{s}$

$$M = \frac{4\pi^2 r^3}{GT^2} = \frac{4\pi^2 \times (1.22 \times 10^9)^3}{6.67 \times 10^{-11} \times (1.38 \times 10^6)^2} \text{ [1]}$$

$$= 5.64 \times 10^{26}\,\text{kg} \text{ [1]}$$

b Volume $V = \dfrac{4}{3}\pi R^3 = \dfrac{4}{3}\pi (6.03 \times 10^7)^3 = 9.18 \times 10^{23}\,\text{m}^3$ [1]

density $= \dfrac{M}{V} = \dfrac{5.64 \times 10^{26}\,\text{kg}}{9.18 \times 10^{23}\,\text{m}^3} = 615\,\text{kg m}^{-3}$ [1]

c Let P be the point along the line between their centres at which the gravitational field strengths are equal and opposite.

For Saturn, $g_s = -\dfrac{GM_s}{r_1^2}$ where r_1 is the distance from the centre of Saturn to point P

For Triton, $g_T = -\dfrac{GM_T}{r_2^2}$ where r_2 is the distance from the centre of Triton to point P

At P, $g_s = g_T$ therefore $-\dfrac{GM_s}{r_1^2} = -\dfrac{GM_T}{r_2^2}$ [1]

Rearranging this equation gives $\dfrac{r_1}{r_2} = \sqrt{\dfrac{M_s}{M_T}} = \sqrt{\dfrac{5.64 \times 10^{26}\,\text{kg}}{1.35 \times 10^{23}\,\text{kg}}} = 64.4$ therefore $r_1 = 64.6 r_2$ [1]

Since $r_1 + r_2 = D$ where D = the distance between their centres = $1.22 \times 10^9\,\text{m}$

Therefore $64.6 r_2 + r_2 = D$ so $65.6 r_2 = D$ [1]

$r_2 = D/65.6 = 1.22 \times 10^9\,\text{m}/65.6 = 1.86 \times 10^7\,\text{m}$ [1]

6 a i $F_{\text{grav}} = Gm^2/D^2 = mv^2/r$ [1] where r = radius of orbit = $D/2$ and v = orbital speed

Therefore $v^2 = GM/2D$ hence $v = \sqrt{\dfrac{GM}{2D}}$ [1]

ii Time period T = circumference/speed = $\pi D/v$

[1] $= \pi D \div \sqrt{\dfrac{GM}{2D}} = \pi \sqrt{(2D^3/GM)}$ [1]

b Squaring and rearranging $T = \pi \sqrt{(2D^3/GM)}$

gives $D^3 = \dfrac{GM}{2\pi^2} \times (80 \times 365 \times 24 \times 3600)^2$ [1]

$= 4.3 \times 10^{37}\,\text{m}^3$ [1]

Therefore $D = 3.5 \times 10^{12}\,\text{m} = 23$ AU [1]

7 For an extension of 180 mm, the spring is at or near its elastic limit and its tension would be about 0.9 N ($= 5.0\,\text{N m}^{-1} \times 0.180\,\text{m}$). [1] Using a mass of weight 0.9 N would give a difference in extension of 180 mm $\times 0.05\,\text{N kg}^{-1}/9.8\,\text{N kg}^{-1} = 0.92$ mm [1].

The extension is too small to measure accurately using the metre ruler. [1]

Using a significantly larger mass would stretch the spring beyond its limit of proportionality and the calculation would not be reliable as the spring constant would change with extension beyond the limit of proportionality. [1]

8 a Closest distance between Jupiter and Mars = $7.8 \times 10^{11}\,\text{m} - 2.3 \times 10^{11}\,\text{m} = 5.5 \times 10^{11}\,\text{m}$ [1]

$g_j = \dfrac{GM}{r^2} = -\dfrac{6.67 \times 10^{-11} \times 1.90 \times 10^{27}}{(5.5 \times 10^{11})^2}$

$= 4.2 \times 10^{-7}\,\text{N kg}^{-1}$ [1]

b $g_s = \dfrac{GM}{r^2} = -\dfrac{6.67 \times 10^{-11} \times 2.0 \times 10^{30}}{(2.3 \times 10^{11})^2}$

$= 2.5 \times 10^{-3}\,\text{N kg}^{-1}$ [1]

Resultant gravitational field strength = $g_s - g_j =$ $2.5 \times 10^{-3}\,\text{N kg}^{-1}$ towards the Sun [1]

Chapter 22

1 **B:** doubling the charge on T doubles the force and doubling the separation makes the force four times smaller. So the effect of both changes is to half the force (i.e., $F \rightarrow 2 \times \frac{1}{4} F = \frac{1}{2} F$) [1]

2 **A:** a small + 'test' charge at P experiences a force F due to X away from X, a force F due to Z away from Z and a force less than F due to Y towards Y. The resultant of the first two forces is greater than F and is along the line PY away from Y. Since this is opposite in direction to the third force and is greater in magnitude, the overall resultant of all 3 forces is along the line PY away from Y. [1]

3 **D:** equate the force expressions for Newton's law of gravitation and for Coulomb's law and make $G\varepsilon_0$ the subject to give $G\varepsilon_0 = Q_1Q_2/m_1m_2$ so the unit of $G\varepsilon_0$ is C^2/kg^2 [1]

4 a $V = kr^{-1}$ where k is constant. Using values for any point on the curve (e.g., $V = 2000$ V m at $r = 0.20$ m) gives $k = Vr = 400$ V m. [1]

Therefore at $r = 0.25$ m, $V = k/r = 400$ V m/0.25 m = 1600 V [1]

b i $V = \dfrac{Q}{4\pi\varepsilon_0 r} = \dfrac{k}{r}$ therefore $\dfrac{Q}{4\pi\varepsilon_0} = k$ so

$Q = 4\pi\varepsilon_0 k = 4\pi \times 8.85 \times 10^{-12}\,\text{F m}^{-1} \times 400\,\text{V m}$ [1] $= 4.4(5) \times 10^{-8}$ C [1]

ii Electric field strength $E = \dfrac{Q}{4\pi\varepsilon_0 r^2} = \dfrac{V}{r}$

At $r = 0.25$ m, $E = \dfrac{V}{r} = \dfrac{+1600\,\text{V}}{0.25\,\text{m}}$ [1] $=$ 6400 V m^{-1} [1]

c i The conduction electrons in the rod are attracted towards the sphere and they therefore move towards end X [1] so end X becomes negatively charged because there are more conduction electrons there than (fixed) protons. [1].

The movement of the conduction electrons towards X means they move away from end Y [1] so there are more (fixed) protons at end Y than there are electrons so end Y becomes positively charged. [1]

ii Surface to X; the line is a curve from 2000 V to a lower potential at X than before [1]

X to Y; the line is flat (because the pd from X to Y is zero) [1] and at a potential between what the potential at X and at Y was. [1]

From Y; the potential decreases beyond from Y [1]

iii The electric field near P is stronger [1] The field will not be radial near P [1]

5 a i Resolving the tension T in the thread into horizontal and vertical components (see Figure 1) gives $F = T \sin \theta$ (horizontally) and $mg = T \cos \theta$ (vertically) [1]

Therefore $F/mg = \sin \theta / \cos \theta = \tan \theta$ so $F = mg \tan \theta$ [1]

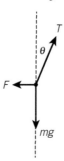

▲ **Figure 1**

ii $F = mg \tan \theta = 0.82 \times 10^{-3}\,\text{kg} \times 9.8\,\text{N kg}^{-1} \times \tan 5.5° = 7.74 \times 10^{-4}$ N [1]

$E = V/d = 2800\,\text{V}/0.035\,\text{m} = 80\,000\,\text{V m}^{-1}$ [1]

Rearranging $F = QE$ gives $Q = \dfrac{F}{E}$

$= \dfrac{7.74 \times 10^{-4}\,\text{N}}{80\,000\,\text{V m}^{-1}}$ [1] $= 9.7 \times 10^{-9}$ C [1]

b θ decreases [1]; This is because the electric field strength $E = V/d$ so E decreases as V is unchanged and d increases [1].

Force F therefore decreases (as $F = QE$) and because $F = mg \tan \theta$, $\tan \theta$ decreases so θ decreases. [1]

6 a Let the charge of A = Q_A and the charge of B = Q_B.

The potential of A at P, $V_A = \dfrac{1}{4\pi\varepsilon_0}\dfrac{Q_A}{r_1}$ where r_1 = AP

The potential of B at P, $V_B = \dfrac{1}{4\pi\varepsilon_0}\dfrac{Q_B}{r_2}$ where

r_2 = BP = 0.25 m − 0.10 m = 0.15 m

$V_A + V_B = 0$ at P therefore $\dfrac{1}{4\pi\varepsilon_0}\dfrac{Q_A}{r_1} + \dfrac{1}{4\pi\varepsilon_0}\dfrac{Q_B}{r_2} = 0$ [1]

So $\dfrac{Q_A}{r_1} = -\dfrac{Q_B}{r_2}$ so $Q_A = -\dfrac{Q_B}{r_2}\times r_1 = -\,0.10\times\dfrac{Q_B}{0.15} =$
$-2\,Q_B/3$ [1]

$Q_A + Q_B = +4.0$ nC so $(-2\,Q_B/3) + Q_B = +4.0$ nC or
$Q_B/3 = +4.0$ nC

Hence Q_B = 12.0 nC, [1]

$Q_A = -2\,Q_B/3 = -8$ nC [1]

b i 3 lines of force including a straight line through P directed from B to A [1]

ii arc through P perpendicular to AB at P [1] bending towards [1] A on both sides of AB

Chapter 23

1 B: Time taken = $Q/I = CV/I$ which is B. [1]

2 C: In a charging circuit, the charge stored increases and the current decreases. So on this basis, C or D could be correct. However, the rate of change in both cases decreases so C is correct. [1]

3 C: The pd across the resistor decreases to $0.6V_B = (V_B - 0.40V_B)$ so the current decreases to 60% in time T. Therefore in time $2T$, the current and therefore the pd across the resistor decreases to 36% (= 60% of 60%) of V_B. So the pd across the capacitor in time $2T$ increases to 64% of the final pd which is $0.64V_B$. [1]

4 a i Q 0.90 mC 1.35 mC 1.80 mC 2.25 mC 2.70 mC [1]

ii Graph−correctly labelled axes [1]; linear scales from 0 covering at least half of each axis[1]; correctly plotted points and best-fit line [1]

iii Large gradient triangle drawn and correct read off to calculate the capacitance [1] Capacitance calculated from graph to give C = 0.90 mF [1]

b [Any 2 of the following 3 marks]

The initial rate of change of pd would be the same. [1]

The voltage would increase at a decreasing rate that becomes zero [1] when the pd is equal to the battery pd [1]

5 a Energy stored = $\frac{1}{2}CV^2 = 0.5 \times 2.2\times 10^{-3}$ F × $(5.5\,\text{V})^2$ [1] = 3.3×10^{-2} J [1]

b i Assuming all the energy stored by the capacitor is transferred to the object's store of gravitational potential energy, $mg\Delta h = 3.3\times 10^{-2}$ J [1]

Therefore 3.8×10^{-3} kg × 9.8 N kg^{-1} × $\Delta h = 3.3\times 10^{-2}$ J [1] which gives $\Delta h = 3.3\times 10^{-2}$ J/0.0373 N = 0.89 m [1]

ii Any 2 of the following 3 points:

Energy is dissipated (or transferred to the surroundings) by

1 the resistance heating effect of the discharge current [1]

2 friction between the moving parts in the motor [1]

3 sound waves created by the rotational motion (of vibrations) of the motor when it rotates [1]

6 a Rearranging $V = V_0 e^{-t/RC}$ gives $V_0/V = e^{t/RC}$ [1]

Taking natural logs on both sides gives ln V = ln $V_0 - t/RC$ which is the equation for a straight line $y = mx + c$ [1] where the gradient $m = -1/RC$ and the y-intercept = ln V_0 [1]

b i Missing ln V values 1.281, 0.993, 0.765, 0.531, 0.300 [1]

Graph; axes labelled correctly [1] suitable scales covering at least half of each graph axis [1], points plotted correctly [1], best-fit line drawn [1] See Figure 2

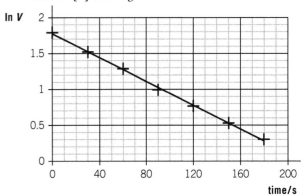

▲ **Figure 2**

ii Large gradient triangle drawn [1] gradient correctly calculated from the height and base of the triangle [1]; gradient −1/RC (= −0.0083 s^{-1}) used to calculate C to give C = 2.6 mF [1] within ± 0.2 mF [1]

iii Time constant = $RC = -1$/gradient = 120 s (or correct calculation of RC) [1]

7 a i The dielectric molecules become polarised by the electric field so each molecule becomes a dipole with positive charge and negative charge at opposite ends of the molecule. [1] As a result, opposite surfaces on the dielectric perpendicular to the field become oppositely charged. [1] The surface nearest the negative plate of the capacitor is positively charged and the opposite surface positively charged. [1]

ii The positive charge that is created on the surface of the dielectric nearest the negative plate allows more electrons to transfer from the battery to the negative plate. [1] The negative charge that is created on the surface of the dielectric nearest the positive plate forces electrons to transfer to the battery from the positive plate. [1] As a result of this movement of electrons, more charge is stored by the capacitor as a result of the dielectric being inserted. [1]

b i The capacitance decreases. [1]

ii The charge stored Q is unchanged and the capacitance C decreases [1] so the potential difference increases (as it is equal to Q/C.) [1]

The energy stored increases [1] because it is equal to ½ QV and Q is unchanged and V increases. [1]

(Alternative for last mark: because work is done to overcome the attraction between the dielectric charge and the charge on each plate [1])

c Rearranging $C = \dfrac{A\varepsilon_0\varepsilon}{d}$

gives $d = \dfrac{A\varepsilon_0\varepsilon}{C} = \dfrac{1.2\,\text{m}^2 \times 8.85 \times 10^{-12}\,\text{F m}^{-1} \times 7}{47 \times 10^{-6}}$ [1]

$= 1.6 \times 10^{-6}\,\text{m}$ [1]

8 a Rearrange $v = V_0 e^{-\frac{t}{RC}}$ to give $\ln V = \ln V_0 - t/RC$ [1]
Therefore $t/RC = \ln V_0 - \ln V = \ln 6.0 - \ln 1.5 = 1.39$ [1]
$t = RC \times 1.39 = 1.39 \times 50\,\text{k}\Omega \times 68\,\mu\text{F} = 4.7\,\text{s}$ [1]

b The time constant needs to be about 4 times longer ($\approx 20\,\text{s}/4.7\,\text{s}$) [1] so the resistor needs to be about 4 times bigger which is about $200\,\text{k}\Omega$. [1]

Chapter 24

1 **A:** For no deflection, $BQv = QE$ therefore $E = Bv$ [1]

2 **B:** The k.e. of the particles $= eV$ therefore $\frac{1}{2}mv^2 = eV$ hence its speed $v = (2eV/m)^{1/2}$. The magnetic force $Bev = mv^2/r$ therefore $r = mv/eB = m(2eV/m)^{1/2}/eB = (2mV/eB^2)^{1/2}$ [1]

3 **C:** X curves in the same direction as the proton so it also has a + charge which means B and D are incorrect. As the radius of curvature $r = mv/QB$ and Q is the same for a pion and a proton, r is proportional to the momentum. Since r is less for X, its momentum is less so A is incorrect. [1]

4 a The magnetic force BQv on the particle is perpendicular to the direction of motion of the particle [1] so no work is done on the particle [1] and so its kinetic energy (and therefore its speed) is constant [1].

b i The radius of curvature of the particle's path in the field, $r = 0.31\,\text{m}/2 = 0.155\,\text{m}$

The mass of the ion $= 20\,\text{u} = 20 \times 1.67 \times 10^{-27}\,\text{kg} = 3.34 \times 10^{-26}\,\text{kg}$ [1]

$Bqv = mv^2/r$ hence $B = \dfrac{mv}{rq} =$

$\dfrac{3.34 \times 10^{-26}\,\text{kg} \times 7.3 \times 10^5\,\text{m s}^{-1}}{0.155\,\text{m} \times 1.6 \times 10^{-19}\,\text{C}}$ [1] $= 0.98\,\text{T}$ [1]

ii $r = mv/BQ$ therefore m is proportional to r [1]

The ratio of distances from $\mathbf{P} = 0.34\,\text{m}/0.31\,\text{m} = 1.1$ therefore the mass of the ions detected $0.34\,\text{m}$ from $\mathbf{P} = 1.1 \times 20\,\text{u} = 22\,\text{u}$ [1]

iii Each neon ion has 10 protons in its nucleus. The neon ions detected at $0.31\,\text{m}$ from P each have 10 neutrons so their mass is $20\,\text{u}$ [1] whereas the ions detected at $0.34\,\text{m}$ each have 12 neutrons so their mass is $22\,\text{u}$. [1]

5 a Length of each turn $l = \pi d$ where d is the diameter of the coil.
$F = BIl\,n = 0.120\,\text{T} \times 0.19\,\text{A} \times \pi \times 0.082\,\text{m} \times 60$ [1] $= 0.35\,\text{N}$ [1]

b i The force of the coil is perpendicular to the field. The a.c. repeatedly reverses its direction. [1] Each time it reverses, the force on the coil reverses so the direction of motion of the coil repeatedly reverses. [1]

ii The magnitude of the force is proportional to the current in the coil. [1] If the rms (or peak) current is increased, the force increases in magnitude so the amplitude of vibration of the coil increases. [1]

6 a Weight of $W = 0.67 \times 10^{-3}\,\text{kg} \times 9.8\,\text{N kg}^{-1} = 6.57 \times 10^{-3}\,\text{N}$

To regain balance, the moment due to the magnetic force = change of moment of W [1] = $6.57 \times 10^{-3}\,\text{N} \times 0.032\,\text{m} = 2.10 \times 10^{-4}\,\text{N m}$ [1]

Force on AB = moment/distance from A to the pivot = $2.10 \times 10^{-4}\,\text{N m}/0.120\,\text{m}$ [1] $= 1.75 \times 10^{-3}\,\text{N}$ [1]

b Magnetic flux density $B = \dfrac{F}{Il} = \dfrac{1.75 \times 10^{-3}\,\text{N m}}{4.5\,\text{A} \times 0.095\,\text{m}}$ [1]

$= 4.1 \times 10^{-3}\,\text{T}$ [1]

7 a i $F = BIl\,n = 0.110\,\text{T} \times 0.67\,\text{A} \times 0.035\,\text{m} \times 160 = 0.41(2)\,\text{N}$ [1]

ii Torque $= Fd = 0.412\,\text{N} \times 0.028\,\text{m}$ [1] $= 1.15 \times 10^{-2}\,\text{N m}$ [1]

b The torque due to W $= 1.4\,\text{N} \times (0.5 \times 0.054\,\text{m}) = 0.0378\,\text{N m}$ [1].

To raise the weight, an equal and opposite torque must be applied by the motor [1].

The torque due the motor must be increased from $0.0115\,\text{N m}$ to $0.0378\,\text{N m}$ which is $3.3 \times$ greater. [1]

The torque is proportional to the motor current so the current must therefore be increased by $\times 3.3$ to $3.3 \times 0.67\,\text{A} = 2.0\,\text{A}$ [1]

Chapter 25

1 **D:** The induced emf V = change of magnetic flux per second $= BAn/t$ therefore $B = Vt/nA$. [1]

2 **C:** D is incorrect because V A^{-1} is the ohm. The unit of magnetic flux is the weber. The tesla equals 1 weber per square metre so A and B are incorrect. C is correct because induced emf = magnetic flux change per second therefore the unit of magnetic flux = the unit of emf × the unit of time = V s. [1]

3 A: The area per unit time swept out by the rod is constant and because the field is uniform, the flux per second swept out by the rod is constant so the induced emf is constant. [1]

4 a Faraday's law tells you that the induced emf = change of flux per second so the product of induced emf and time represents magnetic flux linkage. [1] Therefore, the area under the curve represents the total change of magnetic flux linkage [1]

b No of small squares under the curve = 9 + 35 + 82 + 83 + 23 + 1 = 233 ± 5 [1]

Each large square = 25 small squares represents $10\,mV \times 50\,ms = 500\,\mu Wb = 0.50\,mWb$ [1]

Change of flux linkage = $(233 \div 25) \times 0.50\,mWb = 4.7\,mWb$ [1]

c i Final flux linkage = 0 so initial flux linkage = 4.7 mWb [1]

Area of coil $A = \pi \times (0.025\,m)^2 = 1.96 \times 10^{-3}\,m^2$ [1]

Magnetic flux density

$B = \dfrac{\text{initial flux linkage}}{An} = \dfrac{0.0047\,Wb}{1.96 \times 10^{-3}\,m^2 \times 75}$

$= 0.032\,T$ [1]

ii % uncertainty in coil area = 2 × % uncertainty in diameter = $(0.5/25) \times 100\% = 2.0\%$ [1]

% uncertainty in the magnetic flux estimate ≈ $(5/233) \times 100\% = 2.1\%$ [1]

Overall % uncertainty = 4.1% so uncertainty in V = 4.1% × 4.7 mWb ≈ ±0.2 Wb [1]

5 a i Max flux linkage = $BAn = 0.092\,T \times 2.3 \times 10^{-3}\,m^2 \times 180$ [1] $= 0.038(1)\,Wb$ [1]

ii Peak emf = $BAn\omega = Ban\,(2\pi f) = 0.0381\,Wb \times 2\pi \times 50\,Hz$ [1] = 12 V [1]

b correct shape 'sine wave or negative sine wave [1]; peak values at +12V shown [1]; time period = 20 ms shown or correct time scale shown [1]

c The peak-to-peak voltage would be increased to 1.8 its initial value (= initial value 90 Hz / 50 Hz) [1]

The time period would be reduced to 0.56 its initial value (= initial value 50 Hz / 90 Hz) [1]

6 a i $V_P = V_s \times \dfrac{N_P}{N_s} = 12\,V \times \dfrac{1800}{94}$ [1] = 230 V [1]

ii Peak pd = $\sqrt{2} \times 230\,V = 325\,V$ [1]

b Efficiency = $\dfrac{I_s V_s}{I_P V_P} = \dfrac{50\,W}{0.23\,A \times 230\,V}$ [1] = 0.95 (= 95%) [1]

c Any 2 of the following:

Resistance heating [1] by the current in the coils [1]

Resistance heating [1] by eddy currents in the core [1]

Heating in the core due to [1] by repeated magnetisation and demagnetisation [1]

7 a i The conduction electrons in the disc are forced to move in/across the magnetic field [1] These electrons experience a magnetic force which is towards or away from the centre of the disc (depending on its direction of rotation) [1] If the electrons move to the centre, the centre becomes negative and the rim becomes positive (or if the electrons move to the rim, the rim becomes negative and the centre becomes negative) [1]

OR Each radial length in the disc cuts across the field lines [1] and sweeps out magnetic flux at a constant rate [1] so an emf is induced between the ends of each radial length with the ends near the centre have opposite polarity to the ends at the rim [1]

ii Using the right hand rule; Rim is + and centre – [1]

b flux swept out per turn = $B A$ [1]

number of turns per second = f therefore flux swept out per second = $BA f$ [1]

(magnitude of the) induced emf = flux swept out per second = $BA f$ [1]

c Induced emf = $BA f$ = 0.270 T × π × $(0.015\,m)^2$ × 46 Hz [1] = 8.8 mV [1]

8 Magnetic flux linkage = BAn = 0.13 T × 0.018 m × 0.018 m × 55 [1] = 2.3×10^{-3} Wb [1]

The magnetic flux linkage changes from 2.3 mWb to −2.3 mWb. Therefore the magnitude of the change of magnetic flux linkage = (2.3 mWb) − (−2.3 mWb) = 4.6 mWb [1]

The magnitude of the induced emf = change of magnetic flux linkage per second = $\dfrac{0.0046\,Wb}{0.092\,s}$ [1] = 0.05 V [1]

Chapter 26

1 D: The mass number decreases by 4 for each α decay and doesn't change in a β decay. Since the mass number decreases by 12, there must be 3 α decays. The proton number decreases by 2 for each α decay and increases by 1 in each β decay. Since the overall change of the proton number is a decrease of 2 and the α decays cause a decrease of 6 , there must be 4 β decays to give an overall decrease of 2. Therefore there are 7 decays. [1]

2 B: For N_o atoms initially, when there are N undecayed atoms, the number of decayed atoms is 3 N. Therefore $N_o = N + 3N = 4 N$ or N/N_o = ¼

Time taken for N to decrease from N_o to ¼ N_o = 2 half-lives = 20 hours. [1]

3 B: Since $A = N\lambda$ and $\lambda = \ln 2/T^{1/2}$ then $N = A\,T^{1/2}/\ln 2$ so the ratio of the number of atoms of Q to P = their activity ratio × their half-life ratio

Half-life Q/P = 100/50 = 2 /1

Activity of Q/P after 200 s = $(A_o/4)/(A_o/16)$ = 16/4 = 4/1

Therefore ratio of number of atoms of P to Q = 8 [1]

4 a 11.3, 8.0, 6.2, 4.8, 3.8, 2.6, 2.1 [1]

b See Figure 1

Axes labelled, correctly plotted points [1]

Suitable scales [1]

Best-fit line [1]

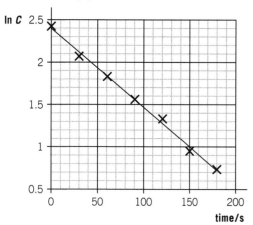

▲ **Figure 1**   [3]

c i Large triangle clearly shown and used [1]

Gradient = −decay constant $\lambda = (2.40 - 1.00)/(150 - 0)\,\text{s}^{-1} = 0.00933\,\text{s}^{-1}$ [1]

Half life = $\ln 2/\lambda = 74\,\text{s}$ [1]

ii The count rate changes if the distance from the source to the tube changes. [1] The distance needs to be constant so the decrease of the count rate is due only to radioactive decay [1]

5 a i Most of the mass of the atom is concentrated in its nucleus which has a much smaller diameter than the atom [1] The nucleus is + charged and it repels and scatters α particles that come close to it [1] Only a small proportion of the incident α particles come close enough to be scattered by more than 90° [1]

ii The probability of an α particle being deflected by a nucleus is very small because most of the atom is empty space. [1] The foil has multiple layers of atoms and the probability of an α particle between scattered more than once is very very small [1] as it is equal to the probability of being scattered once raised to the power n where $n > 1$ [1]

b $E_K = \dfrac{Q_\alpha Q_N}{4\pi\varepsilon_0 d}$ gives $d = \dfrac{Q_\alpha Q_N}{4\pi\varepsilon_0 E_K}$

$= \dfrac{2e \times 29e}{4\pi \times 8.85 \times 10^{-12}\,\text{Fm}^{-1} \times 4.5 \times 10^6\,e}$ [1]

$= \dfrac{58 \times 1.6 \times 10^{-19}\,\text{C}}{4\pi \times 8.85 \times 10^{-12}\,\text{Fm}^{-1} \times 4.5 \times 10^6}$ [1]

$= 1.9 \times 10^{-14}\,\text{m}$ [1]

6 a Rearranging $C = \dfrac{k}{(d + d_0)^2}$ gives $d + d_0 = \dfrac{\sqrt{k}}{\sqrt{C}}$ [1]

Therefore $d = \dfrac{a}{\sqrt{C}} + b$ where $b = -d_0$ and $a = \sqrt{k}$ [1]

b i Values of $1/\sqrt{C}$ correct to 3 sf [1]

Units correct [1]

d/mm	102	136	170	202	245	305	385
C/s⁻¹	218	126	73	57	40	27	15
$1/\sqrt{C}$/ s^{1/2}	0.0678	0.0891	0.117	0.132	0.158	0.192	0.258

ii See Figure 2

Points plotted correctly and axes labelled correctly [1]

Suitable scales [1]

Best fit line [1]

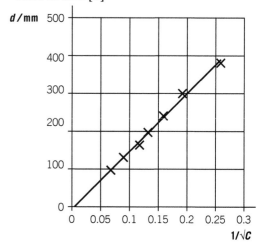

▲ **Figure 8**

c $d_0 = 4$ to $5\,\text{mm}$ [1]

7 a $^{63}_{29}\text{Cu} + ^{1}_{0}\text{n} \rightarrow ^{64}_{29}\text{Cu}$ [1]

b i Beta minus radiation [1]: The copper-64 nucleus is neutron-rich [1] and becomes stable by emitting a β^- particle [1]

ii $^{64}_{29}\text{Cu} \rightarrow ^{64}_{30}\text{Zn}$ [1] $+ ^{0}_{-1}\beta + \bar{v}$ [1]

c If the copper-63 nucleus absorbs a proton, it would become a zinc-64 nucleus [1] which is stable. [1]

8 a The radiation from the source included α particles with a range just less than the tube–source distance at which the count rate dropped. [1] These particles were unable to reach the tube at this distance. [1]

b The source must be emitting either longer-range α radiation and/or β radiation and/or γ radiation. [1]

Placing paper between the source and the tube would cause a further drop in the count rate if the source was emitting longer-range α radiation. [1]

Placing a 5 mm thick aluminium plate instead would cause a drop in the count rate if the source was emitting β radiation. [1] If no drop occurred, the source must be emitting γ radiation. [1]

Chapter 27

1 **D:** The mass defect $= Z m_p + (A - Z) m_n - M = m_n A + (m_p - m_n) Z - M$ therefore $r = m_n$, $s = (m_p - m_n)$ and $t = -1$ so D is correct. [1]

2 **B:** Fission energy/second $= 1200\,MW/0.4 = 3000\,MW$ so number of fission events per second $= 3000\,MW/$ $(200\,MeV \times 1.6 \times 10^{-13}\,J/MeV) = 9.38 \times 10^{19} \approx 10^{20}$ [1]

3 **D:** The total number of protons is unchanged at 92 so Y has 92 − Q protons which means A and B are incorrect. The total number of protons and neutrons decreases by 1 from 235 as 2 fission neutrons are released after a neutron has been absorbed. So Y has 234 − P protons and neutrons. [1]

4 a $^{64}_{29}Cu \rightarrow ^{\ 0}_{+1}\beta + ^{64}_{28}Ni$ [1] $+ \bar{\nu}$ [1] $+ Q$

  b mass difference $\Delta m = 0.66\,MeV/931.5\,MeV\,u^{-1} = 0.00071\,u$ [1]

  mass of nickel nucleus = mass of copper nucleus − mass of positron $- \Delta m$ [1]

  $= 63.91384 - 0.00055 - 0.00071\,u = 63.91258\,u$ [1]

  c a = up quark (*u*); b = down quark (*d*) [1]
  c = a W^+ boson; d = an (electron) neutrino (ν_e) [1]

  d i similarity: both are elementary (or leptons) [1]

   difference: neutrino is uncharged whereas the β^+ particle is charged (or the neutrino is a matter particle whereas the β^+ particle is an antimatter particle) [1]

   ii similarity: both are charged [1]

   difference: the W^+ boson is very short-lived compared with the β^+ particle [1]

5 a i See Topic 27.2, Figure 2: Curve showing an increase up to about $A = 55$ [1] then a more gradual decrease up $A \approx 240$ [1] at BE/nucleon > 6 MeV [1]

   Maximum BE/nucleon between 8 to 9 MeV [1]

   ii The average binding energy per nucleon increases in fusion [1] provided the atomic mass of the fused nucleus is less than about 55. [1] The number of nucleons is unchanged [1] so the total binding energy increases [1]

  b i Mass defect $\Delta m = 2m_p + 2m_n - M_{NUC} = 0.03040\,u$ [1]

   Therefore, binding energy $= 0.03040\,u \times 931.5\,MeV/u = 28.3\,MeV$ [1]

   Binding energy per nucleon $= 28.3\,MeV/4 = 7.1\,MeV$ [1]

   ii Correct position at $A = 4$ [1]

   iii The energy released when protons and neutrons form a cluster (or a nucleus) is equal to the binding energy of the cluster. [1]

   So the energy released when a proton and a neutron form a ^2_1H nucleus is much less than

the energy released when an alpha particle forms from 2 protons and 2 neutrons. [1]

A ^2_1H nucleus formed inside a large nucleus does not gain sufficient energy to leave the large nucleus whereas an alpha particle does gain sufficient energy to leave. [1].

6 a Induced fission is the splitting of a nucleus of a fissile material into two approximately equal smaller nuclei [1] when a neutron from outside the original nucleus collides with it. [1]

  b i When a neutron causes a nucleus to undergo fission, 2 or 3 neutrons are released in the process. [1] These neutrons can go on to produce further fission events in which more neutrons are released that cause further fission events. [1] In this way, a chain of reactions occurs until all the nuclei of the fissile material have undergone fission. [1]

   A minimum mass (the critical mass) is required because if the mass is too low, too many of the fission neutrons escape from the fissile material without causing fission [1] and some are absorbed by other nuclei without fission. [1]

   ii For a chain reaction to occur, at least one fission neutron per fission on average must go on to produce further fission. [1] The rate of fission events is proportional to the mass of the material. [1] A minimum mass (the critical mass) is required because if the mass is too low, too many of the fission neutrons escape from the fissile material without causing fission [1] as the surface area in relation to the mass is too great. [1]

  c i The control rods are used to absorb excess neutrons to ensure a constant rate of fission events occurs. [1]. The depth of the rods in the reactor core is controlled automatically according to the rate at which fission events occur. [1] If the rate increases, the rods are pushed further into the reactor core so they absorb more neutrons, causing the rate of fission events to decrease to where it was. [1] If the rate of fission events decreases, the rods are withdrawn gradually to increase the rate of fission events back to where it was. [1]

   Either from: boron or cadmium [1]

   ii Fission neutrons are released with too much speed to cause further fission of uranium 235 nuclei [1] and their kinetic energy must be reduced considerably before they can cause induced fission. [1] Their kinetic energy is reduced by allowing them to undergo repeated elastic collisions with the atoms of the moderator [1] as some of their kinetic energy is transferred to the moderator atoms in each collision. [1] The amount of kinetic energy transferred in such a collision is greater the closer the mass of the moderator atoms is to the mass of a neutron. [1]

iii Any 2 points: The coolant must be a liquid or a gas that can be pumped through channels in the moderator.[1] It must have a high specific heat capacity [1] and not react chemically with the materials in the reactor core. [1]

The coolant is pumped through pipes from the reactor core to a heat exchanger [1] where it is used to heat water and create steam to drive turbines [1]. After passing through the heat exchanger, the coolant returns through pipes to the reactor core where it is heated up again before returning to the heat exchanger. [1]

7 a i $^{210}_{84}\text{Po} \rightarrow {}^{206}_{82}\text{Pb}$ [1] $+ {}^{4}_{2}\alpha$ [1] $+ Q$

ii mass difference $\Delta m = 209.93675\,\text{u} - 205.92945\,\text{u} - 4.00150\,\text{u} = 0.00580\,\text{u}$ [1]

$Q = 931.5\,\text{MeV/u} \times 0.00580\,\text{u}$ [1] $= 5.4\,\text{MeV}$ [1]

b i The alpha particle and the lead nucleus move away from each other with equal and opposite momentum [1]

Therefore the energy released is shared between them as kinetic energy so the kinetic energy of the alpha particle is less than the energy released. [1]

ii As $E_K = p^2/2m$ and the momentum p is equal in magnitude for the alpha particle and the lead nucleus [1], the ratio of their kinetic energies is the inverse of their mass ratio [1]

Their mass ratio $m_{Pb}/m_\alpha = 206/4 = 51.5$ [1]

Therefore the $E_{KPb}/E_{K\alpha} = m_\alpha/m_{Pb} = 1/51.5$ so $E_{KPb} = E_{K\alpha}/51.5 = 0.0194\,E_{K\alpha}$ [1]

Since $E_{K\alpha} + E_{Pb} = Q = 5.4\,\text{MeV}$ then $E_{K\alpha} + 0.0194\,E_{K\alpha} = 5.4\,\text{MeV}$ [1]

Therefore $E_{K\alpha} = (5.4/1.0194)\,\text{MeV} = 5.3\,\text{MeV}$ [1]

Answers to summary questions

1.1

1 a 19 n, 20 p [1]

   b B (16 neutrons and 14 protons) [1]

   c $^{30}_{16}\text{S}$ [1]

2 a i 8 p, 9 n [1]   **ii** 4 p, 5 n [1]

   b Compared with a $^{9}_{4}\text{Be}$ nucleus, a $^{17}_{8}\text{O}$ nucleus has twice as much charge but less than twice as much mass. [1] So its specific charge $\left(= \dfrac{\text{charge}}{\text{mass}}\right)$ is greater than the specific charge of the $^{9}_{4}\text{Be}$ nucleus. [1]

3 a The charge of the nucleus is +92e. [1] Therefore, the specific charge of this nucleus $= \dfrac{92e}{238m} = \dfrac{92 \times 1.60 \times 10^{-19}\,\text{C}}{238 \times 1.67 \times 10^{-27}\,\text{kg}}$ [1] $= 3.70 \times 10^{7}\,\text{C kg}^{-1}$ [1]

   b The charge of the nucleus is +2e. [1] Therefore, the specific charge of this nucleus $= \dfrac{2e}{9m} = \dfrac{2 \times 1.60 \times 10^{-19}\,\text{C}}{9 \times 1.67 \times 10^{-27}\,\text{kg}}$ [1] $= 2.13 \times 10^{7}\,\text{C kg}^{-1}$ [1]

1.2

1 The strong nuclear force has a range of about 3–4 fm whereas the electrostatic force has an infinite range. [1]

   The strong nuclear force is attractive from 2–3 fm to 0.5 fm and repulsive at < 0.5 fm whereas the electrostatic force between two protons is always repulsive. [1]

2 a A – 4 [1]   **b** Z – 1 [1]

3 a a = 64, b = 30, c = 0 (2 correct [1], all correct [1])

   b a = 224, b = 88, c = 2 (2 correct [1], all correct [1])

1.3

1 a Similarity: same speed in a vacuum. [1] Difference: a radio wave photon has a much longer wavelength/shorter frequency than a light photon. [1]

   b To calculate the wavelength of a 50 keV photon, rearrange the equation $E = \dfrac{hc}{\lambda}$ to give $\lambda = \dfrac{hc}{E}$ Substituting values where $E = 50000 \times e$ $= 50000 \times 1.6 \times 10^{-19}$ J [1] gives $\lambda = \dfrac{hc}{E} = \dfrac{6.63 \times 10^{-34} \times 3.00 \times 10^{8}}{50000 \times 1.60 \times 10^{-19}} = 2.5 \times 10^{-11}$ J (answer to 2 significant figures) [1]

2 a To calculate the energy of a photon of wavelength 30 mm, we can use $E = \dfrac{hc}{\lambda} = \dfrac{6.63 \times 10^{-34} \times 3.00 \times 10^{8}}{30 \times 10^{-3}}$ $= 6.6 \times 10^{-24}$ J microwave [1]

   b i The wavelength is 10^{-8} times smaller than in **a** so the energy is 10^{8} times larger $\left(\text{because } E \propto \dfrac{1}{\lambda}\right)$. Hence, $E = 6.6 \times 10^{-24} \times 10^{8}$ J $= 6.6 \times 10^{-16}$ J [1]

     ii X-rays [1]

3 Using $E = \dfrac{hc}{\lambda}$ with $\lambda = 600$ nm gives $E = \dfrac{6.63 \times 10^{-34} \times 3.00 \times 10^{8}}{600 \times 10^{-9}} = 3.3(1) \times 10^{-19}$ J. [1] In 1 s, the LED emits a maximum of 5.0 mJ. Therefore, the maximum number of 600 nm photons that could be emitted is $\dfrac{5.0 \times 10^{-3}}{3.3(1) \times 10^{-19}\,\text{J}}$ [1] $\approx 1.5 \times 10^{16}$ [1]

1.4

1 1876 MeV (= 2 × 938 MeV) [1]

2 a Total energy = (2 × 0.511 MeV) [1] + 0.250 MeV = 1.272 MeV [1]

   b 0.636 MeV (= 1.272 MeV ÷ 2) [1]

3 a Pair production is a process in which a single photon passing near a nucleus creates a particle and its corresponding antiparticle and ceases to exist. [1]

   b Any two of: **1** A particle and its corresponding antiparticle must be created from a photon. [1] **2** The photon energy must be greater than twice the rest energy of the particle. **3** The interaction must take place near a nucleus. [1]

1.5

1 a a virtual photon [1]   **b** a W⁻ boson [1]

   c a W⁺ boson [1]

2 a See Topic 1.5 Figure 2a (emitted particles both correctly labelled [1], exchange particle = W⁻ [1])

   b See Topic 1.5 Figure 3b (emitted particles both correctly labelled [1])

3 a See Topic 1.5 Figure 4 (ingoing particles correctly labelled [1], outgoing particles correctly labelled [1], exchange particle = W⁺ from p to e⁻ [1])

   b A proton in the nucleus interacts with an inner-shell electron by emitting a W⁺ boson which is absorbed by the electron [1]. The proton changes into a neutron in the process [1] and the electron changes into a neutrino. [1]

2.1

1 a electron, antiproton [1]   **b** K⁰ meson [1]

2 a weak [1]   **b** strong [1]   **c** weak [1]

3 a A muon lasts longer than a π meson so it is less unstable. [1]

   b Similarity: they are both negatively charged. [1] Difference: a muon is less unstable than a π meson or a π⁻ meson decays into a muon. [1]

2.2

1 a i Leptons do not interact through the strong interaction whereas hadrons do. [1]

     ii A baryon decays directly or eventually into a proton whereas meson decay products do not include protons. [1]

b i meson

ii baryon

iii lepton (all 3 correct [2], 2 correct [1])

2 a $K^+ \rightarrow \pi^+ + \pi^0$ [1]

b 219 MeV [1] (= 494 − 140 − 135 MeV) [1]

3 a Energy of each colliding proton = 1.94 GeV
(= 1.00 + 0.94 GeV) [1]

0.12 GeV [1] ((2 × 1.94) − (4 × 0.94) GeV) [1]

b Momentum is always conserved so the total momentum after the collision is not zero as the initial momentum is not zero. [1] The total energy is insufficient to provide the kinetic energy associated with the momentum and the rest energy of the 3 protons and the antiproton after the collision. [1]

2.3

1 Q: −1, +1, 0, −1   L: +1, −1, −1, 0 (all 6 correct [3], 5 correct [2], 4 correct [1])

2 a $\mu^- \rightarrow e^- + \nu_\mu + \bar{\nu}_e$ [2]

b Electron lepton numbers are conserved:
0 = 1 + 0 + −1 [1]

Muon lepton numbers are conserved:
+1 = 0 + +1 +0 [1]

(Electron and muon lepton numbers are both conserved.)

3 a No [1] charge is not conserved in the change. [1]

b No [1] electron lepton number is not conserved because the total is 1 before the change and −1 after the change. [1]

c No [1] electron lepton number is not conserved because the total is +1 before the change and −1 after the change. [1]

d No [1] electron lepton number is not conserved because the total is 0 before the change and −2 (−1 for the positron and −1 for the electron antineutrino) after the change. [1]

Note muon lepton number is conserved because the antimuon and the antimuon neutrino both have a lepton number of −1.

2.4

1 a i $u\bar{d}$, S = 0 [1]

ii $ud\bar{d}$, S = 0 [1]

iii $u\bar{s}$, S = 1 [1]

b A K⁻ meson consists of an up antiquark and a strange quark. [1] So the antiparticle of a K⁻ meson consists of an up quark and a strange antiquark. This combination is a K⁺ meson. [1]

2 a See Topic 2.4 Fig 2a (u and d quarks and W⁻ boson correctly labelled [1], β⁻ and antineutrino correctly shown [1])

b A down quark in a neutron emits a W⁻ boson and changes to an up quark [1] so the neutron changes to a proton. [1] The W⁻ quark decays into a β⁻ particle (i.e., an electron) and an electron antineutrino. [1]

3 a S = −1 so the Σ⁻ contains one strange quark. [1] Because the charge of the Σ⁻ is −1 and the charge of the strange quark is $-\frac{1}{3}$ and there are two more quarks in the Σ⁻ particle, [1] these two quarks must have a combined charge of $-\frac{2}{3}$ so they must be down quarks. Therefore, the Σ⁻ particle must be composed of a strange quark and two down quarks. [1]

b The total initial strangeness is zero because the initial particles are non-strange particles. [1] The strangeness of the K⁺ is +1 by definition. The Σ⁻ particle has strangeness −1. So the total final strangeness is zero. [1] Therefore, as the total initial strangeness is equal to the total final strangeness, strangeness is conserved. [1]

2.5

1 a The charge and the baryon number of an antiproton are both −1. Table 1 shows that both Q and B are conserved because the total of each is the same before and after the reaction. [2]

▼ **Table 1**

	Before	After
	p + p	p + p + p + p̄
Q	1 + 1	1 + 1 + 1 − 1
B	1 + 1	1 + 1 + 1 − 1

b The total rest energy before the change is 2 × 0.94 GeV. The total after the change is 4 × 0.94 GeV. [1] Therefore, the two protons that collide must have at least 2 × 0.94 GeV of kinetic energy before the change. So each of these protons must have 0.94 GeV of kinetic energy. [1]

2 a i $-\frac{1}{3}$ **ii** 0

iii +1 (all 3 correct [2], 2 correct [1])

b i 1   **ii** −1

iii 0 (all 3 correct [2], 2 correct [1])

3 a uds [1] Its strangeness of −1 means it contains 1 strange quark. [1] Since it is uncharged, the total charge of the other 2 quarks present must be equal and opposite to the charge of the strange quark (or −1/3). Therefore, the 2 other quarks must be an up quark and a down quark. [1]

b Table 2 below shows the numbers for charge Q, baryon number B, and strangeness S. The results show that X is uncharged and is not a baryon. [1] It does have non-zero strangeness so it cannot be a lepton. [1] Therefore, it must be a strange uncharged meson. [1] Because its strangeness is +1 and it is uncharged, it must consist of a strange antiquark and a down quark. So it is a K⁰ meson [1]

▼ **Table 2**

	Before		After
	π⁻ + p	→	Λ⁰ + X
Q	−1 + 1		0 + 0
B	0 + 1		1 + 0
S	0 + 0		−1 + 1

3.1

1 The work function is the minimum energy an electron must have to escape from the surface of the metal. [1]

2 When a conduction electron at or near the surface of a metal absorbs a photon from the incident light, it gains energy equal to the photon energy hf. [1] The conduction electron can only escape from the metal if its energy is greater than or equal to the work function ϕ of the metal. [1] Therefore, in order for the electron to escape, $hf \geq \phi$. Therefore, the frequency of the incident radiation must be greater than or equal to $\dfrac{\phi}{h}$. [1]

3 a The work function of the metal is equal to the energy of a photon of wavelength 390 nm. Therefore, the work function is equal to 5.1×10^{-19} J. [1]

   b A photon of wavelength more than 390 nm will have a frequency and therefore an energy less than a photon of wavelength 390 nm. [1] Therefore, a photon of wavelength less than 390 nm will not give a conduction electron enough energy to overcome the work function of the metal and escape. [1]

3.2

1 The stopping potential of a metal is the minimum potential needed to stop photoelectric emission from the metal. [1]

2 a i Photoelectric emission of electrons takes place from the cathode because each emitted electron absorbs a single photon and gains enough energy to leave the cathode. [1] The emitted electrons reach the anode and flow through the ammeter round the circuit. The microammeter current is due to this flow of electrons round the circuit. [1]

   ii The flow of charge Q in 1 second = current × time = $0.85\,\mu A \times 1$ second = $0.85\,\mu C$ [1]

   The number of electrons per second
   $= \dfrac{\text{charge flow } \overline{Q}}{e} = \dfrac{0.85 \times 10^{-6}}{1.60 \times 10^{-19}} = 5.3 \times 10^{12}\,s^{-1}$ [1]

   b The current through the microammeter is proportional to the intensity of the incident light. [1] This is because the number of electrons from the cathode is proportional to the number of photons incident on the cathode. [1] As the light intensity is reduced to zero the current decreases to zero in proportion to the light intensity. [1]

3 a Photon energy $= \dfrac{hc}{\lambda} = \dfrac{6.63 \times 10^{-34} \times 3.00 \times 10^8}{430 \times 10^{-9}}$

   $= 4.63 \times 10^{-19}\,J = \dfrac{4.63 \times 10^{-19}}{1.60 \times 10^{-19}} = 2.89\,eV$ [1]

   b Maximum kinetic energy of a photoelectron $= hf - \phi = 2.89 - 1.30\,eV$ [1] $= 1.59\,eV$ [1]

3.3

1 Ionisation is the process of adding or removing an electron from an uncharged atom so the atom becomes charged. [1] Excitation is a process in which an electron in an atom gains sufficient energy to move away from the nucleus to an electron shell further away from the nucleus. [1]

2 a $2.18 \times 10^{-18}\,J$ [1] $(= 13.6 \times 1.60 \times 10^{-19}\,J)$

   b $1.6\,eV$ [1] $(= 16.2 - 1.0 - 13.6\,eV)$ [1]

3 a The external electron collides with an electron in one of the electron shells of the atom and gives the atomic electron enough kinetic energy [1] to move to an electron shell further away from the nucleus. [1]

   b The slow-moving electron does not have enough kinetic energy to enable any of the atomic electrons to move to an outer shell. So it cannot cause excitation of the atom. [1]

3.4

1 a $\approx 7.2\,MeV$ [1]

   b i 10. [1] 4 from the 7.2 MeV level to each lower level, 3 from the next level down to each lower level, 2 from the next level below to each lower level, and 1 from the 1st excited level to the ground state. (all correct [2], all except one correct [1])

   ii The photon emitted when the atom de-excites to the ground state from the 7.2 MeV energy level. [1]

3.5

1 Line emission spectrum is a spectrum consisting of discrete lines of different wavelengths. It is produced when gas atoms become excited and then de-excite and emit photons. [1] The photons have specific energies because an excited gas atom has different energy levels. [1] Each photon is released when an electron in the excited atom moves to a lower energy level in the atom. [1]

2 a Photon energy $= \dfrac{hc}{\lambda} = \dfrac{6.63 \times 10^{-34} \times 3.00 \times 10^8}{656 \times 10^{-9}}$
   $= 3.03 \times 10^{-19}\,J$ [1]

   b The energy of a 486 nm photon
   $= \dfrac{hc}{\lambda} = \dfrac{6.63 \times 10^{-34} \times 3.00 \times 10^8}{486 \times 10^{-9}} = 4.09 \times 10^{-19}\,J$ [1]
   Therefore, the energy difference ΔE between X and Y $= 4.09 \times 10^{-19}\,J - 3.03 \times 10^{-19}\,J = 1.06 \times 10^{-19}\,J$ [1]

   The wavelength of a photon released in an electron transition from X to Y
   $= \dfrac{h}{\Delta E} = \dfrac{6.63 \times 10^{-34} \times 3.00 \times 10^8}{1.06 \times 10^{-9}} = 1.88 \times 10^{-6}\,m$ [1]

3.6

1 Matter particles have a wave-like nature as well as a particle-like nature. [1] Their wave-like behaviour is characterised by their de Broglie wavelength which is related to their momentum, p, by means of the equation $\lambda = \dfrac{h}{p}$. [1]

2 a de Broglie wavelength $\lambda = \dfrac{h}{mv}$
   $= \dfrac{6.63 \times 10^{-34}}{9.11 \times 10^{-31} \times 2.0 \times 10^7}$ [1] $= 3.64 \times 10^{-11}\,m$ [1]

b speed $v = \dfrac{h}{m\lambda} = \dfrac{6.63 \times 10^{-34}}{1.67 \times 10^{-27} \times 3.64 \times 10^{-11}}$ [1]

$= 1.09 \times 10^4 \, \text{ms}^{-1}$ [1]

4.1

1 a electromagnetic; transverse

b electromagnetic; transverse

c mechanical; longitudinal

d mechanical; longitudinal (all correct [3], 3 correct [2], 2 correct [1])

2 a A longitudinal wave is a wave in which the vibrations of the particles are parallel to the direction in which the wave travels [1]

b Each time the end of the coil moved forwards, the coils at that end are compressed and they push on the nearby coils further along the slinky. [1] When the end of the slinky is pulled backwards, the coils at the end move backwards and create a rarefaction, pulling the nearby coils back which pulls the coils back further away. [1] As a result, the earlier compression is followed by a rarefaction. Repeating the process continually sends a series of compressions and rarefactions along the slinky. [1]

3 a 1 The vibrations of a transverse wave are perpendicular to the direction in which the wave travels whereas the vibrations of a longitudinal wave are parallel to the direction in which the wave travels.

2 Transverse waves can be polarised whereas longitudinal waves cannot. [2]

b i Transverse wave shown and direction correct. [1]

ii X on a crest and direction correctly shown. [1]

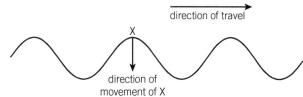

▲ **Figure 1**

4.2

1 a $f = \dfrac{c}{\lambda} = \dfrac{3.00 \times 10^8}{320 \times 10^{-9}} = 9.4 \times 10^{14} \, \text{Hz}$ [1]

b $\lambda = \dfrac{c}{f} = \dfrac{1500}{20 \times 10^3} = 0.075 \, \text{m}$ [1]

2 a The frequency of a progressive wave is the number of complete waves passing a point per second. [1]

b Period $= \dfrac{1}{f} = \dfrac{1}{1.6} = 0.63 \, \text{s}$ (to 2 significant figures) [1]

$\lambda = \dfrac{c}{f} = \dfrac{2.9}{1.6} = 1.8 \, \text{m}$ [1]

3 a i 30° $\left(= \dfrac{360°}{12} \text{ as there are 12 intervals from O to} \right.$ S which are in phase with each other $\left.\right)$

ii $\dfrac{\pi}{6}$ (or 0.52) rad [1]

b π rad $\left(= 6 \text{ intervals} \times \dfrac{\pi}{6} \right)$ [1]

c 1 They have the same amplitude and this is unchanged half a cycle later. [1]

2 Their phase difference is π rad $\left(= 6 \text{ intervals} \times \dfrac{\pi}{6} \right)$ and this is constant so is unchanged. [1]

3 P is at negative maximum displacement and R is at positive maximum displacement. Half a cycle later, P is at positive maximum displacement and R is at negative maximum displacement. [1]

4.3

1 a See Figure 1 [1]

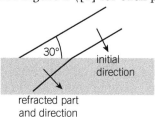

▲ **Figure 1**

2 a/b See Figure 2 ([1] for each part correct)

▲ **Figure 2**

4.4

1 a The principle of superposition states that when two waves meet, the total displacement at a point is equal to the sum of the individual displacements at that point. [1]

b The crests and troughs of a progressive wave travel along the rope whereas the crests and troughs of a stationary wave do not move along the rope. [1]

2 If the frequency is increased, the wavelength of the waves decreases. [1] The points of cancellation and reinforcement therefore become closer together. [1]

3 a Coherent sources emit waves of the same frequency with a constant phase difference. [1]

b The receiver signal is a maximum at points along the line where the microwaves from one slit arrive in phase with the microwaves from the other slit so they reinforce each other. [1] These points are regularly spaced along XY. [1] The receiver signal is a minimum midway between the maxima because at the minima, the microwaves from each slit are out of phase by 180° so they cancel each other out. [1]

4.5

1 a Two progressive waves form a stationary wave pattern when they pass through each other if they have the same frequency and they travel in opposite directions through each other. [1]

   b i $\lambda = 2L = 0.60 \times 2 = 1.20\,m$ (because the string has only 2 nodes which are at its ends. Therefore, the string length $L = \dfrac{\lambda}{2}$ where λ is the wavelength.) [1]

   ii $v = f\lambda = 150 \times 1.20 = 180\,m\,s^{-1}$ [1]

2 a The amplitude of a progressive wave is the same at all positions along the wave. [1] The amplitude of a stationary wave varies with position and is greatest at the antinodes and zero at the nodes. [1]

   b The phase difference between two different points along a progressive wave increases with the distance between the two points from zero to 2π for each cycle of the wave. [1]

   For a stationary wave, the phase difference is π for any two points separated by an odd number of nodes and zero for any two points separated by an even number of nodes. [1]

3 Sound waves travelling along the pipe reflect at the closed end and the reflected waves pass through the sound waves from the loudspeaker. [1] The two sets of waves have the same frequency and therefore form a stationary wave pattern in the pipe which makes the pipe resonate with sound. [1]

   This happens at specific frequencies which produce a node at the closed end of the pipe and an antinode just outside the open end. [1]

4.6

1 a $\lambda = 2L = 2 \times 0.76\,m = 1.52\,m$ [1]

   b Rearranging $f_1 = \dfrac{1}{2L}\sqrt{\dfrac{T}{\mu}}$ gives

   $\mu = \dfrac{T}{(2Lf_1)^2} = \dfrac{64}{(2 \times 0.760 \times 280)^2}$ [1]
   $= 3.5(3) \times 10^{-4}\,kg\,m^{-1}$ [1]

2 As T and μ are unchanged the equation $f_1 = \dfrac{1}{2L}\sqrt{\dfrac{T}{\mu}}$ means that $f \times L = $ constant. [1]

   Therefore, the length L for a frequency of $384\,Hz$ is given by the equation $384L = 280 \times 0.760$ [1]

   Hence, $L = \dfrac{280 \times 0.760}{384} = 0.554\,m$ [1]

3 Mass per unit length $\mu = $ density $\rho \times$ area of cross section $\left(\dfrac{\pi d^2}{4}\right)$ where d is the diameter.

   Therefore, $\dfrac{\mu_x}{\mu_y} = \dfrac{\dfrac{\rho_x \pi d_x^2}{4}}{\dfrac{\rho_y \pi d_y^2}{4}} = \dfrac{8\rho_y}{\rho_y} \times \dfrac{\pi(0.5d_y^2)}{\dfrac{\pi d_y^2}{4}} = 8 \times (0.5)^2$
   $= 2.0$ [1]

   Because tension T and length L are constant, the equation

   $f_1 = \dfrac{1}{2L}\sqrt{\dfrac{T}{\mu}}$ means that $f \times \mu^{\frac{1}{2}} = $ constant [1]

   Therefore, $f_y = f_x \times \sqrt{\dfrac{\mu_x}{\mu_y}} = 100 \times \sqrt{2.0} = 141\,Hz$ [1]

4.7

1 $0.52\,V (= 0.5 \times 5.2\,cm \times 0.2\,V\,cm^{-1})$ [1]

2 Period $T = \dfrac{5.0\,ms\,cm^{-1} \times 6.4\,cm}{4} = 8.0\,ms$ [1]

   Therefore, $f = \dfrac{1}{T} = \dfrac{1}{8.0\,ms} = 130\,N$ (2 s.f) [1]

5.1/5.2

1 $n_1 \sin\theta_1 = n_2 \sin\theta_2$ gives $1\sin 30.0 = 1.52\sin\theta_2$. [1]
   Therefore, $\sin\theta_2 = \dfrac{\sin 30.0}{1.52}$ so $\theta_2 = 19.2°$ [1]

2 $n_1 \sin\theta_1 = n_2 \sin\theta_2$ gives $1.52\sin\theta_1 = 1\sin 70.0$ [1]
   Therefore, $\sin\theta_1 = \dfrac{\sin 70.0}{1.52}$ so $\theta_1 = 38.2°$ [1]

3 a The light ray was not refracted because it entered the block along the normal. [1]

   b At C, $\theta_1 = 35°$. Therefore, $n_1 \sin\theta_1 = n_2 \sin\theta_2$ gives $1.54\sin 35 = 1\sin\theta_2$. Therefore, $\sin\theta_2 = 1.54\sin 35$ [1] so $\theta_2 = 62°$. [1] So the deflection of the light ray $= 62 - 35 = 27°$. [1]

4 $n_1 \sin\theta_1 = n_2 \sin\theta_2$ gives $1.33\sin 40.0 = 1.50\sin\theta_2$. [1]
   Therefore, $\sin\theta_2 = \dfrac{1.33\sin 40.0}{1.50} = 0.570$ so $\theta_1 = 34.7°$. [1]

5.3

1 a $n_1 \sin\theta_1 = n_2 \sin\theta_2$ gives $1.54\sin c = 1\sin 90$. [1]
   Therefore, $\sin c = \dfrac{1}{1.54} = 0.649$ so $c = 40.5°$ [1]

   b $n_1 \sin\theta_1 = n_2 \sin\theta_2$ gives $1.54\sin c = 1.33\sin 90.0$ [1]
   Therefore, $\sin c = \dfrac{1.33}{1.54} = 0.864$ so $c = 59.7°$ [1]

2 a The cladding is designed such that where two fibres are in contact, their cores are not in contact. If the cores were in contact, light would pass between the two fibres, causing signals to be insecure and also light loss. [1] The cladding needs to have a lower refractive index than the core so that total internal reflection of light rays in the core can occur at the boundary. [1] The cores need to be narrow so the difference in the distance travelled by the axial rays and non-axial rays is as little as possible otherwise light pulses would become longer as they travel along the fibre. [1]

   b $n_1 \sin\theta_1 = n_2 \sin\theta_2$ gives $1.51\sin c = 1.42\sin 90.0$ [1]
   Therefore, $\sin c = \dfrac{1.42}{1.51} = 0.940$ so $c = 70.1°$ [1]

3 a A coherent bundle of fibres is a bundle in which the fibre ends are in the same relative positions at each end of the bundle. [1]

   b Two fibre bundles are needed because one bundle is used to send light into the body cavity to illuminate it and the other bundle, the coherent bundle, is used to observe the cavity. [1] A lens over the end of the coherent bundle forms an image of the body cavity on the end of the coherent bundle. The image is observed at the other end of the coherent bundle. [1]

5.4

1 a The fringe spacing increases so the fringes become further apart. [1]

b The fringes would be blue and they would be closer together. [1]

2 Fringe spacing $= \dfrac{4.5\,\text{mm}}{4} = 1.125\,\text{mm}$ [1]

$\lambda = \dfrac{ws}{D} = \dfrac{1.125\times10^{-3}\times0.40\times10^{-3}}{0.85}$ [1] $= 5.30 \times 10^{-7}\,\text{m}$

$= 530\,\text{nm}$ (2 s.f.) [1]

3 The fringes would be a different colour because the wavelength is different. [1] The fringes would be further apart because the fringe spacing is larger for longer wavelengths. [1]

5.5

1 a Two or more light sources are said to be coherent sources if they emit light waves of the same frequency with a constant phase difference. [1] Note: A laser is a coherent light source because it emits light waves in phase with one another.

b A lamp filament emits light waves at random so the light waves from two filament lamps can never have a constant phase difference. [1] Therefore, two filament lamps can never produce a double slit interference pattern, as the points of cancellation and reinforcement would move about at random. [1]

2 $w = \dfrac{\lambda D}{s} = \dfrac{635\times10^{-9}\times1.90}{0.45\times10^{-3}}$ [1] $= 2.7\times10^{-3}\,\text{m}$ [1]

3 The fringes all become brighter and the colour of the central fringe changes from red to white. [1] The fringe separation decreases [1] and each non-central fringe becomes white in the middle and tinged with blue on the side nearest the central fringe and red on the other side. [1]

5.6

1 a The pattern consists of alternate bright and dark fringes. [1] The central bright fringe is brighter and twice as wide as each of the other bright fringes [1] which are less and less bright the further they are from the central fringe. [1]

b See Topic 5.6 Figure 1. (Central max is at least 3 times higher than the 1st max either side [1], central fringe is twice the width of the other bright fringes [1])

2 a The bright fringes become less intense because less light passes through the narrower slit and the diffracted light spreads out more as they diffract more. [1] The increase in diffraction also causes the bright fringes to become wider and further apart. [1]

b The colour of the fringes changes from red to blue and the amount of diffraction decreases as the wavelength of blue light is less than that of red light. [1] As a result of the decrease in diffraction, the bright fringes become closer together and narrower. [1]

5.7

1 a The grating has 600 000 lines per metre. Hence, $d = \dfrac{1}{600\,000}\,\text{m} = 1.67\times10^{-6}\,\text{m}$ [1]

Since $\lambda = d\sin\theta$ for the 1st order beam, $\sin\theta = \dfrac{\lambda}{d}$

$= \dfrac{630\times10^{-9}}{1.67\times10^{-6}} = 0.377$ [1] which gives $\theta =$ 22° (2 s.f.) [1]

b maximum order number $= \dfrac{1.67\times10^{-6}}{630\times10^{-9}} = 2.65$ [1] which rounded down $= 2$ [1]

2 $\lambda = d\sin28°$ for the 1st order beam so the maximum order number $= \dfrac{d}{\lambda}$ rounded down ($= 1 \div \sin28°$)

$= 2.13$ rounded down $= 2$ [1]

The angle of diffraction of the 2nd order beam is given by the equation $2\lambda = d\sin\theta_2$ hence $\sin\theta_2 = 2\lambda/d$ $= 2/2.13 = 0.9389$ [1] which gives $\theta_2 = 70°$ [1]

6.1

1 a See Figure 1. [1]

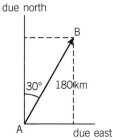

▲ Figure 1

b i Component due East $= 180\sin30 = 90\,\text{km}$

ii Component due North $= 180\cos30 = 156\,\text{km}$ [1]

2 a Resultant $= 11.0 - 5.0 = 6.0\,\text{N}$ in the direction of the 11 N force [1]

b See Figure 2 for a sketch diagram.

Magnitude of the resultant $= \sqrt{F_1^2 + F_2^2}$

$= \sqrt{5.0^2 + 11.0^2} = 12.1\,\text{N}$ [1]

Direction of resultant is given by $\tan\theta = \dfrac{F_2}{F_1}$ which

gives $\tan\theta = \dfrac{11.0}{5.0} = 2.20$ hence $\theta = 66°$ (2 s.f.) [1]

between F_1 and the resultant.

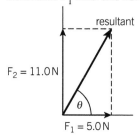

▲ Figure 2

3 See Figure 3 for a sketch diagram.

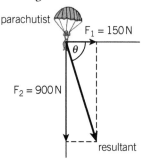

▲ Figure 3

Magnitude of the resultant
$= \sqrt{F_1^2 + F_2^2} = \sqrt{150^2 + 900^2} = 912\text{N}$ [1]

Direction of resultant is given by $\tan\theta = \dfrac{F_2}{F_1}$ [1] which gives $\tan\theta = \dfrac{900}{150} = 6.00$ hence $\theta = 80.5°$
between F_1 (which is horizontal) and the resultant. [1]

6.2

1 **a** Magnitude of resultant $= 6.6\,\text{N} = \sqrt{3.2^2 + 5.8^2}\,\text{N}$ [1]

Angle of resultant below the horizontal
$= 29°$ [1] $\left(= \tan^{-1}\dfrac{3.2}{5.8}\right)$ [1]

b The third force must be equal and opposite to the resultant in **a** and therefore the third force must be 6.6 N in a direction of 29° above the horizontal (i.e. in the opposite direction to the resultant in part **a**). [1]

2 Resolving W parallel and perpendicular to the slope correctly [1] gives:

a horizontally: $F = W\sin\theta = 8.6\sin 40 = 5.5\,\text{N}$ [1]

b vertically: $S = W\cos\theta = 8.6\cos 40 = 6.6\,\text{N}$ [1]

Note: The lines of action of the three forces in Figure 3 have no overall turning effect because they act through the same point.

3 See Figure 1.

Resolving T_1 and T_2 vertically and horizontally gives:

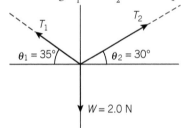

Horizontally: $T_1\cos 35 = T_2\cos 30$ therefore
$T_1 = \dfrac{T_2\cos 30}{\cos 35} = 1.057\,T_2$ [1]

Vertically: $T_1\sin 35 + T_2\sin 30 = W$ therefore
$1.057\sin 35\,T_2 + 0.500\,T_2 = W$ [1]

Therefore, $0.606\,T_2 + 0.500\,T_2 = W$ which gives
$1.106\,T_2 = 2.0$ so $T_2 = \dfrac{2.00}{1.106} = 1.8(1)\,\text{N}$ [1] and
$T_1 = 1.057\,T_2 = 1.9(1)\,\text{N}$ [1]

6.3

1 **a** See Figure 1 (figure not to scale). [1]

▲ **Figure 1**

b Taking moments about the pivot:

Sum of the clockwise moments
$= W \times (0.90 - 0.50) = 0.40\,W$ [1]

Sum of the anticlockwise moments
$= [2.0 \times (0.50 - 0.10)] + [3.0 \times (0.50 - 0.25)]$
$= 0.80 + 0.75$ [1]

Applying the principle of moments gives
$0.40\,W = 0.80 + 0.75 = 1.55$
Hence, $W = \dfrac{1.55}{0.40} = 3.9\,\text{N}$ (to 2 significant figures) [1]

2 See Figure 2 for a sketch of the arrangement. Assume the force F of the adult on the seesaw is vertically downwards.

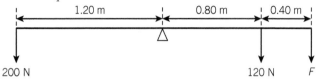

▲ **Figure 2**

Taking moments about the pivot,

Sum of the clockwise moments $= (120 \times 0.80) + 1.20\,F = 96.0 + 1.20\,F$ [1]

Sum of the anticlockwise moments $= (200 \times 1.20) = 240\,\text{N m}$ [1]

Applying the principle of moments gives
$96.0 + 1.20\,F = 240$

Therefore, $1.20\,F = 240 - 96 = 144$ which gives
$F = \dfrac{144}{1.20} = 120\,\text{N}$ (to 2 significant figures) [1]

3 See Figure 3 where F is the force to be calculated [1]

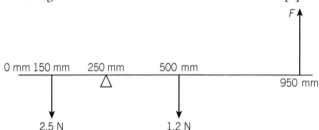

Taking moments about the pivot,

Sum of the clockwise moments
$= 1.2 \times (0.500 - 0.250) = 0.300\,\text{N m}$ [1]

Sum of the anticlockwise moments
$= [2.5 \times (0.250 - 0.150)] + [F \times (0.950 - 0.250)]$
$= 0.250 + 0.70\,F$ [1]

Applying the principle of moments gives
$0.70\,F + 0.250 = 0.300$

Hence, $0.70\,F = 0.300 - 0.250 = 0.050$ therefore
$F = \dfrac{0.050}{0.70} = 0.071\,\text{N}$ (to 2 significant figures) [1]

6.4/6.5

1 See Figure 1 [1]

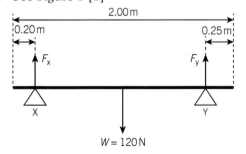

▲ **Figure 1**

Taking moments about Y to find the force F_x at X,

Sum of the clockwise moments $= F_x \times 1.55$ since XY
$= (2.00 - 0.25 - 0.20)$ [1]

Sum of the anticlockwise moments
= 120 × (1.00 − 0.25) = 90 Nm [1]

Applying the principle of moments gives $1.55F_x = 90$

Hence, $F_x = \dfrac{90}{1.55} = 58\,N$ (to 2 significant figures) [1]

To find the force F_y at Y, since $F_x + F_y = W$, then
$F_y = W − F_x = 120 − 58 = 62\,N$ [1]

2 Taking moments about Q to find the force F_p at P, since the perpendicular distances from Q of the forces are 11.0 m for the force at P, 5.5 m for the weight of the span (6400 N) and 7.0 m (= 11.0 − 4.0 m) for the load weight (500 N) then:

Sum of the clockwise moments = $F_p × 11.0$ [1]

Sum of the anticlockwise moments
= (500 × (11.0 − 4.0)) + (6400 × 5.5) = 38 700 Nm

Applying the principle of moments gives
$11.0F_p = 38\,700$ [1]

Hence, $F_p = \dfrac{38\,700}{11.0} = 3520\,N$ (to 3 significant figures) [1]

The total weight on the pillars = 6400 + 500 = 6900 N

To find the force F_q at Q, since $F_p + F_q = 6900$,
then $F_q = W − F_p = 6900 − 3520 = 3380\,N$ [1]

3 a See Figure 2. The force on the board at A acts vertically downward to stop the board turning about P and tipping into the pool. [1]

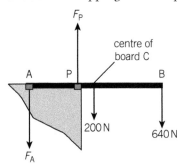

▲ Figure 2

b i The perpendicular distance of the force from P is PB = 1.50.

AP = AB − PB = 2.50 − 1.50 = 1.00 m
PC = AC − AP = 1.25 − 1.00 = 0.25 m

Taking moments about P to find the force at A, F_A, gives:

Sum of the clockwise moments = (200 N × 0.25 m) + (640 N × 1.50 m) = 1010 Nm [1]

Sum of the anticlockwise moments = $F_A × 1.00\,m$ [1]

Applying the principle of moments gives
$F_A = \dfrac{1010\,N\,m}{1.00\,m} = 1010\,N$ downwards. [1]

ii To find the force F_p at P use
$F_p = F_A + 200\,N + 640\,N = 1010 + 840$
= 1850 N [1]

4 A sidewind exerts a horizontal force on the side of the lorry. The line of action of this force is above the ground and so the force has a turning effect about the points where the wheels on the other side of the lorry are in contact with the road. [1] The turning

effect makes the lorry unstable as it could make the lorry topple over. [1]

5 Pulling a drawer out shifts the centre of mass of the cabinet in the direction in which the drawer is pulled out. [1] Pulling a second drawer out would shift the centre of mass even further and may even cause the line of action of the weight of the filing cabinet to act beyond the base of the cabinet. [1] If this happens, the moment of the weight about the edge of the filing cabinet would be the only moment acting on the filing cabinet so the cabinet would topple over. [1]

6 The maximum angle of the slope for no toppling is such that the centre of mass is directly above the lower wheel, as shown in Figure 3. [1] At this position, the angle of inclination $θ$ of the slope is given by $\tan θ = \dfrac{0.5 × 1.7}{0.9} = 0.944$ [1] which gives $θ = 43°$. [1]

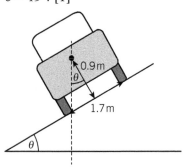

▲ Figure 3

6.6/6.7

1 a See Figure 1. M is the midpoint of the plank. [1]

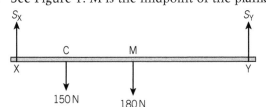

▲ Figure 1

b XY = 4.00 − 0.20 = 3.80 m, XC = 0.90 − 0.10 = 0.80 m

CY = 3.80 − 0.80 = 3.00 m, MY= 2.0 − 0.10 = 1.90 m

Taking moments about Y gives:

Clockwise moments = $(S_x × XY) = 3.80 S_x$ [1]

Anticlockwise moments = (150 N × CY) + (180 N × MY) [1]

= (150 × 3.00) + (180 ×1.90) = 792 Nm

Since the plank is in equilibrium:

the sum of the clockwise moments = the sum of the anticlockwise moments

Therefore, $3.80 S_x = 792$ hence $S_x = \dfrac{792}{3.80} = 208\,N$ [1]

Considering all the force components are vertical therefore

$S_x + S_y = 150 + 180 = 330\,N$
So, $S_y = 330 − 208 = 122\,N$ [1]

2 a See Figure 2a: F = force due to the wall, S = support force due to floor, W = weight [1]

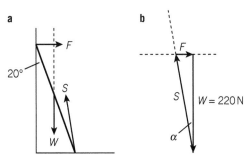

▲ **Figure 2**

b The distance from the top of the ladder to the floor = $L\cos 20$ where L is the length of the ladder. The distance from the bottom of the ladder to the wall = $L\sin 20$.

Taking moments about the bottom of the ladder (length L) gives:

$FL\cos 20$ [1] $= \frac{1}{2}WL\sin 20$ [1] therefore

$$F = \frac{\frac{1}{2}W\sin 20}{\cos 20} = 40\,N$$

Since F and W are perpendicular, $S^2 = F^2 + W^2 = 220^2 + 40^2$ which gives $S = 224\,N$. [1]

Resolving S vertically and horizontally gives $S\cos\alpha = W$ and $S\sin\alpha = F$.

Therefore, $\tan\alpha = \frac{F}{W} = 10.3°$ [1]

Alternative method for part b: S acts through the point where lines of action of W and F intersect on a scale diagram of the ladder at 20° to the wall. The direction of S can therefore be measured from the scale diagram and then used to draw the triangle of forces as in Figure 2b. Using Figure 2b gives $\alpha = 10.5°$ and $S = 224\,N$.

3 a See Figure 3. [1]

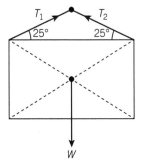

▲ **Figure 3**

b The tension in each section of the cord is the same because their horizontal components are equal and opposite to each other and the angle between each cord and the horizontal is the same (i.e., $T_1\cos 25 = T_2\cos 25$). [1]

The sum of their vertical components is equal and opposite to the weight (i.e., $T_1\sin 25 + T_2\sin 25 = W$). [1]

Hence, $2T_1\sin 25 = W$ so $T_1 = T_2$

$= \frac{W}{2\sin 25} = \frac{21}{2\sin 25} = 25\,N$ [1]

7.1

1 a $28\,m\,s^{-1}$ $\left(= \frac{42\,000\,m}{1500\,s}\right)$ [1]

b Time taken for 2nd part of journey

$= \frac{distance}{speed} = \frac{20\,000\,m}{20\,ms^{-1}} = 1000\,s$ [1]

Average speed $= \frac{total\ distance}{total\ time} = \frac{62\,000\,m}{2500\,s}$ [1]

$= 25\,m\,s^{-1}$ (to 2 significant figures) [1]

2 a Speed $= \frac{2\pi r}{t} = \frac{2\pi \times 6.36 \times 10^6\,m}{24 \times 3600\,s}$ [1] $= 460\,m\,s^{-1}$ (to 2 significant figures) [1]

b i The satellite next passes over the equator in the same direction 2 hours later when it has moved through 360°. [1]

ii In 2 hours, the Earth has rotated through $\frac{1}{12}$th of a rotation hence [1]

$PQ = \frac{1}{12}$th of the circumference $= \frac{1}{12} \times 2\pi r$

$= \pi \times \frac{6360}{6}$ [1] $= 3300\,km$ (to 2 significant figures) [1]

3 a The speed of an object is change of distance per unit time. Velocity is change of displacement per unit time. [1] Displacement is distance in a certain direction. Therefore, velocity is speed in a certain direction. [1]

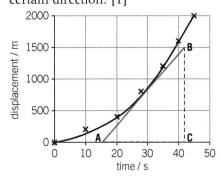

▲ **Figure 1**

b i See Figure 1: [1] suitable scales [1] correctly plotted points [1] best-fit curve.

ii Using the gradient triangle ABC, time taken = AC = 42 − 15 = 27 s, distance moved = BC = 1500 − 0 = 1500 m. [1]

Therefore, speed $= \frac{distance}{time} = \frac{1500\,m}{27\,s} = 56\,m\,s^{-1}$. [1]

7.2

1 $u = 26\,m\,s^{-1}$, $v = 0$, $t = 6.7\,s$

$a = \frac{v-u}{t} = \frac{0-26}{6.7}$[1] $= -3.9\,m\,s^{-2}$ [1]

2 $u = 0$, $v = 1.8 \times 10^7\,m\,s^{-1}$, $t = 51\,ns$

$a = \frac{v-u}{t} = \frac{1.8 \times 10^7 - 0}{51 \times 10^{-9}}$[1] $= 3.5 \times 10^{14}\,m\,s^{-2}$ [1]

3 a i The velocity increased from zero at a decreasing rate. [1]

ii The acceleration decreased gradually from a non-zero value. [1]

b Draw a tangent to the line at 0.5 s and measure the gradient of the tangent which should be $7.4\,cm\,s^{-2}$. [1] The acceleration at 0.5 s is therefore $7.4\,cm\,s^{-2}$ [1]

7.3

1 $u = 0$, $v = 28.0\,\text{m s}^{-1}$, $s = 170$ m, $a = ?$, $t = ?$

  a To find a, use $v^2 = u^2 + 2as$. Rearranging this
   equation gives $2as = v^2 - u^2$
   Therefore, $a = \dfrac{v^2 - u^2}{2s} = \dfrac{28.0^2 - 0^2}{2 \times 170}$ [1] $= 2.3\ \text{m s}^{-2}$ [1]

  b Rearranging $s = \dfrac{(u+v)t}{2}$ to find t gives
   $t = \dfrac{2s}{u+v} = \dfrac{2 \times 170}{0 + 28.0}$ [1] $= 12.1\,\text{s}$ [1]

2 $u = 86\,\text{m s}^{-1}$, $v = 0$, $t = 28\,\text{s}$, $a = ?$, $s = ?$

  a To find s, use $s = \dfrac{(u+v)t}{2}$ to give
   $s = \dfrac{86 + 0}{2} \times 28$ [1] $= 1200$ m [1]

  b To find a, rearrange $v = u + at$ to give
   $a = \dfrac{v - u}{t} = \dfrac{0 - 86}{28}$ [1] $= -3.1\ \text{m s}^{-2}$ [1]

3 The displacement is given by the area under
   the curve. Each block of graph grid represents a
   displacement of 2.0 cm (because the height of each
   block represents a speed of $2.0\,\text{m s}^{-1}$ and the width
   of each block represents a time interval of 1.0 s).
   [1] Counting the number of grid blocks under the
   line from 0 to 2.5 s gives 21.0 ± 0.5 blocks so the
   displacement is 21.0 ± 0.5 cm. [1]

7.4

1 a $u = 0$, $v = ?$, $s = -1.21$ m, $a = -9.81\,\text{m s}^{-2}$, $t = ?$

    i To find v, using $v^2 = u^2 + 2as$ gives
     $v^2 = 0 + (2 \times -9.81 \times -1.21)$ [1] $= 23.7$
     so $v = -4.87\ \text{m s}^{-1} = 4.9\,\text{m s}^{-1}$ to 2 significant
     figures. [1]

    ii To find t, rearranging $v = u + at$ gives
     $t = \dfrac{v - u}{a} = \dfrac{-4.87 - 0}{-9.81}$ [1] $= 0.50\,\text{s}$ [1]

  b $u = ?$, $v = 0$ at maximum height, $s = +0.95$ m,
   $a = -9.81\,\text{m s}^{-2}$, $t = ?$

   To find u, using $v^2 = u^2 + 2as$ gives $0 = u^2 +$
   $(2 \times -9.81 \times 0.95)$ [1] so $u^2 = (2 \times 9.81 \times 0.95)$
   $= 18.6$. Therefore, $u = +4.3\,\text{m s}^{-1}$ to 2 significant
   figures. [1]

2 $u = +2.5\,\text{m s}^{-1}$, $v = ?$, $s = -13$ m, $a = -9.81\,\text{m s}^{-2}$, $t = ?$

   To find t, find v first then find t $\left(\text{rather than}\right.$
   using the quadratic equation $\left. s = ut + \dfrac{1}{2}at^2\right)$.

   Step 1: Using $v^2 = u^2 + 2as$ gives
   $v^2 = 2.5^2 + (2 \times -9.81 \times -13)$ [1] so $v^2 = 261$.
   Therefore, $v = -16.2\,\text{m s}^{-1}$. (− because the
   velocity is downwards) [1]

   Step 2: Rearranging $v = u + at$ gives
   $t = \dfrac{v - u}{a} = \dfrac{-16.2 - 2.5}{-9.81}$ [1] $= 1.9\,\text{s}$ to
   2 significant figures. [1]

3 a Dividing $s = ut + \dfrac{1}{2}at^2$ by t gives $\dfrac{s}{t} = u + \dfrac{1}{2}at$
    Therefore, a graph of $\dfrac{s}{t}$ against t should give a
    straight line with a y-intercept equal to u and a
    gradient $\dfrac{1}{2}a$. [1]
    As $u = 0$, the y-intercept should therefore be zero
    and the line should go through the origin. [1]

  b i A systematic error in a set of measurements
     is where there is a pattern (or a bias) in the
     measurements that causes them to differ from
     the true measurements. [1]

    ii A systematic error in s would give the same
     constant gradient and a non-zero y-intercept
     [1] whereas a systematic error in t would not
     give a constant gradient as well as a non-
     zero intercept. [1] Since the line is straight
     and does not pass through the origin, there
     is likely to have been a systematic error in
     the measurement of s. [1] and it would give
     a different gradient as well as a different
     y-intercept. [1]

7.5/7.6

1 a See Figure 1. A is at the end of the straight
    section of the line. B is where the line ends where
    the gradient is zero. [2]

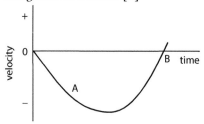

▲ Figure 1

  b The acceleration is constant and negative and equals
   $-g$ from O to A [1] and then it gradually decreases
   to zero [1] and then becomes a deceleration
   reducing the jumper's velocity to zero at B. [1]

2 a i $u = 0$, $v = -2.4\,\text{m s}^{-1}$ (− for downwards),
     $a = -9.81\,\text{m s}^{-2}$, $t = ?$

     To find t, rearranging $v = u + at$ gives
     $t = \dfrac{v - u}{a} = \dfrac{-2.4 - 0}{-9.81}$ [1] $= 0.245\,\text{s} = 0.24\,\text{s}$ to
     2 significant figures [1]

    ii See Figure 2 (straight line from the origin [1],
     velocity and time values marked at 0.245 s [1])

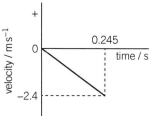

▲ Figure 2

  b Distance fallen = area between the line and the
   time axis in Figure 2
   $= \dfrac{1}{2} \times 2.4 \times 0.245$ [1] $= 0.294$ m [1]
   Note that the equation $v^2 = u^2 + 2as$ can be used
   to calculate the displacement $(= -0.294$ m$)$ which
   gives 0.29 m for the distance fallen to 2 significant
   figures.

3 a $a = +3.7\,\text{m s}^{-2}$, $u = 0$, $t = 30$ s, $v = ?$, $s = ?$

   To calculate v, using $v = u + at$ gives
   $v = 0 + 3.7 \times 30$ [1] $= 110\,\text{m s}^{-1}$ (2 s.f.) [1]

To calculate s, using $s = ut + \frac{1}{2}at^2$ gives

$s = 0 + 0.5 \times 3.7 \times 30^2$ [1] $= 1670$ (2 s.f.) m [1]

b i After the engines were switched off, $a = -9.81\,\text{m s}^{-2}$, $u = +111\,\text{m s}^{-1}$, and $v = 0$ at maximum height. To find t, rearranging $v = u + at$ gives $t = \dfrac{v - u}{a} = \dfrac{0 - 111}{-9.81}$ [1] $= 11$ s (2 s.f.) [1]

ii See Figure 3. The graph has two straight line sections OA and AB. [1] The gradient of section OA is $+3.7\,\text{m s}^{-2}$ and A is the point 30 s from zero where the velocity is $111\,\text{m s}^{-1}$.[1] The gradient of section AB is $-9.81\,\text{m s}^{-2}$ and B is the point where $v = 0$ at 41 s from the start. [1]

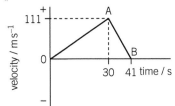

▲ **Figure 3**

c To calculate the velocity just before hitting the ground, consider the motion from 30 s after launch to the ground. For this section of the flight, $a = -9.81\,\text{m s}^{-2}$, $u = +111\,\text{m s}^{-1}$, and $s = -1665\,\text{m}$ (to the ground). Using the equation $v^2 = u^2 + 2as$ gives $v^2 = 111^2 + (2 \times -9.81 \times -1665)$ [1] $= 1.23 \times 10^4 + 3.23 \times 10^4 = 4.50 \times 10^4\,\text{m}^2\,\text{s}^{-2}$. [1] Therefore, $v = -210\,\text{m s}^{-1}$ (2 s.f.). (– for downwards) [1]

Note: The speed just before impact can also be calculated by adding the loss of potential energy per unit mass due to a height difference of 1665 m to the kinetic energy per unit mass at a speed of $111\,\text{m s}^{-1}$ to give the kinetic energy per unit mass and hence the speed just before impact.

7.7/7.8

1 a Consider the vertical motion: $u_y = 0$, $t = 3.1$ s, $a_y = -9.81\,\text{m s}^{-2}$ (– for downwards)

Using $y = \frac{1}{2}a_y t^2$ gives $y = \frac{1}{2} \times -9.81 \times 3.1^2 = -47$ m [1]

Therefore, the clifftop is 47 m above the sea. [1]

b Consider the horizontal motion: $u_x = 16\,\text{m s}^{-1}$, $a_x = 0$, $t = 3.1$ s

Therefore, the horizontal displacement $x = u_x t = 16 \times 3.1 = 50$ m [1]

c At $t = 3.1$ s, vertical component of velocity $v_y = u_x + a_y t = 0 - 9.81 \times 3.1$[1] $= -30.4\,\text{m s}^{-1}$ (– for downwards) [1]

Therefore, impact speed $\sqrt{v_x^2 + v_y^2} = \sqrt{16^2 + 30.4^2} = 34$ m s⁻¹ [1]

2 a Consider the vertical motion: $u_y = 0$, $t = 7.5$ s, $a_y = -9.81\,\text{m s}^{-2}$ (– for downwards)

Using $y = \frac{1}{2}a_y t^2$ gives $y = \frac{1}{2} \times -9.81 \times 7.5^2$ [1] $= -280$ m to 2 significant figures.

Therefore, the aircraft is 280 m above the sea. [1]

b Assuming the drag force is negligible, the horizontal component of the parcel's velocity will be equal to the aircraft's velocity so the parcel will be directly below the aircraft as it falls. [1]

If the drag force is not negligible, the horizontal component of the drag force will reduce the horizontal component of the parcel's velocity [1] so its horizontal displacement at the point of impact will be less than that of the aircraft and the horizontal position of the parcel will fall behind the plane's horizontal position. [1]

3 a Horizontal component $u_x = 29.0 \cos 15.0°$ $= 28.0\,\text{m s}^{-1}$

Vertical component $u_y = 29.0 \sin 15.0° = 7.51\,\text{m s}^{-1}$ [1]

b Consider the vertical motion: $u_y = 7.51\,\text{m s}^{-1}$, $y = -2.60$ m, $a_y = -9.81\,\text{m s}^{-2}$

(– for downwards)

i To calculate the vertical component of velocity, v_y, using $v_y^2 = u_y^2 + 2a_y y$, gives:

$v_y^2 = 7.51^2 + (2 \times -9.81 \times -2.60)$ [1] $= 56.4 + 51.0 = 107.4\,\text{m}^2\,\text{s}^2$

Therefore, $v_y = -10.4\,\text{m s}^{-1}$[1]

ii To calculate the time taken t rearranging $v_y = u_y + a_y t$ gives:

$t = \dfrac{v_y - u_y}{a_y} = \dfrac{-10.4 - 7.51}{-9.81}$ [1] $= 1.83$ s [1]

c Consider the horizontal motion: $u_x = 28.0\,\text{m s}^{-1}$, $a_x = 0$, $t = 1.83$ s

Therefore, the horizontal displacement $x = u_x t = 28.0 \times 1.83 = 51.2$ m [1]

8.1

1 a $u = 0$, $v = 12\,\text{m s}^{-1}$, $t = 50$ s

acceleration $a = \dfrac{v - u}{t} = \dfrac{12 - 0}{50} = 0.24\,\text{m s}^{-2}$ [1]

b $F = ma = 26\,000 \times 0.24 = 6200$ N (2 s.f.) [1]

c Weight $= mg$

Ratio of resultant force:weight $= \dfrac{ma}{mg} = \dfrac{0.24}{9.81}$ $= 0.024$ to 2 significant figures [1]

2 a $u = 62\,\text{m s}^{-1}$, $v = 0$, $s = 1600$ m

Rearranging $v^2 = u^2 + 2as$ gives

$a = \dfrac{v^2 - u^2}{2s} = \dfrac{0 - 62^2}{2 \times 1600}$ [1] $= -1.2\,\text{m s}^{-2}$ [1]

b $F = ma = 4000 \times -1.2 = -4800$ N

Therefore, the decelerating force = 4800 N [1]

3 a $F = 6000$ N, $m = 0.027$ kg

acceleration $a = \dfrac{F}{m} = \dfrac{6000}{0.027} = 2.2 \times 10^5\,\text{m s}^{-2}$ to 2 significant figures [1]

b i Average speed $= \dfrac{\text{distance}}{\text{time}} = \dfrac{310\,\text{m}}{4.2\,\text{s}}$ $(= 73.8\,\text{m s}^{-1}) = 74\,\text{m s}^{-1}$ to 2 significant figures. [1]

ii During the contact time t assume the ball accelerated from rest to $74\,\text{m s}^{-1}$ with an acceleration of $2.2 \times 10^5\,\text{m s}^{-2}$.

Rearranging $v = u + at$ gives $t = \dfrac{v - u}{a} = \dfrac{74 - 0}{2.2 \times 10^5}$ [1] $= 3.4 \times 10^{-4}$ s [1]

c **1** The golf ball is assumed to have negligible vertical motion which is not the case as the ball leaves the golf club with a vertical component of velocity as well as a horizontal component. [1] The average speed calculation assumes the vertical component of its velocity is negligible. The actual contact time is therefore longer than the calculated value in part **b**. [1]

2 Air resistance is assumed to be negligible which is not the case as air resistance on the golf ball reduces its speed and its range. [1] Without air resistance the average speed would have been greater which would therefore give a longer contact time. [1]

8.2

1 **a** Weight = mg = 1050 × 9.81 = 10 300 N [1]

b $T - mg = ma$ where T is the thrust [1]

$T = mg + ma$ = 10 300 + (1050 × 4.80) = 15 300 N [1]

2 **a** At constant velocity, the tension T in the cable is equal and opposite to the weight.

Therefore, $T = mg$ = 1200 × 9.81 = 11 800 N [1]

b $T - mg = ma$ [1] therefore $T = mg + ma$

Since the lift is accelerating downwards, then $a = -0.50\,\text{m s}^{-2}$. [1]

Therefore, $T = mg + ma$ = 11 800 + (1200 × −0.50) = 11 200 N [1]

3 **a** $u = 0$, $s = 5.0\,\text{m}$, $t = 3.0\,\text{s}$

To calculate a, substitute $u = 0$ into $s = ut + \frac{1}{2}at^2$

to give $s = \frac{1}{2}at^2$ and rearrange the equation to give

$a = \dfrac{2s}{t^2} = \dfrac{2 \times 5.0}{3.0^2}$ [1] = 1.1 m s⁻² (2 s.f.) [1]

b **i** Component of weight acting down the slope = $mg\sin\theta$ = 36 × 9.81 × sin 20 = 120N [1]

ii The resultant force on the skateboarder = $mg\sin\theta - F_0$ where F_0 is the frictional force.

Therefore, $mg\sin\theta - F_0 = ma$ [1] hence $F_0 = mg\sin\theta - ma$ = 120 − (36 × 1.1) = 80 N [1]

8.3

1 **a** Terminal speed = $\dfrac{\text{distance}}{\text{time}} = \dfrac{0.32\,\text{m}}{7.1\,\text{s}}$ = 0.045 m s⁻¹ [1]

b At terminal speed, the drag force = weight of the object [1]

= mg = 0.045 kg × 9.81 N kg⁻¹ = 0.44 N [1]

2 **a** See Figure 1 (OW correct [1]. Beyond W, decrease at decreasing gradient to constant speed [1])

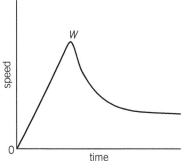

▲ **Figure 1**

b After the ball is released, until it reaches the water surface, it has a constant acceleration equal to g because the air resistance on it is negligible and it accelerates due to the force of gravity on it. When it enters the water, because it is moving fast, it experiences a drag force which is greater than its weight [1] so the resultant force on it is in the opposite direction to its velocity and it decelerates and slows down. [1] As it slows down the drag force on it decreases so the resultant force on it decreases [1] and its deceleration decreases gradually to zero [1] when the drag force is equal to its weight and it moves at terminal speed. [1]

3 **a** maximum acceleration = $\dfrac{\text{maximum engine force}}{\text{mass}} = \dfrac{3800}{22000}$ = 0.17 m s⁻² [1]

b At top speed, the total resistive force = its maximum engine force = 3800N [1]

8.4

1 **a** **i** Thinking distance = initial speed × reaction time = 20 m s⁻¹ × 0.70 s = 14 m (2 s.f.) [1]

ii $u = 20\,\text{m s}^{-1}$, $v = 0$, $a = -5.1\,\text{m s}^{-2}$, braking distance = s to be calculated

To calculate s, rearrange $v^2 = u^2 + 2as$ to give

$s = \dfrac{v^2 - u^2}{2a} = \dfrac{0 - 20^2}{2 \times -5.1}$ [1] = 39 m (2 s.f.) [1]

b Distance from the stopped vehicle to the pedestrian = 58.0 − (14.0 + 39.2) = 4.8 m [1]

2 **a** $u = 28\,\text{m s}^{-1}$, $v = 0$, $s = 71\,\text{m}$

To calculate s, rearrange $v^2 = u^2 + 2as$ to give

$a = \dfrac{v^2 - u^2}{2s} = \dfrac{0 - 28^2}{2 \times 71}$ [1] = −5.5 m s⁻² [1]

b Resultant force = mass × acceleration = 15 000 × −5.5 = 83 000 N

The braking force is 83 000 N [1]

(assuming the resultant force is due to the frictional force only).

3 **a** Stopping distance is the sum of the thinking distance and the braking distance. The braking force on a vehicle on a wet road must be reduced compared with on a dry road otherwise the vehicle will skid. [1] So the braking distance for a vehicle on a wet road is longer than on a dry road for the same initial speed. Therefore, the stopping distance is longer than on a dry road. [1]

b Braking force = $0.57mg$

Deceleration = $\dfrac{\text{braking force}}{\text{mass}} = \dfrac{0.57mg}{m}$ [1] = $0.57g$

= 0.57 × 9.81 = 5.59 m s⁻² [1]

$u = 31\,\text{m s}^{-1}$, $v = 0$, $a = -5.59\,\text{m s}^{-2}$ (− for deceleration)

To calculate s, rearrange $v^2 = u^2 + 2as$ to give

$s = \dfrac{v^2 - u^2}{2a} = \dfrac{0 - 31^2}{2 \times -5.59}$ [1] = 86 m [1]

8.5

1 $u = 12\,\text{m s}^{-1}$, $v = 18\,\text{m s}^{-1}$, $s = 3.0\,\text{m}$

impact time $t = \dfrac{2s}{u + v} = \dfrac{2 \times 3.0}{12 + 18}$ [1] = 0.20 s [1]

acceleration $a = \dfrac{v - u}{t} = \dfrac{18 - 12}{0.20}$ = 30 m s⁻² [1]

impact force $F = ma$ = 1100 × 30 = 33 000 N [1]

2 $u = 3.0\,\text{m s}^{-1}$, $v = 0$, $t = 0.40\,\text{s}$

acceleration $a = \dfrac{v - u}{t} = \dfrac{0 - 3}{0.40} = -7.5\,\text{m s}^{-2}$ [1]

impact force $F = ma = 900 \times 7.5 = 6800\,\text{N}$ [1]

3 a $u = 23\,\text{m s}^{-1}$, $v = 0$, $s = 4.8 + 0.5 = 5.3\,\text{m}$

impact time $t = \dfrac{2s}{u + v} = \dfrac{2 \times 5.3}{23 + 0}$ [1] $= 0.46\,\text{s}$ [1]

acceleration $a = \dfrac{v - u}{t} = \dfrac{0 - 23}{0.46} = -50\,\text{m s}^{-2}$ [1]

impact force $F = ma = 62 \times 50 = 3100\,\text{N}$ [1]

b As the car braked, the passenger would move at a speed of $23\,\text{m s}^{-1}$ until she collided with the inside of the car. [1] The impact time in this collision would be much shorter than the impact time when she was wearing the seat belt. [1] So the force of the impact without the seat belt on would be much larger than the force on her with the seat belt on. [1]

9.1/9.2

1 Change of momentum $= \Delta(mv) = 25\,000 \times 160 = 4.0 \times 10^6\,\text{N s}$ [1]

Force $= \dfrac{\Delta(mv)}{\Delta t} = \dfrac{4.0 \times 10^6}{58} = 69000\,\text{N}$ [1]

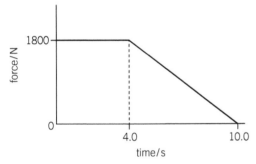

▲ **Figure 1**

2 a See Figure 1 [1]

b Change of momentum = area under line

$= (1800\,\text{N} \times 4.0\,\text{s}) + (0.5 \times 1800\,\text{N} \times 6.0\,\text{s})$

$= 12\,600\,\text{N s}$ [1]

Change of velocity $= \dfrac{\text{change of momentum}}{\text{mass}}$

$= \dfrac{12\,600}{1300} = 9.7\,\text{m s}^{-1}$ [1]

Final velocity $= 20 - 9.7 = 10.3\,\text{m s}^{-1}$ [1]

3 Change of momentum = final momentum − initial momentum

$= (1500 \times 6.5) - (1500 \times 2.0) = 6750\,\text{N s}$ [1]

Force $= \dfrac{\text{change of momentum}}{\text{contact time}} = \dfrac{6750\,\text{N s}}{0.55\,\text{s}}$

$= 12\,000\,\text{N}$ [1]

4 The normal component of its momentum is $+ mu \cos\theta$ before the impact and $-mu \cos\theta$ after the impact.

Change of momentum $= (-mu \cos\theta) - (mu \cos\theta)$

$= -2mu \cos\theta$ [1]

$= -2 \times 5.0 \times 10^{-26}\,\text{kg} \times 420\,\text{m s}^{-1} \times \cos 60$

$= 2.1 \times 10^{-23}\,\text{kg m s}^{-1}$ [1]

Force $= \dfrac{\text{change of momentum}}{\text{contact time}} = \dfrac{2.1 \times 10^{-23}\,\text{kg m s}^{-1}}{0.22 \times 10^{-9}\,\text{s}}$

$= 9.5 \times 10^{-14}\,\text{N}$ [1]

9.3/9.4/9.5

1 Let m = mass of B.

Total initial momentum $= (1.50 \times 0.35) + (m \times -0.25) = 0.525 - 0.25m$ [1]

Total final momentum $= (m + 1.50) \times 0.050 = 0.050m + 0.075$ [1]

Therefore, $0.050m + 0.075 = 0.525 - 0.25m$ [1]

Rearranging this equation gives:

$0.25m + 0.05\,m = 0.525 - 0.075$ which gives

$0.30\,m = 0.450$ so $m = \dfrac{0.450}{0.30} = 1.5\,\text{kg}$ [1]

2 Let v = the velocity of Y after the collision.

Total initial momentum $= (0.60 \times 3.0)$

$= 1.80\,\text{kg m s}^{-1}$ [1]

Total final momentum $= (0.60 \times 0.80) + (0.40 \times v)$

$= 0.48 + 0.40v$ [1]

Therefore, $0.40v + 0.48 = 1.80$

Rearranging this equation gives $0.40v = 1.80 - 0.48$

$= 1.32$ [1]

Therefore, $v = \dfrac{1.32}{0.40} = 3.3\,\text{m s}^{-1}$ in the same direction as the direction of A initially. [1]

3 a Let v = the velocity of P after the collision.

Total initial momentum $= (3000 \times 1.2) + (2000 \times -2.5) = -1400\,\text{kg m s}^{-1}$ [1]

Total final momentum $= (3000 \times v) + (2000 \times 0.50) = 3000v + 1000$ [1]

Therefore, $3000v + 1000 = -1400$ [1]

Rearranging this equation gives $3000v = -2400$

Therefore, $v = \dfrac{-2400}{3000} = -0.80\,\text{m s}^{-1}$

The velocity of P after the collision $= 0.80\,\text{m s}^{-1}$ in the opposite direction to its initial direction. [1]

b Kinetic energy E_K of P before the collision

$= \dfrac{1}{2} \times 3000 \times 1.2^2 = 2160\,\text{J}$

E_K of Q before the collision $= \dfrac{1}{2} \times 2000 \times -2.5^2 = 6250\,\text{J}$

Total E_K before collision $= 2160 + 6250 = 8410\,\text{J}$ [1]

E_K of P after the collision $= \dfrac{1}{2} \times 3000 \times -0.80^2 = 960\,\text{J}$

E_K of Q after the collision $= \dfrac{1}{2} \times 2000 \times 0.5^2 = 250\,\text{J}$

Total E_K after collision $= 960 + 250 = 1210\,\text{J}$ [1]

Total E_K after collision < total E_K before the collision, so the collision is inelastic. [1]

4 a Let V = recoil velocity of the gun.

Total final momentum $= (2.4 \times 150) + (900 \times V) = 360 + 900V$

Total initial momentum $= 0$ [1]

Therefore, $360 + 900V = 0$ which gives $V = \dfrac{-360}{900} = -0.40\,\text{m s}^{-1}$

The recoil velocity of the gun $= 0.40\,\text{m s}^{-1}$ [1]

b Kinetic energy of shell $= \dfrac{1}{2} \times 2.4 \times 150^2 = 27\,000\,\text{J}$

Kinetic energy of gun $= \dfrac{1}{2} \times 900 \times -0.40^2 = 72\,\text{J}$ [1]

Kinetic energy of gun/total kinetic energy
= 72/(27 000 + 72) = 0.0027

Therefore, kinetic energy of the gun as a
percentage of the total kinetic energy
= 0.0027 × 100% = 0.27% [1]

10.1/10.2

1 a Work done = $Fs\cos\theta$ = 14 × 3.0 cos 60 = 21 J [1]

b $W = \frac{1}{2}F\Delta L$ = 0.5 × 20 × 0.45 = 4.5 J [1]

2 a KE at ground = initial KE + loss of GPE

$= \frac{1}{2}mu^2 + mg\,\Delta h$

= (0.5 × 0.048 × 16.0²) + (0.048 × 9.81 × 23.0) [1]

= 6.1 + 10.8 = 16.9 J [1]

b Let v = speed just before impact so $\frac{1}{2}mv^2 = E_K$
(= 16.9 J)

Rearranging to find v gives

$v = \sqrt{\frac{2E_K}{m}} = \sqrt{\frac{2 \times 16.9}{0.048}}$ [1] = 28 m s⁻¹ (2 s.f.) [1]

3 a Loss of potential energy = $mg\,\Delta h$ = 80 × 9.81 × 40
= 31 J [1]

b Gain of kinetic energy = $\frac{1}{2}mv^2$ = 0.5 × 80 × 14²
= 7.8 kJ [1]

c Work done to overcome resistive forces = loss of
GPE – gain of KE = 31.4 – 7.8 = 23.6 kJ [1]

average resistive force =

$\frac{\text{work done}}{\text{distance moved}} = \frac{23.6 \text{ kJ}}{560 \text{ m}}$ = 42 N [1]

10.3/10.4

1 a Rearranging power $P = Fv$ where F is the force
and v is the velocity gives

$F = \frac{P}{v} = \frac{240\,000 \text{ W}}{31 \text{ m s}^{-1}}$ [1] = 7.7 kN [1]

b Useful energy transferred per second = 240 kJ s⁻¹
= 37% of energy per second supplied

Energy supplied per second = 240 000 ÷ 0.37
= 650 kJ s⁻¹ [1]

Energy wasted per second = 650 – 240
= 410 kJ s⁻¹ [1]

2 a Power $P = kv^3$ where k is the constant of
proportionality

Therefore, $k = \frac{P}{v^3}$ [1] $= \frac{1.4}{12^3}$ MW (m s⁻¹)⁻³ [1]

Hence, at $v = 8$ m s⁻¹, $P = kv^3 = \frac{1.4}{12^3} \times 8.0^3 =$
0.41 MW [1]

b Kinetic energy/second of the wind

$= \frac{\text{output power}}{\text{efficiency}} = \frac{1.4 \text{ MW}}{0.48}$ [1] = 2.9 MW [1]

3 Mass of water passing through per second= vol/sec ×
density = 52 × 1000 = 52 000 kg s⁻¹ [1]

Loss of potential energy each second
= $mg\,\Delta h$ (where m = 52 000 kg)
= 52000 × 9.81 × 350 = 179 MJ s⁻¹ [1]

Output power = 28% of 179 MJ s⁻¹ = 50 MW [1]

11.1

1 a Volume = 20 × 10⁻³ × 60 × 10⁻² × 75 × 10⁻²
= 9.0 × 10⁻³ m³ [1]

b Density $= \frac{\text{mass}}{\text{volume}} = \frac{22.4 \text{ kg}}{9.0 \times 10^{-3} \text{ m}^3}$ = 2500 kg m⁻³
(2 s.f.) [1]

2 a Mass of liquid = 248 g – 85 g = 163 g = 0.163 kg [1]

b Depth H of liquid in tin = 85 mm
Volume of liquid $= \frac{1}{4}\pi d^2 \times H$ = 0.25 × π ×
(50 × 10⁻³)² × 85 × 10⁻³ = 1.67 × 10⁻⁴ m³ [1]

c Density $= \frac{\text{mass}}{\text{volume}} = \frac{0.163 \text{ kg}}{1.67 \times 10^{-4} \text{ m}^3}$ = 980 kg m⁻³ [1]

3 a i Volume of the wire $= \frac{1}{4}\pi d^2 \times L$ = 0.25 × π ×
(0.220 × 10⁻³)² × 2415 × 10⁻³ [1]

= 9.18 × 10⁻⁸ m³ [1]

ii Density of metal $= \frac{\text{mass}}{\text{volume}} = \frac{0.730 \times 10^{-3} \text{ kg}}{9.18 \times 10^{-8} \text{ m}^3}$

= 7950 kg m⁻³ [1]

b % uncertainties: mass $= \frac{0.005}{0.730} \times 100\%$ = 0.7%;

length $= \frac{5}{2415} \times 100\%$ = 0.2%; [1]

diameter $= \frac{0.005}{0.220} \times 100\%$ = 2.27% therefore

% uncertainty in the area of cross-section
= 2 × 2.27% = 4.54%. [1] Total % uncertainty
= 0.7 + 0.2 + 4.54 = 5.44% [1]

Therefore, uncertainty in the density value =
5.44% of 7950 = 430 kg m⁻³ [1]

11.2

1 a Rearranging $F = k\Delta L$ gives $\Delta L = \frac{F}{k} = \frac{8.0 \text{ N}}{40 \text{ Nm}^{-1}}$
= 0.20 m [1]

b Energy stored $= \frac{1}{2}F\Delta L$ = 0.5 × 8.0 × 0.20 = 0.80 J [1]

2 a i Since the springs are identical and in parallel,
the tension T in each spring is the same and
equal to half the weight supported.
So T = 0.80 N. [1]

ii Extension ΔL of each spring = 274 – 250
= 24 mm [1]

Tension T in each spring = $k\Delta L$ where k is the
spring constant of each spring.

Hence, $k = \frac{T}{\Delta L} = \frac{0.80 \text{ N}}{0.024 \text{ m}}$ [1] = 33.3(3) N m⁻¹
= 33 N m⁻¹ to 2 significant figures [1]

b Total energy stored $= \frac{1}{2}k\Delta L^2$ where the effective
spring constant k = 2 × 33.3(3) = 66.7 N m⁻¹ and
ΔL = 0.024 m. [1]

Therefore, the total energy stored $= \frac{1}{2} \times 66.7$
× 0.024² [1] = 0.019 J. [1]

3 a The tension T in each spring is the same because
they are in series and is equal to the weight
supported. Hence, T =1.6 N. [1]

b For P, ΔL = 280 –250 = 30 mm.[1] Therefore,

$k_P = \frac{T}{\Delta L} = \frac{1.6 \text{ N}}{0.030 \text{ m}}$ = 53 N m⁻¹ [1]

For Q, ΔL = 300 – 250 = 50 mm. [1] Therefore,

$k_P = \frac{T}{\Delta L} = \frac{1.6 \text{ N}}{0.050 \text{ m}}$ = 32 N m⁻¹ [1]

11.3/11.4

1 Area of cross-section $A = \frac{1}{4}\pi d^2 = 0.25\pi \times$
$(0.26 \times 10^{-3})^2 = 5.3 \times 10^{-8}\,\text{m}^2$ [1]
$E = \frac{FL}{A\Delta L} = \frac{48 \times 1.524}{5.3 \times 10^{-8} \times 6.6 \times 10^{-3}}$ [1] $= 2.1 \times 10^{11}\,\text{Pa}$ [1]

2 a Area of cross-section $A = \frac{1}{4}\pi d^2 = 0.25\pi \times$
$(0.22 \times 10^{-3})^2 = 3.8 \times 10^{-8}\,\text{m}^2$ [1]
Rearranging
$E = \frac{FL}{A\Delta L}$ gives $\Delta L = \frac{FL}{AE} = \frac{32 \times 1.500}{3.8 \times 10^{-8} \times 2.1 \times 10^{11}}$ [1]
$= 6.0 \times 10^{-3}\,\text{m}$ [1]

   b Elastic energy stored $= \frac{1}{2}T\Delta L = 0.5 \times 32 \times$
$6.0 \times 10^{-3} = 0.096\,\text{J}$ [1]

3 a Rearranging $E = \frac{F}{A\Delta L}$ where F is the compressive
force gives stress
$= \frac{F}{A} = \frac{E\Delta L}{L} = \frac{2.1 \times 10^{11} \times 0.13 \times 10^{-3}}{0.050}$ [1]
$= 5.5 \times 10^8\,\text{Pa}$ [1]

   b Compressive force = stress $\times A = 5.5 \times 10^8 \times$
3.4×10^{-4} [1] $= 187\,\text{kN}$ [1]
Elastic energy stored $= \frac{1}{2}F\Delta L = 0.5 \times 187 \times 10^3 \times$
$0.13 \times 10^{-3} = 12\,\text{J}$ to 2 significant figures [1]

12.1/12.2

1 a $\Delta Q = I\Delta t = 1.2 \times 10^{-3} \times 300\,\text{s} = 0.36\,\text{C}$ [1]

   b no. of electrons $= \frac{\Delta Q}{e} = \frac{0.36\ \text{C}}{1.60 \times 10^{-19}\ \text{C}}$
$= 2.3 \times 10^{18}$ (2 s.f.) [1]

2 a Current $I = \frac{P}{V} = \frac{800\ \text{W}}{230\ \text{V}} = 3.5\,\text{A}$ (2 s.f.) [1]

   b Energy transferred in 1 minute $= Pt$
$= 800\,\text{W} \times 60\,\text{s} = 48\,000\,\text{J}$ [1]

3 a Rearranging $\Delta Q = I\Delta t$ gives
$\Delta t = \frac{\Delta Q}{I} = \frac{3.6 \times 10^6\ \text{C}}{13} = 2.8 \times 10^5\,\text{s}$ (2 s.f.) [1]

   b Rearranging $V = \frac{W}{Q}$ gives energy delivered
$= QV = 3.6 \times 10^6\,\text{C} \times 12\,\text{V} = 4.3 \times 10^7\,\text{J}$ (2 s.f.) [1]
(or energy delivered $= IV\Delta t = 13\,\text{A} \times 12\,\text{V} \times$
$2.77 \times 10^5\,\text{s} = 4.3 \times 10^7\,\text{J}$ (2 s.f.))

12.3

1 a $V = IR = 0.970\,\text{A} \times 0.156\,\Omega = 0.151\,\text{V}$ [1]

   b Energy per second dissipated $= IV$
$= 0.970\,\text{A} \times 0.151\,\text{V} = 0.146\,\text{W}$ (3 s.f.) [1]

2 Area of cross-section of wire
$A = 35.1 \times 10^{-3}\,\text{m} \times 0.62 \times 10^{-3}\,\text{m} = 2.18 \times 10^{-5}\,\text{m}^2$ [1]
Resistivity $\rho = \frac{RA}{L} = \frac{0.156\ \text{V} \times 2.18 \times 10^{-5}\ \text{m}^2}{0.382\ \text{m}}$
$= 8.90 \times 10^{-6}\,\Omega\,\text{m}$ [1]

3 a Area of cross-section $= \frac{1}{4}\pi d^2$
$= 0.25\pi \times (1.28 \times 10^{-3})^2 = 1.29 \times 10^{-6}\,\text{m}^2$ [1]

Rearranging,
$\rho = \frac{RA}{L}$ gives $R = \frac{\rho L}{A} = \frac{1.70 \times 10^{-8}\,\Omega\,\text{m} \times 25.0\ \text{m}}{1.29 \times 10^{-6}\ \text{m}^2}$
[1]

   $= 0.329\,\Omega$ [1] (to 2 significant figures)

   b Energy dissipated per second $P = I^2R = 13.0^2$
$\times 0.329\,\Omega = 55.6\,\text{W}$ [1]

12.4

1 a See Figure 1. [1]

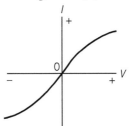

▲ Figure 1

   b The resistance increases as the current increases
in either direction. [1] The increase of resistance
is because the filament becomes hotter as the
current increases and the positive ions in the
filament vibrate more. [1] The conduction
electrons need to do more work because they
collide more with the positive ions [1] so a
greater potential difference is needed to maintain
the same current. [1]

2 a i $I = \frac{V}{R} = \frac{4.5\ \text{V}}{100\,000} = 4.5 \times 10^{-5}\,\text{A}$ [1]

      ii $I = \frac{V}{R} = \frac{4.5\ \text{V}}{400} = 1.1 \times 10^{-2}\,\text{A}$ (2 s.f.) [1]

   b The number of charge carriers in the LDR
increases if the incident light intensity is
increased. [1] Therefore, there are fewer charge
carriers in the LDR in darkness than in daylight
so the current for the same pd is smaller in
darkness than in daylight. Hence, the resistance
in darkness is larger than in daylight. [1]

3 a See Figure 1. [1]

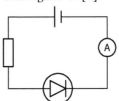

▲ Figure 1

   b When the diode conducts, the pd across it does
not change when the current increases. The pd
across the resistor is therefore unchanged if the
current changes. [1] The ammeter reading would
decrease because the resistor has the same pd
across its terminals and its resistance is greater so
the current in the circuit would be less. [1]

13.1

1 a See Figure 1. [1]

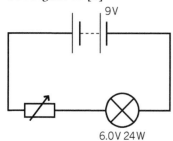

▲ **Figure 1**

b i 4.0 A [1]

ii Pd across variable resistor = 9.0 − 6.0 = 3.0 V [1]

iii Power supplied by battery = IV = 4.0 A × 9.0 V
= 36 W [1]

2 a See Figure 2. [1]

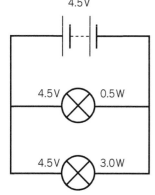

▲ **Figure 2**

b $I = \dfrac{P}{V} = \dfrac{0.5 \text{ W}}{4.5 \text{ V}} = 0.11$ A for the 0.5 W torchbulb [1]

$I = \dfrac{P}{V} = \dfrac{3.0 \text{ W}}{4.5 \text{ V}} = 0.67$ A for the 3.0 W torchbulb [1]

c Battery current = 0.11 + 0.67 = 0.78 A [1] hence energy per second supplied by the battery = IV
= 0.78 A × 4.5 V = 3.5 W [1]

Energy supplied per second to the two torchbulbs
= 3.0 W + 0.5 W = 3.5 W = energy per second supplied by the battery. [1]

3 a See Figure 3. [1]

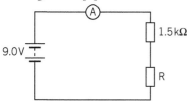

▲ **Figure 3**

b i Pd across 1.5 kΩ resistor (= current × resistance) = 2.0 mA × 1.5 kΩ = 3.0 V [1]

ii Pd across resistor R = 9.0 − 3.0 = 6.0 V [1]

iii Resistance of $R = \dfrac{V}{I} = \dfrac{6.0 \text{ V}}{2.0 \text{ mA}} = 3.0$ kΩ [1]

13.2

1 a See Figure 1. [1]

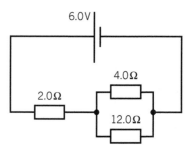

▲ **Figure 1**

b i Resistance of the 2 parallel resistors

$= \left[\dfrac{1}{4.0} + \dfrac{1}{12.0} \right]^{-1} = 3.0$ Ω [1]

Total circuit resistance = 3.0 + 2.0 = 5.0 Ω [1]

ii Battery current = $\dfrac{\text{cell emf}}{\text{total circuit resistance}}$

$= \dfrac{6.0 \text{ V}}{5.0 \text{ Ω}} = 1.2$ A [1]

c Pd across 2.0 Ω resistor = IR = 1.2 × 2.0 = 2.4 V therefore pd across parallel resistors

= 6.0 − 2.4 = 3.6 V [1]

Power supplied to 4.0 Ω resistor = $\dfrac{V^2}{R} = \dfrac{3.6^2}{4.0} = 3.2$ W
(2 s.f.) [1]

2 a See Figure 2 (all correct [2], correct circuit with incorrect labels [1])

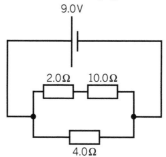

▲ **Figure 2**

b i Resistance of the 2 series resistors
= 2.0 Ω + 10.0 Ω = 12.0 Ω [1]

Total circuit resistance = $\left[\dfrac{1}{4.0} + \dfrac{1}{12.0} \right]^{-1}$

= 3.0 Ω [1]

ii Battery current = $\dfrac{\text{cell emf}}{\text{total circuit resistance}}$

$= \dfrac{9.0 \text{ V}}{3.0 \text{ Ω}} = 3.0$ A [1]

c Power supplied to 4.0 Ω resistor

$= \dfrac{V^2}{R} = \dfrac{9.0^2}{4.0} = 20.3$ W [1]

Current through the 2.0 Ω and 10.0 Ω resistors

$= \dfrac{V}{R} = \dfrac{9.0 \text{ V}}{12.0 \text{ Ω}} = 0.75$ A

Power supplied to 2.0 Ω resistor = I^2R
= 0.75² × 2.0 = 1.1 W (2 s.f.) [1]

Power supplied to 10.0 Ω resistor = I^2R
= 0.75² × 10.0 = 5.6 W (2 s.f.) [1]

Note: The power supplied by the battery = IV
= 3.0 A × 9.0 V = 27 W. This is equal to the total power supplied to the resistors (= 20.25 W + 1.125 W + 5.625 W as required by conservation of energy).

3 $P = 3000\,\text{W}$, $V = 230\,\text{V}$. Rearranging $P = \dfrac{V^2}{R}$ gives

$R = \dfrac{V^2}{P} = \dfrac{230^2}{3000}$ [1] $= 18$ (2 s.f.) [1]

13.3

1 a Total resistance $= 4.5 + 0.5 = 5.0\,\Omega$ [1]

 b Battery current $= \dfrac{V}{R} = \dfrac{9.0\,\text{V}}{5.0\,\Omega} = 1.8\,\text{A}$ [1]

 c Pd across cell terminals $= IR = 1.8\,\text{A} \times 4.5\,\Omega$

 $= 8.1\,\text{V}$ [1] (or $\varepsilon - Ir = 9.0 - (1.8 \times 0.5) = 8.1\,\text{V}$)

2 a Current $= \dfrac{\text{emf}}{\text{total circuit resistance}} = \dfrac{2.0}{3.5 + 0.5}$ [1]

 $= 0.50\,\text{A}$ [1]

 b Power delivered to 3.5 Ω resistor $= I^2R$

 $= 0.50^2 \times 3.5 = 0.88\,\text{W}$ (2 s.f.) [1]

 c Power wasted $= I^2r = 0.50^2 \times 0.5 = 0.13\,\text{W}$ (2 s.f.) [1]

3 a Using $\varepsilon = V + Ir$ with $V = 1.05\,\text{V}$ and $I = 0.25\,\text{A}$ gives

 $\varepsilon = 1.05 + 0.25r$ [1]

 Using $\varepsilon = V + Ir$ with $V = 1.35\,\text{V}$ and $I = 0.10\,\text{A}$ gives

 $\varepsilon = 1.35 + 0.10r$ [1]

 Therefore, $1.35 + 0.10r = 1.05 + 0.25r$ which gives

 $0.25r - 0.10r = 1.35 - 1.05$

 Therefore, $r = \dfrac{1.35 - 1.05}{0.25 - 0.10} = \dfrac{0.30}{0.15} = 2.0\,\Omega$ [1]

 b Using $\varepsilon = V + Ir$ with $V = 1.05\,\text{V}$, $I = 0.25\,\text{A}$ and $r = 2.0\,\Omega$ gives:

 $\varepsilon = 1.05 + (0.25 \times 2.0\,\Omega) = 1.55\,\text{V}$ [1]

13.4

1 a See Figure 1. [1]

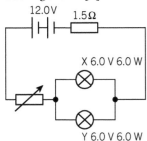

X 6.0 V 6.0 W

Y 6.0 V 6.0 W

▲ **Figure 1**

 b i Current in each lamp $= \dfrac{P}{V} = \dfrac{6.0\,\text{W}}{6.0\,\text{V}} = 1.0\,\text{A}$ [1]

 ii Battery current $I = I_X + I_Y = 1.0 + 1.0 = 2.0\,\text{A}$ [1]

 iii Battery emf = pd across variable resistor + pd across lamps + pd across internal resistance

 Pd across internal resistance $= Ir = 2.0 \times 1.5 = 3.0\,\text{V}$ [1]

 Pd across lamps $= 6.0\,\text{V}$

 Therefore, pd across variable resistor $= 12.0\,\text{V} - 6.0\,\text{V} - 3.0\,\text{V} = 3.0\,\text{V}$ [1]

13.5

1 a See Figure 1. [1]

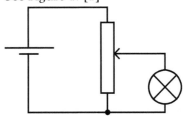

▲ **Figure 1**

 b A potential divider can be adjusted to give a range of pd from zero to a maximum equal to the pd of the supply voltage. Therefore, the brightness of the lamp can be adjusted from zero to a maximum by adjusting the potential divider. [1] A variable resistor in series with the lamp gives a range of current between a minimum when the variable resistor has maximum resistance to a maximum current when the variable resistor has zero resistance. [1] Therefore, the brightness of a lamp cannot be reduced to zero using a variable resistor. [1]

2 a i Total circuit resistance $= 8.0\,\Omega + 12.0\,\Omega = 20.0\,\Omega$ [1]

 Current in each resistor

 $= \dfrac{\text{battery pd}}{\text{total circuit resistance}} = \dfrac{6.0\,\text{V}}{20.0\,\Omega} = 0.30\,\text{A}$ [1]

 ii Let V_1 = pd across the 8.0 Ω resistor and V_2 = pd across the 12.0 Ω resistor. The battery pd of 6.0 V is shared between the 8.0 Ω resistor and the 12.0 Ω resistor in proportion to their resistances. Therefore, $\dfrac{V_2}{V_1} = \dfrac{12.0}{8.0} = 1.5$ so $V_2 = 1.5V_1$. [1]

 So the battery pd $= V_1 + V_2 = V_1 + 1.5V_1 = 2.5V_1$

 Hence, $2.5V_1 = 6.0\,\text{V}$ so $V_1 = \dfrac{6.0}{2.5} = 2.4\,\text{V}$ and $V_2 = 1.5V_1 = 1.5 \times 2.4 = 3.6\,\text{V}$ [1]

 (OR pd across each resistor = current × resistance)

 Therefore, pd across 8.0 Ω resistor $= 0.30\,\text{A} \times 8.0\,\Omega = 2.4\,\text{V}$ [1]

 (and pd across 12.0 Ω resistor $= 0.30\,\text{A} \times 12.0\,\Omega = 3.6\,\text{V}$)

 b i The total circuit resistance $= 16.0 + 8.0 = 24.0\,\Omega$. The ratio of the pd across the 16 Ω resistor to the cell pd $= 16.0/24.0 = 0.667$. [1] Therefore, the pd across the 16 Ω resistor $= 0.667 \times$ the cell pd $= 0.667 \times 4.5\,\text{V} = 3.0\,\text{V}$. [1]

(Alternative method: current in the circuit
= 4.5 V / (16.0 Ω + 8.0 Ω) = 0.188 A

Therefore, pd across 16 Ω resistor
= 0.188 A × 16.0 Ω = 3.0 V)

ii The total circuit resistance = 6.0 + 8.0 = 14.0 Ω.
The ratio of the pd across the 6 Ω resistor to
the cell pd = 6.0/14.0 = 0.429. [1] Therefore,
the pd across the 6 Ω resistor = 0.429 × the cell
pd = 0.429 × 4.5 V = 1.9(3) V. [1]

(Alternative method: current in the circuit
= 4.5 V / (6.0 Ω + 8.0 Ω) = 0.321 A

Therefore, pd across 6 Ω resistor
= 0.321 A × 6.0 Ω = 1.9(3) V)

3 a See Figure 2. [1]

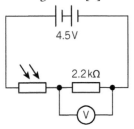

▲ **Figure 2**

b i Pd across LDR = 4.5 V − 1.2 V = 3.3 V [1]

ii Current through 2.2 kΩ resistor
$= \dfrac{V}{R} = \dfrac{1.2\ V}{2.2\ k\Omega} = 0.55$ mA [1]

LDR resistance $= \dfrac{V}{I} = \dfrac{3.3\ V}{0.55\ mA} = 6.0$ kΩ [1]

c When the LDR is exposed to daylight, its resistance
decreases so the share of the battery pd across the
LDR decreases. [1] Therefore, the pd across the
resistor increases. [1]

17.1

1 a Circumference of wheel = π × diameter
= π × 0.60 m = 1.885 m

Number of turns $= \dfrac{\text{distance}}{\text{circumference}} = \dfrac{100\ m}{1.885\ m}$
= 53.1 = 53 (2 s.f.) [1]

b Time taken to travel 100 m = 100 m/25 m s⁻¹ = 4.0 s

Frequency f = number of turns/time taken =
53.1/4.0 s = 13.3 Hz = 13 Hz (2 s.f.) [1]

c Angular speed $\omega = 2\pi f = 2\pi \times 13.3$ Hz = 83.6 rad s⁻¹ =
84 rad s⁻¹ (2 s.f.) [1]

2 a Frequency $f = \dfrac{1}{\text{time for 1 rotation}} = \dfrac{1}{(93 \times 60\ s)}$
$= 1.8 \times 10^{-4}$ Hz [1]

b Speed $v = \omega r = 2\pi f r = 2\pi \times 1.8 \times 10^{-4}$ Hz ×
$(6400 + 400) \times 10^3$ m ≈ 7.7 km s⁻¹ [1]

3 a Angular speed $\omega = 2\pi f = 2\pi \times 0.95$ Hz =
6.0 rad s⁻¹ [1]

b The two gear wheels move at the same speed, v,
where they are in contact. [1]

Therefore $v = \omega_1\, r_1 = \omega_2\, r_2$, so $\omega_2 \times 2.5$ mm =
6.0 rad s⁻¹ × 8.2 mm [1]

Hence $\omega_2 = (6.0\ \text{rad s}^{-1} \times 8.2)/2.5$ mm = 19.7 rad s⁻¹

Rearranging $\omega = 2\pi f$ gives $f = \dfrac{\omega}{2\pi} = \dfrac{19.7\ \text{rad s}^{-1}}{2\pi}$
= 3.1 Hz [1]

c i $\Delta\theta = \omega\, \Delta t = 19.7$ rad s⁻¹ × 0.02 s =
0.39(4) rad [1]

ii speed of tooth, $v = \omega_2\, r_2 = 19.7$ rad s⁻¹ × 2.5 mm =
49 mm s⁻¹

distance moved $\Delta s = v\, \Delta t = 49$ mm s⁻¹ × 0.02 s
= 0.98 mm

(or $\Delta s = r\, \Delta\theta = 2.5$ mm × 0.39(4) rad
= 0.98 mm) [1]

17.2

1 a Angular speed $\omega = 2\pi/T = 2\pi/2.3$ s = 2.7(3) rad s⁻¹
[1]

b Speed $v = \omega r = 2.73$ rad s⁻¹ × 1.1 m = 3.0 m s⁻¹ [1]

c Centripetal force $= m\, \omega^2\, r = 28$ kg × (2.73 rad s⁻¹)² ×
1.1 m = 230 N [1]

2 Centripetal force $= \dfrac{mv^2}{r} = \dfrac{22\,\text{kg} \times (0.78\,\text{ms}^{-1})^2}{2.8\,\text{m}} = 4.8$ N [1]

3 a Angular speed $= 2\pi/T = 2\pi/12$ s = 0.52(4) rad s⁻¹

Speed $v = \omega r = 0.524$ rad s⁻¹ × 2.5 m = 1.3(1) m s⁻¹
[1]

b Centripetal force $= m\, \omega^2\, r = 61$ kg ×
(0.524 rad s⁻¹)² × 2.5 m = 42 N [1]

17.3/17.4

1 a Centripetal acceleration $a = \dfrac{v^2}{r} = \dfrac{(22\,\text{ms}^{-1})^2}{82\,\text{m}}$
$= 5.9$ m s⁻² [1]

b Support force $S = mg + ma$ (as $S - mg = ma$) [1]

Therefore $S = (1200\,\text{kg} \times 9.81\,\text{m s}^{-2}) + (1200\,\text{kg} \times 5.9\,\text{m s}^{-2}) = 1.9 \times 10^4\,\text{N}$ [1]

2 Angular speed $\omega = 2\pi/T = 2\pi/7.0\,\text{s} = 0.90\,\text{rad s}^{-1}$ [1]

For a passenger at the top, the centripetal acceleration $= \omega^2 r = (0.90\,\text{rad s}^{-1})^2 \times (16\,\text{m}) = 13.0\,\text{m s}^{-2}$

Support force $S = ma - mg$ (as $S + mg = ma$) [1]

Therefore $S = (48\,\text{kg} \times 13.0\,\text{m s}^{-2}) - (48\,\text{kg} \times 9.81\,\text{m s}^{-2}) = 153\,\text{N} = 150\,\text{N}$ (2 s.f.) [2]

3 Centripetal acceleration $a = \dfrac{v^2}{r} = \dfrac{(2.8\,\text{m s}^{-1})^2}{3.3\,\text{m}} = 2.38\,\text{m s}^{-2}$ [1]

Support force $S = mg + ma$ (as $S - mg = ma$)

Therefore $S = (23\,\text{kg} \times 9.81\,\text{m s}^{-2}) + (23\,\text{kg} \times 2.38\,\text{m s}^{-2}) = 280\,\text{N}$ [2]

18.1/2

1 a i 0.771 s [1]

ii 1.30 Hz [1]

b Range of timings = ±0.2 s, mean timing = 15.4 s [1]

% uncertainty in frequency = % uncertainty in timings = range of timings/mean value ×100% [1] = (±0.2 s/15.4 s) × 100% = 1.3% [1]

2 a $\omega = 2\pi/T = 2\pi/0.77\,\text{s} = 8.2\,\text{rad s}^{-1}$ [1]

acceleration $a = -\omega^2 x = -(8.2\,\text{rad s}^{-1})^2 \times 0.025\,\text{m} = -1.7\,\text{m s}^{-2}$ [1]

b $x = 0$ so $a = -\omega^2 x = 0$ [1]

c As $x = -25\,\text{mm}$, from part **a** $a = (-\omega^2 x = -(8.2\,\text{rad s}^{-1})^2 \times -0.025\,\text{m}) = +1.7\,\text{m s}^{-2}$ [1]

18.3

1 a Angular frequency $\omega = 2\pi/T = 2\pi/2.3\,\text{s} = 2.73\,\text{rad s}^{-1} = 2.7\,\text{rad s}^{-1}$ (2 s.f.) [1]

b Rearranging $a = -\omega^2 x$ gives $x = -a/\omega^2$ therefore at maximum displacement, $x = +A$ when $a = -7.2\,\text{m s}^{-2}$, $-A = -7.2\,\text{m s}^{-2}/(2.73\,\text{rad s}^{-1})^2 = -0.97\,\text{m}$. Therefore $A = 0.97\,\text{m}$. [1]

2 a Sketch graph as SB 18.3 Figure 3 top graph; sine wave [1]

time for 1 cycle shown as 2.3 s [1]

b $x = A \sin(\omega t) = 0.97\,\text{m} \times \sin(2.73\,\text{rad s}^{-1} \times 2.0\,\text{s}) = -0.71\,\text{m}$ [1]

3 a After one half cycle, its displacement $x = -55\,\text{m}$ (because it is at the opposite extreme to its initial displacement) [1]

Its angular frequency $\omega = 2\pi/T = 2\pi/0.27\,\text{s} = 23.3\,\text{rad s}^{-1}$ [1]

Therefore at $x = -55\,\text{mm}$, its acceleration $= -\omega^2 x = -(23.3\,\text{rad s}^{-1})^2 \times -0.055\,\text{m} = 30\,\text{m s}^{-2}$ [1]

Therefore the magnitude of its acceleration $= 30\,\text{m s}^{-2}$ [1]

b At time $t = 0.20\,\text{s}$, its displacement $x = A \cos \omega t = 55\,\text{mm} \times \cos(23.3\,\text{rad s}^{-1} \times 0.20\,\text{s}) = -2.9\,\text{mm}$ [1]

18.4

1 a i Extension of spring at equilibrium is $\Delta L_0 = 79\,\text{mm} = 0.079\,\text{m}$ [1]

ii Spring constant $k = \dfrac{mg}{L_0} = \dfrac{0.20 \times 9.8}{0.079} = 25\,\text{N m}^{-1}$ [1]

b $T = 2\pi\sqrt{\dfrac{0.20\,\text{kg}}{25\,\text{N m}^{-1}}} = 0.56\,\text{s}$ [1]

2 a From the equation $a = -250x$, $\omega_2{}^2 = 250\,\text{rad}^2\,\text{s}^{-2}$ [1]

so $T = 2\pi/(250\,\text{rad}^2\,\text{s}^{-2})^{1/2} = 0.40\,\text{s}$ [1]

b From the equation $T = 2\pi\sqrt{m/k}$, $T \propto m^{1/2}$ so $m \propto T^2$ [1]

therefore $m_2/m_1 = T_2{}^2/T_1{}^2 = (0.40/0.56)^2 = 0.51$ [1] which gives $m_2 = 0.51 m_1 = 0.51 \times 0.20\,\text{kg} = 0.10\,\text{kg}$ [1]

3 a $T = 2\pi\sqrt{\dfrac{L}{g}} = 2\pi\sqrt{\dfrac{0.250\,\text{m}}{9.81\,\text{m s}^{-2}}} = 1.00(3)\,\text{s}$ [1]

b For $L = 255\,\text{mm}$, $T = 2\pi\sqrt{\dfrac{L}{g}} = 2\pi\sqrt{\dfrac{0.255\,\text{m}}{9.81\,\text{m s}^{-2}}} = 1.01(3)\,\text{s}$ [1]

Therefore the uncertainty in $T = \pm 0.01\,\text{s}$ which is ±1% of T [1]

Alternative method

% uncertainty in $L = \pm(\Delta L/L) \times 100\% = \pm(5\,\text{mm}/250\,\text{mm}) \times 100\% = \pm 2\%$ [1]

As $T \propto L^{1/2}$, % uncertainty in $T = \frac12 \times$ % uncertainty in $L = 0.5 \times \pm 2\% = \pm 1\%$ [1]

18.5

1 a $T = 2.2\,\text{s}$ so angular frequency $\omega = 2\pi/T = 2\pi/2.2\,\text{s} = 2.86\,\text{rad s}^{-1}$ [1]

b Rearranging $T = 2\pi\sqrt{m/k}$ gives $k = 4\pi^2 (m/T^2)$ [1]

Therefore $k = 4\pi^2 (0.150\,\text{kg}/(2.2)^2) = 1.22\,\text{N m}^{-1}$ [1]

c Amplitude $A = 30\,\text{mm}$ so maximum speed $= \omega A = 2.86\,\text{rad s}^{-1} \times 0.030\,\text{m} = 0.09\,\text{m s}^{-1}$ [1]

2 a Total energy $= \frac12 kA^2 = 0.5 \times 1.22\,\text{N m s}^{-1} \times (0.030\,\text{m})^2 = 0.55\,\text{mJ}$ [1]

b PE curve parabola from origin [1]

PE at $\pm A$ = . max PE [1]

KE curve inverted PE curve [1] with max KE (= min PE) at $x = 0$ and zero KE at $\pm A$ [1]

3 a Its maximum speed v_{max} is given by $\frac12 m v_{max}{}^2 = mgh$ where $h = 10.0\,\text{mm}$ [1]

Therefore $v_{max} = (2gh)^{1/2} = 0.44(3)\,\text{m s}^{-1}$ [1]

b i $T = 1.50\,\text{s}$ so angular frequency $\omega = 2\pi/T = 2\pi/1.50\,\text{s} = 4.19\,\text{rad s}^{-1}$ [1]

Maximum speed $= \omega A$ therefore $A = $ Maximum speed$/\omega = 0.44(3)\,\text{m s}^{-1}/4.19\,\text{rad s}^{-1} = 0.106\,\text{m}$ [1]

ii At $x = 0.5A$, speed
$$v = \omega \sqrt{A^2 - x^2} = 4.19\sqrt{0.106^2 - 0.053^2}\ \text{m}$$
$$= 0.385\ \text{m s}^{-1}\ [1]$$

18.6

1 a The object and the spring are in resonance at this frequency [1]

because the periodic force is in phase with the velocity of the object at this frequency [1]

b The object gains energy from the work done by the periodic force [1]

and its amplitude increases until the resistive forces due to damping becomes equal and opposite to the periodic force. [1]

2 a Because the time period T is proportional to the square root of the object's mass, increasing the mass by a factor of 4 increases the time period by $\times 2$ ($= \sqrt{4}$) [1]

So the natural frequency of oscillation is halved and resonance occurs at $0.5 f_o$ [1]

b If a stiffer spring is used, the spring constant k is larger therefore the time period T is reduced as T is proportional to $1/\sqrt{k}$. [1]

Therefore the natural frequency of the system is increased so resonance takes place at a higher frequency [1]

3 a Rearranging $T = 2\pi\sqrt{m/k}$ gives $m = kT^2/4\pi^2$ where $T = 1/f = 1/1.8\ \text{Hz} = 0.56\ \text{s}$ [1]

Therefore $m = 25\ \text{N m}^{-1} \times (0.56\ \text{s})^2/4\pi^2$ [1] $= 0.20\ \text{kg}$ [1]

b Using two springs in parallel doubles the spring constant to $50\ \text{N m}^{-1}$ (see Topic 11.2). [1]

As T is proportional to $1/\sqrt{k}$ and f is proportional to $1/T$, then f is proportional to $\sqrt{k}$. [1]

As k is doubled, the frequency of oscillation is increased by $\times \sqrt{2}$ to $1.8\sqrt{2}\ \text{Hz}$ which equals $2.5(5)\ \text{Hz}$ [1]

19.1

1 a The internal energy of an object is the sum of the kinetic and potential energies of its molecules' positions due to their random motion and positions. [1]

b Its internal energy increases by $100\ \text{J}$ ($= 400\ \text{J}$ gained due to work done $- 300\ \text{J}$ lost due to energy transfer from it by heating) [1]

2 a The average kinetic energy decreases and the average potential energy is unchanged. [1]

b The average potential energy decreases and the average kinetic energy is unchanged as the substance transfers energy by heating to its surroundings. [1]

3 When the rubber band is released, the potential energy of its molecules decreases as they move closer together due to the forces (or bonds) between them [1]. These forces pull the molecules together and they collide with each other and vibrate more [1]. So the work done by these forces reduces their potential energy and increases their kinetic energy and the temperature of the rubber increases [1]. Energy transfer to the surroundings takes place and the rubber band loses internal energy and cools down to the same temperature as its surroundings. [1]

19.2

1 Energy needed to heat the tank $= mc\,\Delta T = 20\ \text{kg} \times 390\ \text{J kg}^{-1}\text{K}^{-1} \times (60°C - 15°C) = 3.51 \times 10^5\ \text{J}$ [1]

Energy needed to heat the tank $= mc\,\Delta T = 430\ \text{kg} \times 4200\ \text{J kg}^{-1}\text{K}^{-1} \times (60°C - 15°C) = 8.127 \times 10^7\ \text{J}$

Total energy needed $= 8.16 \times 10^7\ \text{J} = 8.2 \times 10^7\ \text{J}$ (2 s.f.) [1]

2 Energy to heat the aluminium $=$ PE loss therefore $mc\Delta T = mgh$ [1]

Therefore temperature increase $\Delta\theta = gh/c = (9.81\ \text{m s}^{-1} \times 8000\ \text{m})/900\ \text{J kg}^{-1}\text{K}^{-1} = 87\ \text{K}$ [1]

3 Mass flow in 1 second, $m =$ density $\times$ volume flow in 1 second $= 1000\ \text{kg m}^{-3} \times 3.8 \times 10^{-5}\ \text{m}^3 = 0.038\ \text{kg}$ [1]

Energy supplied in 1 second $E = 5000\ \text{J} = mc\,\Delta T$ [1]

therefore $\Delta T = E/mc = 5000\ \text{J}/(0.038\ \text{kg} \times 4200\ \text{J kg}^{-1}\text{K}^{-1}) = 31\ \text{K}$ [1]

19.3

1 When a liquid cools, the molecules move slower as the temperature decreases [1] so the force bonds are able to pull the molecules closer and closer together [1]. When the temperature decreases to the melting point, the molecules hold each other in fixed positions and the liquid solidifies at its melting point. [1]

2 Energy transferred to cool the water to $0°C = mc\,\Delta T = 0.095\ \text{kg} \times 4200\ \text{J kg}^{-1}\text{K}^{-1} \times 12\ \text{K} = 4790\ \text{J}$ [1]

Energy transferred to freeze the water $= mL = 0.095\ \text{kg} \times 3.40 \times 10^5\ \text{J kg}^{-1} = 32300\ \text{J}$ [1]

Energy transferred per second $= (32300\ \text{J} + 4790\ \text{J})/600\ \text{s} = 62\ \text{J s}^{-1}$ [1]

3 a Energy used to boil $0.015\ \text{kg}$ of water away $= mL = 0.015\ \text{kg} \times 2.25 \times 10^6\ \text{J kg}^{-1} = 33750\ \text{J}$ [1]

Energy supplied by kettle $= 3000\ \text{W} \times 20\ \text{s} = 60\,000\ \text{J}$

% energy supplied that is used to boil the water $= 33750\ \text{J}/60\,000\ \text{J} \times 100\% = 56\%$ [1]

b Energy gained by the water in the beaker $= m_w c\,\Delta T = 0.11\ \text{kg} \times 4200\ \text{J kg}^{-1}\text{K}^{-1} \times (55°C - 20°C) = 1.62 \times 10^4\ \text{J} = 1.6 \times 10^4\ \text{J}$ (2 s.f.) [1]

Energy transferred to the water when the steam condensed at $100°C = mL = m \times 2.25 \times 10^6\ \text{J kg}^{-1}$ where m is the mass of the condensed steam [1]

Energy transferred to the water when the condensed steam cooled from 100°C to 55°C = $mc\,\Delta T = m \times 4200\,J\,kg^{-1}\,K^{-1} \times (100°C - 55°C) = m \times 1.89 \times 10^5\,J\,kg^{-1}$ [1]

Assuming no energy is transferred to the surroundings,
energy gained

by the water = energy transferred from the steam and the condensed steam

Therefore $(m \times 2.25 \times 10^6\,J\,kg^{-1}) + (m \times 1.89 \times 10^5\,J\,kg^{-1}) = 1.62 \times 10^4\,J$ [1]

$(2.25 \times 10^6\,J\,kg^{-1} + 1.89 \times 10^5\,J\,kg^{-1}) \times m = 1.62 \times 10^4\,J$

$2.44 \times 10^6\,J\,kg^{-1} \times m = 1.62 \times 10^4\,J$

$m = 1.62 \times 10^4\,J\,/2.44 \times 10^6\,J\,kg^{-1} = 6.6 \times 10^{-3}\,kg$ [1]

20.1

1 Stage 1; Use $p_2 V_2 = p_1 V_1$ to give, after compression, pressure $p_2 = p_1 V_1/V_2 = 120\,kPa \times 8.0 \times 10^{-4}\,m^3 /5.0 \times 10^{-4}\,m^3 = 192\,kPa$ [1]

Stage 2; Use $p_2/T_2 = p_1/T_1$ to give, after compression, [1]

pressure $p_2 = p_1 T_2/T_1 = 192\,kPa \times (20 + 273)\,K\,/ (50 + 273)\,K = 174\,kPa = 170\,kPa$ (2 s.f.) [1]

2 a Pressure $p_2 = p_1 T_2/T_1 = 100\,kPa \times 173\,K\,/373\,K = 46\,kPa$ [2]

b Graph; straight line with a +gradient; line shown between 173 and 373 K with one temp and pressure values shown correctly; correct gradient or 2nd values shown correctly for either 173 or 373 K pressure value [3]

3 New volume $V_2 = V_1 T_2/T_1 = 0.085\,m^3 \times 328\,K\,/288\,K = 0.097\,m^3$ [2]

20.2

1 a Rearrange $pV = nRT$ to give $n = pV/RT = 120\,kPa \times 0.020\,m^3/8.31\,J\,mol^{-1}\,K^{-1} \times (273 + 20) = 0.99\,mol$ [1]

Mass = $0.99\,mol \times 0.028\,kg = 0.028\,kg$ [1]

b Pressure $p_2 = p_1 T_2/T_1 = 120\,kPa \times 278\,K\,/293\,K = 114\,kPa$ [1]

2 Rearrange $pV = nRT$ to give $n = pV/RT = 101\,kPa \times 2.2 \times 10^{-5}\,m^3/8.31\,J\,mol^{-1}\,K^{-1} \times (273 + 12) = 9.38 \times 10^{-4}\,mol$ [1]

Mass = $9.38 \times 10^{-4}\,mol \times 0.032\,kg\,mol^{-1} = 3.0 \times 10^{-5}\,kg$ [1]

3 a Rearrange $pV = nRT$ for $1.0\,m^3$ to give $n = pV/RT = 100\,kPa \times 1.0\,m^3/8.31\,J\,mol^{-1}\,K^{-1} \times 300\,K = 40.1\,mol$ [1]

Number of molecules in n moles = $n\,N_A = 40.1\,mol \times 6.02 \times 10^{23}\,mol^{-1} = 2.41 \times 10^{25}$ [1]

Volume per molecule = $1/2.41 \times 10^{25} = 4.1 \times 10^{-26}\,m^3$ [1]

b Assume each molecule occupies a cube of length d so $d^3 = 4.1 \times 10^{-26}\,m^3$ [1]

The average spacing $d = (4.1 \times 10^{-26}\,m^3)^{1/3} = 3.4 \times 10^{-9}\,m$. [1]

20.3

1 The pressure increases because reducing its volume means the gas molecules travel less distance on average between impacts at the walls. [1]

Therefore, there are more impacts per second so the pressure is greater. [1]

2 Mass of a hydrogen molecule = $0.002\,kg\,mol^{-1}/6.02 \times 10^{23}\,mol^{-1} = 3.32 \times 10^{-27}\,kg$ [1]

Mean kinetic energy $\frac{1}{2}mc_{rms}^2 = \frac{3}{2}kT = 1.5 \times 1.38 \times 10^{-23}\,J\,K^{-1} \times 293\,K = 6.07 \times 10^{-21}\,J$ [1]

Therefore rms speed = $(2 \times$ mean $KE/m)^{1/2} = (2 \times 6.07 \times 10^{-21}\,J/3.32 \times 10^{-27}\,kg)^{1/2} = 1910\,m\,s^{-1} = 2000\,m\,s^{-1}$ (1 s.f.) [1]

3 a Rearrange $pV = nRT$ for $1.0\,m^3$ to give $n = pV/RT = 120\,kPa \times 0.018\,m^3/8.31\,J\,mol^{-1}\,K^{-1} \times 323\,K = 0.805\,mol = 0.81\,mol$ to (2 s.f.) [1]

b Internal energy = $\frac{3}{2}nRT = 1.5 \times 0.805 \times 8.31\,J\,mol^{-1}\,K^{-1} \times 323\,K = 3.2 \times 10^3\,J$ [2]

4 The mean KE is the same for the CO_2 molecules and the nitrogen molecules (as they are at the same temperature). [1]

Therefore, $\frac{1}{2}mc_{rms}^2$ for CO_2 molecules = $\frac{1}{2}mc_{rms}^2$ for nitrogen molecules [1]

$$\frac{c_{rms} \text{ for } CO_2}{c_{rms} \text{ for nitrogen}} = \sqrt{\frac{\text{mass of a nitrogen molecule}}{\text{mass of a } CO_2 \text{ molecule}}}$$

$$= \sqrt{\frac{\text{molar mass of nitrogen}}{\text{molar mass of } CO_2}}$$

$$= \sqrt{\frac{0.044\,kg}{0.028\,kg}} = 1.25 \quad [2]$$

5 a Rearrange $pV = nRT$ to give $n = pV/RT = 1.00 \times 10^5\,Pa \times 0.037\,m^3/(8.31\,J\,mol^{-1}\,K^{-1} \times 300\,K) = 1.48\,mol = 1.5\,mol$ (2 s.f.) [1]

b For 1 mole, $pV = \frac{1}{3}Nmc_{rms}^2 = \frac{1}{3}mc_{rms}^2$ where M = molar mass [1]

So $c_{rms}^2 = 3pV/M = 3 \times 1.00 \times 10^5\,Pa \times 0.037\,m^3/0.032\,kg = 3.47 \times 10^5\,m^2\,s^{-2}$ [1]

Therefore $c_{rms} = 590\,m\,s^{-1}$ [1]

c Mass of gas = $1.48\,mol \times 0.032\,kg\,mol^{-1} = 0.0474\,kg$ [1]

Density = mass/volume = $0.0474\,kg/0.037\,m^3 = 1.28\,kg\,m^{-3} = 1.3\,kg\,m^{-1}$ (2 s.f.) [1]

6 Rearrange $pV = nRT$ for $V = 1.0\,m^3$ to give $n = pV/RT = 1.01 \times 10^5\,Pa \times 1.0\,m^3/(8.31\,J\,mol^{-1}\,K^{-1} \times 288\,K) = 42.2\,mol$. [1]

Mass of gas = $42.2\,mol \times 0.029\,kg\,mol^{-1} = 1.22\,kg$ [1]

Density = mass/volume = $1.22\,kg/1.0\,m^3 = 1.2(2)\,kg\,m^{-3}$ [1]

21.1

1 a $1.6\,N\,kg^{-1}$ [1]

b 400 N [1]

2 **a** In a uniform gravitational field, the gravitational field has the same magnitude and direction at all points in the field [1]

b 3.5 N [1]

3 Weight on Mars = (650 N/9.81 N kg^{-1}) × 3.70 N kg^{-1} [1] = 245 N [1]

21.2

1 **a** 7.4 KJ [1]

b 490 J kg^{-1} [1]

2 **a** The gravitational potential at a point is the work done per unit mass to move a small object from infinity to that point. [1]

b Work done $W = mgh$ therefore $\Delta V = W/m = gh$ [1] = 9.81 N kg^{-1} × 20.0 m = 196 J kg^{-1} [1]

3 **a** For $\Delta x = +500$ km (+ for upwards), $\Delta V = +4.6$ MJ kg^{-1} [1]

$g = -\Delta V/\Delta x = -(4.6 \text{ MJ kg}^{-1})/500 \text{ km}$ [1] = (−)9.2 N kg^{-1} [1]

b $\Delta E_\text{p} = m\Delta V = 300 \text{ kg} \times (+)4.6 \text{ MJ kg}^{-1}$ [1] $= 1.38 \times 10^9$ J [1]

c Maximum kinetic energy at ground = loss of potential energy $= 1.38 \times 10^9$ J [1]

Therefore maximum speed v_max at the ground is such that $\frac{1}{2} m v_\text{max}^2 = 1.38 \times 10^9$ J [1]

$v_\text{max}^2 = 2 \times 1.38 \times 10^9 \text{ J}/300 \text{ kg} = 9.20 \times 10^6 \text{ m}^2\text{s}^{-2}$

so $v_\text{max} = 3.03$ km s^{-1} [1]

21.3

1 $F = \dfrac{6.67 \times 10^{-11} \times 7.4 \times 10^{22} \times 2000}{(3.7 \times 10^8)^2} = 0.072$ N [1]

2 Least distance apart $= 2.3 \times 10^{11}$ m $- 1.5 \times 10^{11}$ m $= 8.0 \times 10^{10}$ m [1]

$F = \dfrac{6.67 \times 10^{-11} \times 6.4 \times 10^{23} \times 6.0 \times 10^{24}}{(8.0 \times 10^{10})^2} = 4.0 \times 10^{16}$ N

[1]

3 Let M_E = mass of the Earth, M_S = mass of the Sun, M_M = mass of the Moon

Force on the Earth due to the Sun $= \dfrac{GM_\text{E}M_\text{S}}{r_\text{ES}^2}$ where

r_ES = the distance between the centres of the Earth and the Sun

Force on the Earth due to the Moon $= \dfrac{GM_\text{E}M_\text{M}}{r_\text{EM}^2}$

where r_EM = the distance between the centres of the Earth and the Moon

$\dfrac{\text{The gravitational force on the Earth due to the Sun}}{\text{The gravitational force on the Earth due to the Moon}}$

$= \dfrac{GM_\text{E}M_\text{S}}{r_\text{ES}^2} \div \dfrac{GM_\text{E}M_\text{M}}{r_\text{EM}^2}$ [1]

$= \dfrac{GM_\text{E}M_\text{S}}{r_\text{ES}^2} \times \dfrac{r_\text{EM}^2}{GM_\text{E}M_\text{M}} = \dfrac{M_\text{S}}{r_\text{ES}^2} \times \dfrac{r_\text{EM}^2}{M_\text{M}}$ [1]

$= \dfrac{2.0 \times 10^{30}}{(1.5 \times 10^{11})^2} \times \dfrac{(3.8 \times 10^8)^2}{7.4 \times 10^{32}} = 173 = 170$ (2 s.f.) [1]

21.4

1 **a** $g = \dfrac{GM}{r^2} = \dfrac{g_\text{s}R^2}{r^2} = \dfrac{9.81 \times (6400 \times 10^3)^2}{(42\,000 \times 10^3)^2}$ [1]

$= 0.23$ N kg^{-1} [1]

b **i** $V = -\dfrac{GM}{10} = -\dfrac{g_\text{s}R^2}{r} = -gR = -0.23 \text{ N kg}^{-1} \times 4.2 \times 10^7$ m [1] $= -9.7$ MJ kg^{-1} [1]

ii $E_\text{p} = mV = 200 \text{ kg} \times -9.7 \text{ MJ kg}^{-1} = -1940 \text{ MJ}$ $= 1900$ MJ (2 s.f.) [1]

2 **a** $g = \dfrac{GM}{r^2} = -\dfrac{6.67 \times 10^{-11} \times 6.4 \times 10^{23}}{(3.4 \times 10^6)^2} = 3.7$ N kg^{-1} [1]

b $v_\text{esc} = \sqrt{2\,g_\text{S}R} = \sqrt{2 \times 3.7 \times 3.4 \times 10^6} = 5000$ m s^{-1} [2]

21.5

1 **a** Rearranging $\dfrac{r^3}{T^2} = \dfrac{GM}{4\pi^2}$ gives $r^3 = \dfrac{GMT^2}{4\pi^2}$ where

$T = 90 \times 60 \text{ s} = 5400$ s

$r^3 = \dfrac{GMT^2}{4\pi^2} = \dfrac{6.67\ 0 \times 10^{-11} \times 6.0 \times 10^{24} \times (5400)^2}{4\pi^2}$ [1]

$= 2.96 \times 10^{20}$ m^3

so $r = 6.7 \times 10^6$ m [1]

Height above the ground = 6.7×10^6 m $- 6.4 \times 10^6$ m $= 300$ km [1]

b Centripetal acceleration $= v^2/r = GM/r^2$ [1] so

$v = \sqrt{\dfrac{GM}{r}} = \sqrt{\dfrac{6.67 \times 10^{-11} \times 6.0 \times 10^{24}}{6.7 \times 10^6}}$ [1]

$= 7.7$ km s^{-1} [1]

2 **a** Rearranging $\dfrac{r^3}{T^2} = \dfrac{GM}{4\pi^2}$ gives $T^2 = \dfrac{4\pi^2 r^3}{GM}$ where

$r = 7600$ km

$T^2 = \dfrac{4\pi^2 r^3}{GM} = \dfrac{4\pi^2 \times (7.6 \times 10^6)^3}{6.67 \times 10^{-11} \times 6.0 \times 10^{24}}$ [1]

$= 4.33 \times 10^7$ s^2 so $T = 6600$ s [1]

b Angular speed $= 2\pi/T = 9.5 \times 10^{-4}$ rad s^{-1} [1]

22.1/22.2

1 **a** The field lines are directed from the plate to Q [1]

b **i** Vertically down [1]

ii Downwards at about 45° to the vertical. [1]

2 **a** No of electrons $n = Q/e = 2.5 \text{ nc}/1.6 \times 10^{-19}$ C $= 1.56 \times 10^{10}$ [1]

Surface area $A = 4\pi R^2 = 4\pi \times (0.15 \text{ m})^2 = 0.283 \text{ m}^2$ $= 2.83 \times 10^5$ mm^2 [1]

No of electrons per mm^2 = n/A = $1.56 \times 10^{10}/2.83 \times 10^5$ mm$^2 = 5.5 \times 10^4$ mm^{-2} [1]

b $t = Q/I = 2.5 \text{ nC}/0.2 \text{ mA} = 12.5 \text{ μs} = 13$ μs (2 s.f.) [1]

3 **a** $F = QE = 3 \times 1.6 \times 10^{-19}$ C $\times 900\,000$ V m^{-1} [1] $= 4.3 \times 10^{-13}$ C [1]

b $V = E\,d = 900\,000$ V m^{-1} × 0.050 m [1] = 45 kV [1]

22.3

1 a $W = F\Delta x = QE \times \Delta x = 1.6 \times 10^{-19}\,\text{C} \times 8000\,\text{V m}^{-1}$
$\times 0.028\,\text{m}$ [1] $= 3.6 \times 10^{-17}\,\text{J}$ [1]

b $\Delta V = E\,\Delta x = 8000\,\text{V m}^{-1} \times 0.028\,\text{m} = 224\,\text{V}$ [1]
Potential at Y $= +400\,\text{V} + 224\,\text{V} = +624\,\text{V}$ [1]

2 a $W = Q\Delta V = +2e\,(6000\,\text{V} - 1000\,\text{V})$ [1]
$= 2 \times 1.6 \times 10^{-19}\,\text{C} \times 5000\,\text{V} = 1.6 \times 10^{-15}\,\text{J}$ [1]

b Average potential gradient $= \Delta V/\Delta x = 5000\,\text{V}/0.036\,\text{m}$
[1] $= 1.4 \times 10^{5}\,\text{V m}^{-1}$ [1]

3 a Graph should be a straight line with a negative
gradient [1] from $y = +4200\,\text{V}$ at $x = 0$ at L to
$y = 0\,\text{V}$ at $x = 80\,\text{mm}$ at m [1]

b i Potential gradient $= \Delta V/\Delta x = (4200 - 0)\,\text{V}/$
$(0 - 0.080\,\text{m}) = -5.3 \times 10^{4}\,\text{V m}^{-1}$ [1]

ii Electric field strength $= -$potential gradient $=$
$+5.3 \times 10^{4}\,\text{V m}^{-1}$ [1] from L to M [1]

22.4/22.5

1 $F = \dfrac{1}{4\pi\varepsilon_o}\dfrac{Q_1\,Q_2}{r^2}$

$= \dfrac{1}{4\pi(8.85\times10^{-12}\,\text{F m}^{-1})} \times \dfrac{(2\times1.6\times10^{-19}\,\text{C})^2}{(5.6\times10^{-7}\,\text{m})^2}$ [1]

$= 2.9 \times 10^{-15}\,\text{m}$ [1]

2 The initial force $F_1 = \dfrac{1}{4\pi\varepsilon_o}\dfrac{Q_X\,Q_Y}{r^2}$ where

$r = 1.6 \times 10^{-2}\,\text{m}$

The force increases to $F_2 = 7.8 \times 10^{-5}\,\text{N}$ from
$5.1 \times 10^{-5}\,\text{N}$, which is an increase of $\times 1.53$
$(= 7.8\times10^{-5}\,\text{N}/5.1\times10^{-5}\,\text{N})$ [1]. So the charge on X
must have increased by $\times 1.53$ [1]

$\dfrac{(Q_X + q)}{Q_X} = 1.53$ where q is the increase of charge so

$(Q_X + q) = 1.53Q_X$ therefore $Q_X = q/0.53 = 2.1\,\text{nC}/0.53$
$= 4.0\,\text{nC}$ [1]

Rearrange $F_1 = \dfrac{1}{4\pi\varepsilon_o}\dfrac{Q_X\,Q_Y}{r^2}$ to give

$Q_Y = \dfrac{4\pi\varepsilon_o r^2 F_1}{Q_X}$

$= \dfrac{4\pi\times8.85\times10^{-12}\,\text{F m}^{-1}\times(1.6\times10^{-2}\,\text{m})^2\times(5.1\times10^{-5}\,\text{N})}{4.0\times10^{-9}\,\text{C}}$

[1]
$= 3.6 \times 10^{-10}\,\text{C} = 0.36\,\text{nC}$ [1]

3 a Potential energy $= \dfrac{1}{4\pi\varepsilon_o}\dfrac{Qq}{r}$

$= \dfrac{1}{4\pi\times8.85\times10^{-12}\,\text{F m}^{-1}} \times$

$\dfrac{(+1.6\times10^{-19}\,\text{C})(+1.6\times10^{-19}\,\text{C})}{0.11\times10^{-9}}$ [1] $= 2.1 \times 10^{-18}\,\text{J}$ [1]

b The distance from the midpoint to each charge $=$
$0.055\,\text{nm}$

Electric field strength at the midpoint

due to the proton $E_1 = \dfrac{1}{4\pi\varepsilon_o}\dfrac{+e}{r^2}$

$= \dfrac{+1.6\times10^{-16}\,\text{C}}{4\pi\times8.85\times10^{-12}\,\text{F m}^{-1}\times(0.055\times10^{-9}\,\text{m})^2}$

$= 4.8 \times 10^{11}\,\text{V m}^{-1}$ away from the proton [1]
Electric field strength at the midpoint due to the

electron $E_2 = \dfrac{1}{4\pi\varepsilon_o}\dfrac{+e}{r^2} = 4.8 \times 10^{11}\,\text{V m}^{-1}$ towards
the electron [1]

Resultant electric field strength $=$
$2 \times 4.8 \times 10^{11}\,\text{V m}^{-1} = 9.6 \times 10^{11}\,\text{V m}^{-1}$ towards the
electron [1]

4 a i Rearranging $V = \dfrac{1}{4\pi\varepsilon_o}\dfrac{Q}{r}$ gives $Q = 4\pi\varepsilon_o rV =$

$4\pi\times8.85\times10^{-12}\,\text{F m}^{-1}\times80\,\text{mm}\times(+6200\,\text{V})$ [1]
$= +55\,\text{nC}$ [1]

ii $V \propto 1/r$ and r is increased by $\times 12.5$
$(= 1000\,\text{mm}/80\,\text{mm})$ [1]
Therefore at $1000\,\text{mm}$, $V = +6.2\,\text{kV}/12.5$
$(= 496\,\text{V}) = 500\,\text{V}$ to 2 significant figures. [1]

b $E = Q/4\pi\varepsilon_o r^2$ and $V = Q/4\pi\varepsilon_o r$ Therefore
$E = V/r = 6200\,\text{V}/0.080\,\text{m}$ [1] $= 77500\,\text{V m}^{-1}$
$= 7.8\,\text{kV m}^{-1}$ to 2 significant figures. [1]

22.6

1 Any 2 of the following:

* field lines are parallel in a uniform field but not
in a radial field

* field lines of an object's gravitational field are
always towards the object whereas the direction
of the field lines in an electric field depends on
whether the charge is + or −

* the field strength in a uniform field is the same at
all points in the field which is not so for a radial field

* the equipotentials in a radial field are spherical
(or circular) whereas in a uniform field the
equipotentials are flat surfaces (or straight lines) [2]

2 Both field strength rows correct [1] Both potential
rows correct [1]

	Z	Y	k	n
Gravitational field strength	g	M	G	−2
Electric field strength	E	Q	$\dfrac{1}{4\pi\varepsilon_o}$	−2
Gravitational potential	V	M	G	−1
Electric potential	V	Q	$\dfrac{1}{4\pi\varepsilon_o}$	−1

3 a Rearranging $F = \dfrac{Q}{4\pi\varepsilon_o r^2}$ gives $\dfrac{1}{4\pi\varepsilon_o} = F r^2 Q^{-1}$ [1]

$= \text{kg m s}^{-2}\,\text{m}^2\,\text{A}^{-1}\,\text{s}^{-1} = \text{kg m}^3\,\text{s}^{-3}\,\text{A}^{-1}$ [1]

b $\sqrt{(2ER)}$ [1]

23.1/23.2/23.3

1 a $Q = C V = 47\,\mu F \times 6.0\,V = 2.8 \times 10^{-4}\,C$ [1]

b $Q = I t$ gives $t = \dfrac{Q}{I} = \dfrac{2.8 \times 10^{-4}\,C}{1.5 \times 10^{-6}\,C} = (187\,s =)$
190 s to 2 significant figures [1]

2 a $Q = C V = 22\,\mu F \times 3.0\,V = 66\,\mu C$ [1]

b i Charge transferred from the 22 µF capacitor to the 10 µF capacitor [1]. (As the total charge is conserved), the 10 µF capacitor gains charge and the 22 µF capacitor loses an equal amount of charge. [1]

ii The pd across the 10 µF capacitor = $q/10\,\mu F$ where q is the charge transferred [1]

The pd across the 22 µF capacitor = $(66\,\mu C - q)/22\,\mu F$ [1]

The pd across the 2 capacitors is the same therefore $q/10\,\mu F = (66\,\mu C - q)/22\,\mu F$ [1]

$2.2q = 66\,\mu C - q$ which gives
$3.2q = 66$ so $q = 20.6\,\mu C$

Therefore the pd across the 10 µF capacitor = $20.6\,\mu C/10\,\mu F = 2.1\,V$ to 2 significant figures [1]

3 a i $Q = C V = 10\,\mu F \times 6.0\,V = 60\,\mu C$ [1]
$E = \tfrac{1}{2}\,C V^2 = 0.5 \times 10\,\mu F \times (6.0\,V)^2$ (or $\tfrac{1}{2}\,QV$) = 180 µJ [1]

ii Energy supplied by the battery = $QV = 60\,\mu C \times 6.0\,V$ = 360 µJ [1]

b The difference is because energy is dissipated due to resistance heating when charge flows round the circuit. [1] The energy supplied by the battery is the sum of the energy stored in the capacitor and the energy dissipated due to resistance heating in the circuit. [1]

4 a i Time constant = $RC = 200\,k\Omega \times 47\,\mu F = 9.4\,s$ [1]

ii The time constant is the time taken to discharge to 37% of the initial pd which is 2.2 V in this case. [1] So the capacitor takes about 10 s (just over 9.4 s) for the pd to decrease to 2 V [1]

b $V = V_0 e^{-\frac{t}{RC}} = 6.0\,V \times e^{-\frac{5.0\,s}{9.4\,s}}$ [1]
$= 6.0\,V \times 0.59 = 3.5\,V$ [1]
Current = $V/R = 3.5\,V/200\,k\Omega = 18\,\mu A$ [1]

5 a i Current = $V/R = 5.0\,V/20\,k\Omega = 2.5 \times 10^{-4}\,A$ [1]

ii The pd across the resistor = $5.0\,V - 3.0\,V = 2.0\,V$ [1] so the current = resistor pd/resistance = $2.0\,V/20\,k\Omega$ = $1.0 \times 10^{-4}\,A$ [1]

b The time for the pd to increase to 3.0 V = time for the current to decrease to $1.0 \times 10^{-4}\,A$ from $2.5 \times 10^{-4}\,A$ [1]
Rearranging $I = I_0 e^{-\frac{t}{RC}}$ gives $\ln (I/I_0) = -t/RC$ [1]
Therefore $t = -RC \ln (I/I_0) = -20\,k\Omega \times 68\,\mu F \times$
$\ln \dfrac{1.0 \times 10^{-4}\,A}{2.5 \times 10^{-4}\,A}$ = 1.2 s to 2 significant figures [1]

23.4

1 a i The capacitance decreases [1] because it is inversely proportional to the distance between the plates. [1]

ii The pd between the plates increases. [1] This is because the charge Q is unchanged (as the capacitor plates are isolated) and the capacitance C decreases so $V = Q/C$ increases. [1]

b Work is done to overcome the force of attraction between the plates when the plates are moved apart [1]. So the energy stored increases by an amount equal to the work done. [1]

2 The molecules are polarised by the electric field to become dipoles (if they are not already polarised). [1] The dipoles rotate to align with the field as their positive pole is attracted to the negative plate and their negative pole is attracted to the positive plate. [1] The effect is to create a layer of positive charge against the negative plate and a layer of negative charge against the positive plate. [1] This 'polarisation' charge allows more electrons to move onto the negative plate from the battery and off the positive plate back to the battery. [1]

24.1

1 a $B = \dfrac{F}{Il} = \dfrac{0.028\,N}{4.3\,A \times 0.080\,m}$ [1] = 0.081 N [1]

b Eastwards and horizontal [1]

2 Arrangement as Figure 1 [1]
magnetic field IN

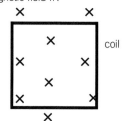

coil

▲ **Figure 1**

Magnitude of the force on each side = $BIln$ = $0.055\,T \times 5.9\,A \times 0.060\,m \times 50$ [1] = 0.97 N [1]

Direction of the force on each side is perpendicular to the side and horizontal [1]

3 Force per unit length = $B I = 6.6 \times 10^{-5}\,T \times 65\,A$ [1]
$= 4.3 \times 10^{-3}\,N\,m^{-1}$ [1]

24.2

1 a $F = B e v = 0.150\,T \times 1.6 \times 10^{-19}\,C \times 8.0 \times 10^6\,s^{-1}$
$= 1.9 \times 10^{-13}\,N$ [1]

b The force is zero because the direction of motion of the electrons is parallel to the magnetic field. [1]

2 a $B q v = m v^2/r$ hence $r = \dfrac{m v}{B q}$ [1]
$= \dfrac{1.67 \times 10^{-27}\,kg \times 3.2 \times 10^7\,m\,s^{-1}}{0.085\,T \times 1.6 \times 10^{-19}\,C}$ [1] = 3.9 m [1]

b From $r = \dfrac{m\,v}{B\,q}$, the radius of curvature is

proportional to $\dfrac{v}{B}$ [1]

The ratio $\dfrac{\text{the new magnetic flux density } B_2}{\text{the initial magnetic flux density } B_1} = 0.5$

The ratio $\dfrac{\text{the new speed } v_2}{\text{the initial speed } v_1} = \sqrt{\dfrac{\text{the new KE}}{\text{the initial KE}}}$

$= 0.71$ [1]

Therefore $\dfrac{\text{the new radius of curvature } r_2}{\text{the new radius of curvature } r_1} =$

$\dfrac{v_2}{B_2} \div \dfrac{v_1}{B_1} = \dfrac{v_2}{v_1}\sqrt{\dfrac{B_2}{B_1}} = 0.71 \div 0.5 = 1.4$

The new radius of curvature is $1.4 \times$ the initial radius of curvature [1]

3 a $B\,q\,v = q\,E$ gives $B = \dfrac{E}{v} = \dfrac{8.2 \times 10^4\,\text{V}}{9.5 \times 10^5\,\text{ms}^{-1}}$ [1]

$= 0.086\,\text{T}$ [1]

b From $r = \dfrac{mv}{Bq}$,

specific charge $\dfrac{q}{m} = \dfrac{v}{Br}$ [1]

$= \dfrac{9.5 \times 10^5\,\text{ms}^{-1}}{0.91\,\text{T} \times 0.185\,\text{m}} =$ [1]

$= 5.6 \times 10^6\,\text{C kg}^{-1}$ [1]

25.1

1 a An emf is induced in the rod because it cuts across the magnetic field lines. [1] The rod is in a complete circuit with the meter so there is an induced current in the circuit while the rod is moving [1]

b Either move the rod faster (or use a stronger magnet) [1]

2 The induced current is upwards from Y to X [1]. See Figure 1 below for the application of the right-hand rule in this situation.

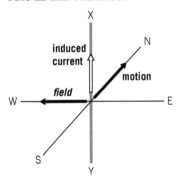

▲ **Figure 1**

Conduction electrons in the rod move from south to north due to the movement of the rod. The magnetic field exerts a force on the moving electrons and forces them towards Y away from X [1]

3 The magnetic flux linkage in the coil decreases to zero and this induces an emf in the coil which causes a current in the circuit. [1] Once the coil has been removed from the field (and stays outside the field) no further change of flux occurs so the meter reading is zero. [1]

When the coil is returned slowly to the magnetic field, the flux linkage through it increases so an emf is induced with opposite polarity to when it was removed so the meter deflects in the opposite direction. [1] The change takes longer so the induced current is less than when the coil was removed. [1]

25.2

1 a Area swept out each second by the rod $= 0.063\,\text{m} \times 0.012\,\text{m} = 7.56 \times 10^{-4}\,\text{m}^2$ [1]

Magnetic flux swept out each second $= B \times A/t = 0.072\,\text{T} \times 7.56 \times 10^{-4}\,\text{m}^2$ [1] $= 5.4 \times 10^{-5}\,\text{Wb}$ [1]

b Induced emf = flux change per second $= 5.4 \times 10^{-5}\,\text{Wb} \div 1.0\,\text{s} = 5.4 \times 10^{-5}\,\text{V}$ [1]

2 a The magnetic flux linkage in the solenoid due to the bar magnet increases as the magnet approaches the solenoid [1] so an emf is induced in the solenoid [1]. The solenoid and the meter form a complete circuit so the induced emf causes an induced current in the circuit. [1]

b The meter deflects in the opposite direction as the magnet is being moved away. This is because the magnetic flux linkage in the solenoid due to the bar magnet decreases as the magnet approaches the solenoid [1] so an emf is induced in the solenoid and this causes an induced current in the circuit. [1]. The current is in the opposite direction because the induced current creates a magnetic field which opposes the magnet moving away from the coil [1]

25.3/4

1 a Maximum flux linkage $= B\,A\,n$
$= 0.110\,\text{T} \times 0.022\,\text{m} \times 0.018\,\text{m} \times 120$ [1]
$= 5.2(3) \times 10^{-3}\,\text{Wb}$ [1]

b Maximum emf $= B\,A\,n\,\omega = 5.2(3) \times 10^{-3}\,\text{Wb} \times 2\pi \times 50\,\text{Hz}$ [1] $= 1.6(4)\,\text{V}$ [1]

2 When the motor is spinning, an emf is induced in the coil that opposes its motion. The faster the motor spins, the greater this back emf is. [1] When the load is increased, the coil's frequency of rotation is reduced so the back emf is reduced. [1]

The current in the coil = (the supply pd − the back emf) ÷ coil resistance so the current increases because the back emf decreases. [1]

3 a rms current = power ÷ pd = 1000 W ÷ 230 V = 4.3(5) A [1]

b The peak current $= \sqrt{2} \times 4.35\,\text{V} = 6.1\,\text{A}$ [1]

c The peak power = peak current × peak pd = 6.1 A × (230√2) V [1] = 2000 W [1]

25.5

1 When the device is switched on, the magnetic flux in the core is due to the difference between the flux due to the primary coil and the flux due to the secondary coil. [1]

Either

When the device is switched off, the current in the secondary coil becomes zero and no power is transferred to the secondary coil from the primary coil. [1] The primary current decreases because less power is supplied to the primary coil. [1]

or

The magnetic flux in the core increases as it is now due only to the primary coil as there is no longer any oppositely-directed flux due to the secondary coil. [1] So the back emf in the primary coil increases and reduces the primary current. [1]

2 a The current in the coils has a heating effect due to their resistance. [1]

Eddy currents induced in the core have a heating effect due to the resistance of the core [1]

Energy is wasted each time the core is magnetised and demagnetised. [1]

b Any 4 out of the following 5 points;-

Power = current × pd. [1] By increasing the grid voltage, less current is needed to transfer the same amount of power. [1] With less current, less power is wasted due to resistance heating by the current in the grid cables. [1] So less power needs to be supplied to the grid cables to deliver the same amount of power to the load connected to the cables. [1] The system efficiency is equal to the power delivered to the load ÷ the power supplied to the cables. [1]

3 a $N_S = N_P \times \dfrac{V_S}{V_P} = 2400 \times \dfrac{6.0\,\text{V}}{230\,\text{V}}$ [1] = 63 turns [1]

b Assume the efficiency = 100% so $I_S V_S = I_P V_P$ [1]

$I_P = \dfrac{I_S V_S}{V_P} = \dfrac{24\,\text{W}}{230\,\text{V}}$ [1] = 0.10(4) A [1]

26.1

1 a The atom is mostly empty space [1] and most of its mass is concentrated in a nucleus which is much smaller than the size of the atom [1]. The nucleus is positively charged because it repels and deflects α particles which are positively charged [1]

b Any two points

Particles moving at different speeds moving along the same initial path would be deflected by different amounts. [1]

The detector reading at any given angle of deflection (other than 180°) would be due to particles moving along different paths at different speeds. [1]

The nuclear model of the atom could not have been proved from the measurements. [1]

2 a See Figure 1

Initial and final directions at 90°

with N in top RH quadrant[1]

P in LH bottom quadrant at midpoint of curve

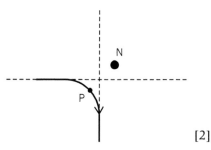

[2]

▲ **Figure 1**

b Any 3 points

E_K decreases to a minimum at P then increases after P [1]

E_P increases to a maximum at P then decreases after P [1]

Initial E_K = Final E_K [1]

At any point $E_K + E_P$ = initial E_K [1]

3 $E_K = \dfrac{Q_\alpha Q_N}{4\pi\varepsilon_0 d}$ gives $d = \dfrac{Q_\alpha Q_N}{4\pi\varepsilon_0 E_K}$

$= \dfrac{2e \times 13e}{4\pi \times 8.85 \times 10^{-12}\,\text{F m}^{-1} \times 5.1 \times 10^6 e}$ [1]

$= \dfrac{26 \times 1.6 \times 10^{-19}\,\text{C}}{4\pi \times 8.85 \times 10^{-12}\,\text{F m}^{-1} \times 5.1 \times 10^6}$ [1]

$= 7.3 \times 10^{-15}\,\text{m}$ [1]

26.2/3

1 a A proton in the nucleus changes into a neutron. [1]

The number of protons in the nucleus decreases by 1 and the number of neutrons increases by 1. [1]

b The count rate C is proportional to the intensity I. From the inverse-square law equation I is proportional to $1/r^2$, so $C = k/r^2$

$k = C r^2 = 25.2\,\text{s}^{-1} \times (0.075\,\text{m})^2$ [1]

At 160 mm, $C = 25.2\,\text{s}^{-1} \times (0.075\,\text{m})^2/(0.160\,\text{m})^2$ [1] = 5.5 counts per second [1]

2 a $^{210}_{85}\text{At} + ^{0}_{-1}e$ [1] $\rightarrow ^{210}_{84}\text{Po} + v_e$ [1]

b $^{210}_{84}\text{Po} \rightarrow ^{4}_{2}\alpha + ^{206}_{82}\text{Pb}$ [1]

26.4

1 α radiation is highly-ionising radiation [1] and when it passes through the body from a source inside the body it creates ions in the body [1] that can kill or damage body cells.

α radiation reaching the body from a source outside the body is stopped by the outer layers of the skin so is much less damaging than if the source is in the body. [1]

2 a Provided the lead lining is thick enough [1], it absorbs all the radiation from the source (including γ radiation) inside the box so no radiation penetrates to the outside of the box. [1]

b The dose of ionising radiation received by the user depends on how long the user is exposed to the radiation. [1] By making the exposure time as short as possible, the risk to the user is reduced. [1]

3 Any 2 points

Ionisation in a living cell can kill cells by destroying cell membranes [1] or damage DNA molecules in cells causing them to divide and grow and create tumours [1] or cause a mutation in a sex cell which may be passed on to future generations. [1]

26.5/6

1 a 8 hours = 4 half-lives so the number of atoms = $(0.5)^4$ [1] × the initial number = 1.1×10^{14} [1]

b 3 hours = 1.5 half-lives so the number of atoms = $(0.5)^{1.5}$ [1] of the initial number = 6.4×10^{14} [1]

2 a The decay constant of an isotope is the probability per unit time of a nucleus (of the isotope) decaying [1]

b i decay constant $\lambda = \ln 2/T_{1/2}$
$= \ln 2/(1620 \times 365 \times 24 \times 3600\,\text{s})$
$= 1.36 \times 10^{-11}\,\text{s}^{-1}$ [1]
number of atoms per kg $= 6.02 \times 10^{23}/0.226\,\text{kg}$
$= 2.67 \times 10^{24}\,\text{kg}^{-1}$ [1]
initial activity $= \lambda N = 1.36 \times 10^{-11}\,\text{s}^{-1} \times 2.67 \times 10^{24} = 3.6 \times 10^{13}\,\text{Bq}$ [1]

ii $m = m_o \text{e}^{-\lambda t}$ where $\lambda t = 1.36 \times 10^{-11}\,\text{s}^{-1} \times 3000 \times 365 \times 24 \times 3600\,\text{s} = 1.29$ [1]

Therefore $m = 1.0\,\text{kg} \times \text{e}^{-1.29} = 0.28\,\text{kg}$ [1]

OR 3000 yrs = 3000/1620 half-lives = 1.85 half-lives [1] therefore $m = 1.0\,\text{kg} \times 0.5^{1.85} = 0.28\,\text{kg}$. [1]

iii Number of atoms remaining $= 2.67 \times 10^{24}\,\text{kg}^{-1} \times 0.28\,\text{kg} = 7.5 \times 10^{23}$ [1]

3 a Rearranging $A = A_0 \text{e}^{-\lambda t}$ gives $\lambda t = \ln (A_0/A)$ [1]
$= \ln (40\,\text{kBq}/36\,\text{kBq}) = \ln 1.11 = 0.105$ [1]
$\lambda = 0.105/24\,\text{h} = 0.105/(24 \times 3600\,\text{s}) = 1.2(2) \times 10^{-6}\,\text{s}^{-1}$ [1]

b $T_{1/2} = \ln 2/\lambda = 0.693/1.2(2) \times 10^{-6}\,\text{s}^{-1}$ [1]
$= 5.7 \times 10^5\,\text{s}$ [1]

c $A = A_0 \text{e}^{-\lambda t}$ where $\lambda t = 1.2(2) \times 10^{-6}\,\text{s}^{-1} \times 120 \times 3600\,\text{s} = 0.527$ [1]
Therefore $A = 40\,\text{kBq} \times \text{e}^{-0.527}$ [1] $= 24\,\text{kBq}$ [1]

26.7

1 a For each U-238 nucleus now present, 6.4 U-238 nuclei decayed and formed Pb-206 nuclei. So the proportion of U-238 now to when the rock formed is 1 : 7.4 [1]

b $\lambda = \ln 2/(4500 \times 10^6 \times 365 \times 24 \times 3600\,\text{s}) = 4.88 \times 10^{-18}\,\text{s}^{-1}$ [1]

Rearranging $N = N_o \text{e}^{-\lambda t}$ gives $t = \ln (N_o/N) \div \lambda$ [1]
$= [\ln (7.4/1)] \div 4.88 \times 10^{-18}\,\text{s}^{-1}$ [1]
$= 4.10 \times 10^{17}\,\text{s} = 1.3 \times 10^{10}$ years [1]

2 a Number of Pu-238 nuclei in 0.20 g $= (6.02 \times 10^{23}/238\,\text{g}) \times 0.20\,\text{g} = 5.1 \times 10^{20}$ [1]
$\lambda = \ln 2/(88 \times 365 \times 24 \times 3600\,\text{s})$
$= 2.50 \times 10^{-10}\,\text{s}^{-1}$ [1]
Activity $= \lambda N = 2.50 \times 10^{-10}\,\text{s}^{-1} \times 5.1 \times 10^{20}$
$(= 1.28 \times 10^{11}\,\text{Bq}) = 1.3 \times 10^{11}\,\text{Bq}$ [1]

b Energy release per second $= AE = 1.28 \times 10^{11}\,\text{Bq} \times 5.6 \times 1.6 \times 10^{-13}\,\text{J}$ [1] $= 0.11\,\text{J}\,\text{s}^{-1}$ [1]

26.8

1 a See Topic 26.8 Figure 1 straight line from origin to approx $Z = A = 20$ [1] ; line curves towards N axis and finishes about $Z = 80$, $N = 120$ [1]

b i near $Z = 80$, $N = 120$ and above [1]

ii near $N = Z$ curve on N-axis side [1]

iii near $N = Z$ curve on Z-axis side [1]

2 a Emission of 8 α particles (reduces A by 32 to 206 and) reduces Z by 16 to 76. [1] Each β⁻ emission increases Z **by 1** so there must be 6 β⁻ emissions to increase Z from 76 to 82. [1]

b Each α emission moves the nuclide towards the N-axis and away from the N-Z curve of stable nuclei [1]. β⁺ emission takes place in nuclei that are on the other side of the $N=Z$ plot to where the nuclide has moved to. [1] α emission increases the neutron to proton ratio which means β⁻ emission not β⁺ emission could then take place to move back towards stability [1]

3 a See Figure 2

[1] for top and bottom energy level values shown

[1] for 1.35 MeV level with energy shown

[1] for 0.47 MeV level with energy shown

[1] for all 4 correct transitions shown

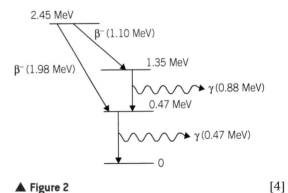

▲ **Figure 2** [4]

b $^{24}_{10}\text{Ne} \rightarrow ^{0}_{-1}\beta + ^{24}_{11}\text{Na}$ [1] $+ \nu_e$ [T]

26.9

1 $\lambda = hc /E = (6.63 \times 10^{-34}\,\mathrm{J\,s} \times 3.0 \times 10^8\,\mathrm{m\,s^{-1}})/$
$(310 \times 1.6 \times 10^{-13}\,\mathrm{J}) = 4.0 \times 10^{-15}\,\mathrm{m}$ [1]
$R \sin \theta_{\min} = 0.61\lambda$ gives $R = 0.61\lambda /\sin \theta_{\min}$
$= 0.61 \times 4.0 \times 10^{-15}\,\mathrm{m}/\sin 39°$ [1] $= 3.9 \times 10^{-15}\,\mathrm{m}$ [1]

2 The angle of diffraction was between 37° and 41°. Since sin 37° = 0.6018, sin 39° = 0.6293, and sin 40° = 0.6561, the % uncertainty in sin θ = (0.6561 – 0.6018/2) ÷ 0.6293 × 100% = 4.3%. [1] Therefore % uncertainty in R = 4.3% so uncertainty in R = ±0.043 × 3.9 fm = ±0.2 fm. [1]

3 $R = r_o A^{1/3} = 1.05\,\mathrm{fm} \times 28^{1/3} = 3.2\,\mathrm{fm}$ [1]
volume = $4\pi R^3/3 = 4\pi (3.2 \times 10^{-15}\,\mathrm{m})^3/3 = 1.37 \times 10^{-43}\,\mathrm{m^3}$
mass = 28 u = 28 × $1.661 \times 10^{-27}\,\mathrm{kg}$ = $4.65 \times 10^{-26}\,\mathrm{kg}$ [1]
density = mass /volume = $4.65 \times 10^{-26}\,\mathrm{kg}$ ÷ $1.37 \times 10^{-43}\,\mathrm{m^3}$
= $3.4 \times 10^{17}\,\mathrm{kg\,m^{-3}}$ [1]

27.1

1 a $^{14}_{6}\mathrm{C} \rightarrow {}^{14}_{7}\mathrm{N}$ [1] $+ {}^{0}_{-1}\beta + \upsilon$ [1] $+ Q$

b mass defect Δm = 13.999 95 u – 13.999 23 u – 0.000 55 u = 0.000 17 u [1]
Energy released Q = 931.5 MeV/u × 0.000 17 u [1] = 0.16 MeV [1]

2 a $^{37}_{18}\mathrm{Ar} + {}^{0}_{-1}e \rightarrow {}^{37}_{18}\mathrm{Cl}$ [1] $+ \upsilon$ [1] $+ Q$ [2]

b mass defect Δm = 36.956 89u + 0.000 55 u – 36.956 57 = 0.000 87 u [1]
Energy released Q = 931.5 MeV/u × 0.000 87 u [1] = 0.81 MeV [1]

3 a $^{241}_{95}\mathrm{Am} \rightarrow {}^{237}_{93}\mathrm{Np}$ [1] $+ {}^{4}_{2}\alpha$ [1] $+ Q$

b mass of $^{241}_{95}\mathrm{Am}$ = 241.056 69 – (95 × 0.000 55)
= 241.004 44 u [1]
mass of $^{237}_{93}\mathrm{Np}$ = 237.048 03 – (93 × 0.000 55) = 236.996 88 u [1]
mass defect Δm = 241.004 44 u – 236.996 88 u – 4.001 50 u = 0.006 06 u [1]
Energy released Q = 931.5 MeV/u × 0.006 06 u [1] = 5.6 MeV [1]

27.2

1 a The binding energy of a nucleus is the work done to separate its protons and neutrons from each other completely. [1]
The greater the binding energy of a nucleus is, the more stable it is. [1]

b The binding energy of the nucleus increases because it releases energy (or it is less unstable or more tightly bound after the emission) [1]. So the binding energy per nucleon of the nucleus increases. [1]

2 a Mass defect $\Delta m = 2m_p + m_n - M_{\mathrm{NUC}} = 0.00726$ u [1]
Therefore, binding energy = 0.00726 u × 931.5 MeV/u = 6.76 MeV [1]
Binding energy per nucleon = 6.76 MeV/3 = 2.3 MeV [1]

b Mass defect $\Delta m = 92m_p + 146m_n - M_{\mathrm{NUC}} = 1.885$ u [1]
Therefore, binding energy = 1.885 u × 931.5 MeV/u = 1756 MeV [1]
Binding energy per nucleon = 1756 MeV/238 = 7.4 MeV [1]

3 a $^{2}_{1}\mathrm{H} + {}^{1}_{1}p$ [1] $\rightarrow {}^{3}_{2}\mathrm{He}$ [1] $+ Q$

b Mass difference Δm = 2.013 55 u + 1.007 28 u – 3.014 93 u = 0.005 90 u [1]
Energy released Q = 931.5 MeV/u × 0.005 90 u [1] = 5.5 MeV [1]

27.3

1 a The strong nuclear force between the protons and neutrons in the nucleus holds them together [1] because it is an attractive force and prevents the coulomb repulsion force between the protons pushing the protons out of the nucleus. [1]

b The work done by the strong nuclear force when the nucleus is formed increases the binding energy of the protons and neutrons when they come together. [1] The mass of the protons, and neutrons is therefore reduced because some of their mass is converted to binding energy [1] in accordance with the equation $E = mc^2$ [1]

2 Mass before = 234.943 42 + 1.008 67 = 235.952 09 u [1]
Mass after = 139.891 92 + 92.893 13 + (3 × 1.008 67) = 235.811 06 u [1]
Mass difference = 0.141 u [1]
Energy released = 931.5 MeV/u × 0.141 u = 131 MeV [1]

3 a $^{3}_{1}\mathrm{H} + {}^{2}_{1}\mathrm{H}$ [1] $\rightarrow {}^{4}_{2}\mathrm{He} + {}^{1}_{0}\mathrm{n}$ [1] $+ Q$

b Mass before = 3.015 50 + 2.013 55 = 5.029 05 u [1]
Mass after = 4.001 50 + 1.008 67 = 5.010 17 u [1]
Mass difference = 0.018 88 u [1]
Energy released = 931.5 MeV/u × 0.018 88 u = 17.6 MeV [1]

27.4

1 a The moderator reduces the average speed (or kinetic energy) of the fission neutrons [1] so they can produce further fission. [1]

b The fission neutrons repeatedly collide with the moderator atoms. [1] The collisions are elastic [1] and in each collision some of the kinetic energy of the fission neutron is transferred to the moderator atom. [1] The percentage of the kinetic energy transferred to the moderator atom from the fission neutron depends on the mass of the moderator atom. [1] The closer the atomic mass of the moderator is to the mass of the neutron, the greater the percentage of the kinetic energy transferred to the moderator atom. [1]

2 a The control rods absorb neutrons. [1] For energy to be released at a constant rate, the depth of the control rods in the reactor core is adjusted (automatically) [1] so that an average of one fission neutron per fission event causes a further fission event. [1]

b The control rods are pushed/inserted fully into the reactor core [1] so the rate at which they absorb neutrons in increased. [1] As a result, the number of fission neutrons in the reactor core decreases [1] and the decrease continues until fission stops. [1]

3 The fuel rods consist of the uranium isotopes $^{235}_{92}$U (which is fissile) and $^{238}_{92}$U [1]. Both isotopes emit α particles and these particles are unable to pass through the metal walls of the fuel. [1]

The used fuel rods contain many different neutron-rich fission fragments which emit β⁻ radiation and γ radiation, [1] both of which can pass through the metal walls [1]. In addition, the used fuel rods contain highly-radioactive plutonium $^{239}_{94}$Pu formed as a result of $^{238}_{92}$U nuclei absorbing neutrons then decaying by β⁻ emission. [1] So more radiation passes through the metal walls of the fuel rods after they have been used than before

Astrophysics

1.1

1 a $v = -0.240$ m, image height = 54 mm [2]

b virtual, upright, enlarged [1]

1.2/1.5

1 a **i** 8.0 [1] **ii** 1.20° [1]

b A parabolic mirror focuses all rays parallel to the axis to a single point whereas a convex lens does not [1] A mirror reflects light whereas a lenses refracts it so chromatic aberration cannot occur with a mirror. [1]

c 6.3×10^{-5} rad [1] $= 3.6 \times 10^{-3}$ degrees [1]

2.1/2.2

1 a The absolute magnitude, M, of a star is its apparent magnitude, m, if it was at a distance of 10 parsecs from Earth. [1]

b $m - M = 5 \log (d / 10)$ gives $M = +4.8 - 5 \log (34/10)$ [1] $= +2.1$ [1]

2.3

1 a P has the greater output power. [1] It is a red giant and will become a white dwarf before it becomes invisible. [1] Q is a main sequence star [1] that will become a red giant then a white dwarf. [1]

b Their magnitude difference is 8 so the power output of P is 1600 times ($= 2.51^8$) greater than that of Q. [1] Its power output per unit area is 0.008 ($=(3000/10\,000)^4$) times that of P. [1] Therefore the area of P is approx $1600 \div 0.008 \times$ that of Q = ×200 000 greater [1] so its diameter about 450 × greater than that of Q. [1]

2.4

1 a The brightness of a star suddenly increases [1] by about 20 magnitudes [1] and then it decreases much more slowly. [1]

b Nuclear fusion ceases in a red giant star which has much more mass than the Sun. [1] The core collapses until its protons and electrons are all converted to neutrons and cannot be compressed any further. [1] The collapsing layers outside the core rebound from the neutron core and are thrown into space. [1]

3.1

1 $v = c \, \Delta\lambda / \lambda = 3.0 \times 10^8 \, \text{m s}^{-1} \times 0.072 \, \text{nm} / 620 \, \text{nm}$ [1] $= 35 \, \text{km s}^{-1}$ [1];
Circumference $= v \, T = 35 \, \text{km s}^{-1} \times 2.5 \, \text{yr} \times 3.15 \times 10^7 \, \text{s yr}^{-1} = 2.76 \times 10^9 \, \text{km}$ [1]
so its radius = circumference $/ 2\pi = 4.4 \times 10^8 \, \text{km}$ [1]

3.2/3.3

1 a $m - M = 5 \log \dfrac{d}{10}$ therefore $\log (d/10) = 40/5$ [1]
$= 8$ [1] hence $d = 10^9 \, \text{pc}$ ($= 1000 \, \text{Mpc}$) [1]

b Red shift estimates of the distance to the more distant type 1a supernovae disagree with distance estimates using their apparent magnitude. [1] The discrepancy is explained by assuming they are accelerating. [1]

3.4

1 Depth of the dip / maximum intensity = 0.10. Therefore the area of the planet disc = 0.10 × the area of the star's disc. [1] Since the area of a disc is proportional to the square of its radius, the radius of the planet [1] = 0.32 (=$0.10^{1/2}$) × the radius of the star. [1]

Medical physics

1.1

1 See Figure 1 [1] for each curve (correct shape, peak at about the correct wavelength) [1] for correct relative peak heights

1.2/1.3

1 a Correctly labelled figure [1]. A diverging (concave) lens of focal length equal to the distance to the eye's far point is placed in front of the eye [1] Parallel rays of light from a distant point object are made to diverge by the correcting lens so they appear to come from the eye's far point [1] and therefore can be focused on the retina. [1]

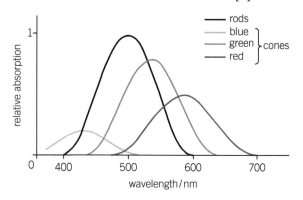

▲ **Figure 1**

b $f = -3.00\,\mathrm{m}$ [1] so $P = \dfrac{1}{f} = -0.333\,\mathrm{D}$ [1]

c $P = \dfrac{1}{f} = \dfrac{1}{+0.25\,\mathrm{m}} - \dfrac{1}{0.50\,\mathrm{m}}$ [1] $= +2.0\,\mathrm{D}$ [1]

2.1

1 They amplify the vibrations [1], they filter out unwanted vibrations [1] and they protect the ear from excessive vibrations. [1]

2 The bones of the middle ear act as levers to increase the force of the vibrations [1] and they cause the vibrations to act on a smaller area (the oval window) than the ear drum [1] So the force per unit area (i.e., pressure) they exert on the oval window is greater than the pressure of the sound waves on the ear drum [1]

2.2

1 a $10\log\left(\dfrac{I}{I_o}\right) = 115\,\mathrm{dB}$ [1]

therefore $I = 10^{11.5}I_o = 1.0 \times 10^{-0.5}\,\mathrm{W\,m^{-2}}$ [1] $= 0.32\,\mathrm{W\,m^{-2}}$ [1]

b $10\log\left(\dfrac{I}{I_o}\right) = 95$ dB gives $I = 10^{9.5}I_o$ [1]. For 2

machines, $I = 2.0 \times 10^{9.5}I_o$ [1]

so the intensity level $= 10\log(I/I_o) = 10\log(2.0 \times 10^{9.5}) = 98$ dB [1]

3.1

1 a i See Figure 1 [1] for correct shape of figure, [1] for correct scales.

ii [1] for correctly positioned labels P, Q, R, S, and T. P is when the atria contract [1]; QRS is when the ventricles contract and the atria relax [1]; T is when the ventricles relax [1].

b Any 2 characteristics from Table 1. [1] for each characteristic, [1] for correct reason

4.1

1 a It absorbs ultrasonic waves created at the rear surface of the disc as these would otherwise cancel each other out. [1] It also damps the disc vibrations rapidly at the end of each pulse before the next pulse is produced. [1]

b An air-skin boundary reflects almost 100% of the incident ultrasonic energy [1]. The probe is applied to the body via a suitable gel or a water bag because most of the ultrasound energy is not reflected and enters the body. [1]

c Acoustic impedances: organ $1.691 \times 10^6\,\mathrm{kg\,m^{-2}\,s^{-1}}$, soft tissue $1.627 \times 10^6\,\mathrm{kg\,m^{-2}\,s^{-1}}$ [1] reflection coefficient 0.019(3) [1]

4.2

1 a The incoherent bundle is used to illuminate the cavity which the endoscope has been inserted into. [1] The coherent bundle is used to see an image of an object inside the body [1] by transmitting light from an image formed by a lens on its end in the body. [1]

b They need to be as thin as possible so the image is as detailed as possible [1] and so the endoscope is flexible and does not lose light where it is bent. [1]

c For the core-cladding boundary, $\sin i_c = \dfrac{1.52}{1.60}$ which gives $i_c = 71.8°$. [1]

Therefore $\sin i_{max} = n_1 \sin(90 - i_c) = 1.60 \times \sin(90 - 71.8)$ [1]

which gives $i_{max} = 30.0°$ [1]

The viewing angle is therefore 60.0°. [1]

4.3

1 a Pulses of radio waves are used in an MR scan to excite hydrogen nuclei [1] systematically across the body using a scanning magnetic field [1]. When the excited nuclei de-excite they emit radio waves [1] that are detected and used to locate the excited nuclei [1] to enable an image of the hydrogen content of the scanned area to be built up. [1]

b The relaxation time of the detected signal is the time the signal takes to decrease. [1] This depends on the type of tissue the excited hydrogen molecules are in [1] and therefore can be used to identify different types of tissue. [1]

5.1

1 a Continuous curve with a peak and tending to zero at long wavelength [1] zero intensity at minimum wavelength [1] intensity spikes [1]

b $\lambda_{MIN} = \dfrac{hc}{eV} = \dfrac{6.63 \times 10^{-34}\,J\,s \times 3.0 \times 10^{8}\,m\,s^{-1}}{1.6 \times 10^{-19}\,C \times 60000\,V}$ [1]

$= 0.021\,nm$ [1]

5.2

1 a $\mu = \dfrac{\ln 2}{X_{\frac{1}{2}}} = \dfrac{\ln 2}{0.015\,m} = 46(.2)\,m^{-1}$ [1] $\mu_m = \mu/\rho =$

$46.2\,m^{-1}/1900\,kg\,m^{-3}$

$= 0.024$ [1] $m^{2}\,kg^{-1}$ [1] (answer given to 2 s.f.)

b $I = I_o\,e^{-X_{\frac{1}{2}}} = I_o e^{-(46.2\,m^{-1} \times 0.004\,m)}$ [1] $= 0.83\,I_o$ [1]

Therefore the % reduction of intensity $= 0.17 \times 100\% = 17\%$ [1]

5.3

1 a A collimator grid [1] prevents X-rays scattered in the patient from reaching the film or detector. [1] Scattered X-rays reaching the unexposed parts of film would reduce the contrast between these parts and the exposed parts. [1]
A contrast medium in an organ [1] absorbs X-rays that enter it and prevents X-rays reaching the film [1] so increasing the contrast between the image of the organ and its surroundings [1].

b The intensifying screen is intended to allow the exposure of the patient to X-rays to be reduced without affecting the image. [1] The screen contains a double-sided film between two layers of fluorescent material in a light-proof wrapper. [1] Each X-ray photon causes many light photons to be emitted [1] so affecting the film much more than if the fluorescent materials had not be present. So the exposure to X-rays can be reduced without affecting the image on the film. [1]

6.1

1 a Rearranging $\dfrac{1}{T_{eff}} = \dfrac{1}{T_b} + \dfrac{1}{T_p} =$ gives

$T_{eff} = \left(\dfrac{1}{24}\,hr + \dfrac{1}{12}\,hr\right)^{-1}$ [1] $= 8.0\,hr$ [1]

b It is a gamma emitter with a half-life of 6 hours [1] which decays to an isotope which has a very long half-life or can be prepared on site. [1]

6.2

1 a Can reach deep tumours [1] less lead shielding needed round the source [1]

2 X-ray photons would need to pass through normal tissues before reaching the tumours whereas the radioisotope attaches itself to the tumour. [1] The X-rays would need to be directed more widely to destroy the tumours whereas the radioisotope targets the tumours. [1]

Engineering physics

1.1

1 a Initial angular speed $\omega_o = 2.0 \times \dfrac{2\pi}{60} = 0.21\,rad\,s^{-1}$

angular speed after 75.0 s, $\omega = 17.0 \times \dfrac{2\pi}{60} = 1.78\,rad\,s^{-1}$ [1]

angular acceleration, $\alpha = \dfrac{(\omega - \omega_o)}{t} = \dfrac{1.78 - 0.21}{75.0}$

$= 2.09 \times 10^{-2}\,rad\,s^{-2}$ [1]

b Angular displacement $\theta = \dfrac{1}{2}(\omega + \omega_o)t = 0.5 \times$

$(0.21\,rad\,s^{-1} + 1.78\,rad\,s^{-1}) \times 75.0\,s$ [1]

$= 74.6\,rad$ [1]

Number of turns $= 74.6/2\pi = 11.9$ [1]

1.2

1 a i Angular displacement for 24.6 turns $= 24.6 \times 2\pi\,rad = 155\,rad$ [1]

Using $\theta = \dfrac{1}{2}(\omega + \omega_o)t$ with $\omega_o = 0$ gives

$\omega = \dfrac{2\theta}{t} = \dfrac{2 \times 155\,rad}{38.2\,s}$

$= 8.11\,rad\,s^{-1}$ [1]

Its average angular acceleration

$\alpha_1 = \dfrac{(\omega - \omega_o)}{t} = \dfrac{8.11\,rad\,s^{-1} - 0}{38.2} = 0.212\,rad\,s^{-2}$ [1]

ii During deceleration, its angular displacement for 97.0 turns $= 97.0 \times 2\pi\,rad = 609\,rad$ [1]
Its initial angular velocity $\omega_o = 8.11\,rad\,s^{-1}$, its final angular velocity $\omega = 0$
Therefore its average angular deceleration

$\alpha_2 = \dfrac{(\omega - \omega_o)}{t} = \dfrac{0 - 8.11\,rad\,s^{-1}}{92.0}$

$= -0.0882\,rad\,s^{-2}$ [1]

b During deceleration, the frictional torque $T_F = I\alpha_2 = 290\,kg\,m^2 \times 0.0882\,rad\,s^{-2} = 25.6\,N\,m$ [1]

During acceleration, the resultant torque $T = I\alpha_1 = 290\,kg\,m^2 \times 0.212\,rad\,s^{-2} = 61.5\,N\,m$

Therefore the applied torque during acceleration
= 61.5 Nm + 25.6 N m = 87.1 Nm [1]

(Note: the resultant torque = the applied torque –
the frictional torque)

1.3

1 a 40 J [1]

 b 0.2 J [2]

 c i 39.8 J [1]

 ii 1.3×10^{-2} kg m^2 [3]

1.4

1 a 0.64 Nm s [2]

 b 0.015 kg m^2 [4]

2.1/2 2

1 a 180 J into the gas. [1]

 b 180 J of work is done on the gas. [1]

2 a 107 kPa, [3] 289 K [2]

 b approx average pressure = 130 kPa, increase of
 volume = 0.0122 m^3 [1]

 estimated work done = 130 kPa × 0.0122 m^3 [1] ≈
 1.6 kJ [1]

 (Note – use this method to check a more accurate
 graphical estimate)

2.3

1 a 130 kW [2]

 b 32% [3]

2.4

1 a 0.73 [2]

 b i 31 MJ kg^{-1} × 9.4 kg [1] / 3600 s [1]

 ii 43 kW [1]

2.5

1 a 5 [2]

 b 480 W [1]

2 a 15.1 kJ [2]

 b 2.5 [2]

Turning points in physics

1.1

1 a Every atom contains electrons. [1] The atoms of
 an element are different from the atoms of any
 other element. [1] When one or more electrons
 are removed from an atom, the electrons
 removed are the same [1] but the atom becomes
 a positive ion which is therefore different for
 each type of gas. [1]

 b 2.5×10^7 m s^{-1} [2]

1.2/1.3

1 a 3.85×10^7 m s^{-1} [2]

 b 1.74×10^{11} C kg^{-1} [2]

1.4

1 a 6.20×10^{-15} kg [3]

 b 4.81×10^{-19} C [2]

2.1

1 a When a corpuscles passes from glass into air, its
 component of velocity parallel to the normal
 decreases and is component perpendicular to the
 normal is unchanged. [1] So the direction of its
 velocity is further from the normal [1] and the
 light ray therefore is refracted away from the
 normal. [1]

 b According to Newton's theory, only two
 bright fringes should be seen corresponding to
 corpuscles passing through each of the two slits.
 [1] The fringe pattern always has more than
 two fringes and Huygens' theory explains this
 by considering the two slits as coherent emitters
 of waves. [1] A bright fringe is seen at regularly
 spaced positions where the waves from each
 slit arrive in phase [1] and a dark fringe is seen
 between the bright fringes where the waves
 arrive out of phase by 180° [1].

2.2

1 a The radio waves from the transmitter are
 polarised. [1] If the aerial is turned through 90°
 from its position at maximum signal strength, the
 strength of the detected signal decreases to zero
 [1] as the orientation of the aerial relative to the
 plane of polarisation changes. [1]

 b $c = 4Df_0N = 4 \times 8500$ m × 12.0 Hz × 720 [1] =
 2.9×10^8 m s^{-1} [1]

2.3

1 a 4.3×10^{-19} J [2]

 b 9.6×10^{-20} J [3]

2.4

1 a 3.5×10^7 m s^{-1} [2], 2.0×10^{-11} m [2]

 b More detail would be seen as the electrons would
 be diffracted less at each lens [1] because the
 electrons would not be slowed down as much
 by the sample [1] so their de Broglie wavelength
 would be less. [1]

3.1

1 a The path difference of the light waves from the beam splitter to the observer via each of the two reflectors is an odd number of half wavelengths [1] so the waves are out of phase by 180° on reaching the observer. [1] So the two waves cancel each other and the observer sees a dark fringe. [1]

b The experimenters expected the fringes to shift. [1]

c The fringes did not shift so the ether does not exist. [1]

3.2

1 a 96 ns [2]

b 852 MeV [3]

Electronics

1.1

1 a As the input voltage increases, the gate–source pd increases and when this pd exceeds the MOSFET threshold voltage, a conducting channel between the drain and the source increases [1] so there is a current between the drain and the source and through the relay coil. [1] The relay coil is therefore magnetised and the electromagnet in the relay attracts and closes the relay switch. [1]

b See Figure 2. Only one curve is required [1] and it should be above the cut-off region [1]. P is left-hand end of the horizontal section of the curve. [1]

1.2

1 a resistor current = (9.0 V – 6.0 V) / 60 W [1] = 50 mA [1]

b diode current = resistor current – lamp current = 50 mA – 30 mA = 20 mA [1]

1.3

1 a 3.4 V [2]

b 6.2 V [2]

1.4

1 a The perpendicular component of the magnetic flux density relative to the device. [1]

b i 7.1 μT [2]

ii 45° [1]

2.1

1 a 25 mV [1]

b 101 001 00 [2]

2.2

1 a 2.4 V [2]

b 9.8 V [2]

3.1

1 a 45 pF [2]

b **The Q factor** is the ratio of the resonant frequency to the bandwidth [1] where the bandwidth is the difference between the frequencies either side of the peak at √2 of the maximum pd. [1]

3.2

1 a The output pd of an op-amp saturates if the pd between its two input terminals exceeds about 150 μV. [1] If one of the two inputs of an op-amp is earthed and the output pd is not saturated, the potential at this other input must be less than 150 μV and is said to be at virtual earth. [1]

b i When the buzzer is off, the output pd is at −15 V so the pd at P must be greater than the pd at Q. [1] When the thermistor's temperature decreases, its resistance increases so the pd at P decreases. [1] When this pd decreases below the pd at Q, the output pd switches to +15 V and the buzzer switches on. [1]

ii The resistance of the variable resistor needs to be increased [1] (so the thermistor temperature does not have to increase as much before the pd at P becomes less than the pd at Q).

3.3

1 a See Figure 1 input pd, output pd amplified [1] in phase [1] saturated [1]

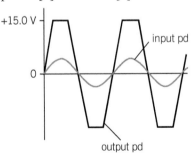

▲ **Figure 1**

b 6.5 V [2]

4.1

1 a [4]

A	B	$\overline{A}$	$\overline{B}$	$\overline{A} \cdot \overline{B}$	$C = \overline{\overline{A} \cdot \overline{B}}$
0	0	1	1	1	0
1	0	0	1	0	1
0	1	1	0	0	1
1	1	0	0	0	1

b The column for C is the same as A + B. Therefore

$\overline{\overline{A} \cdot \overline{B}} = A + B$ [1]

4.2

1 a The inputs of a 2-input AND gate are connected to the Q_3 and Q_1 outputs of the binary counter and the AND gate output is connected to the reset terminal of the counter. [1] Each time Q_3 Q_2 Q_1 reaches 0100 which is decimal 4, the next pulse causes Q_3 Q_2 Q_1 to become 0101 [1] so the AND gate output becomes 1 thus resetting the counter to 0000. [1]

b Each clock pulse causes the Q output of each flip-flop to be shifted to the next Q output. Assuming all the Q outputs are initially 0, the input at the 1st D input is 1 and all the other inputs are zero. [1] The 1st pulse shifts the 1 to the 1st Q output and the second pulse shifts it to the 2nd Q output. [1] There is still a 1 at the 1st D input so this is shifted to the 1st Q output by the second pulse. The third pulse shifts the 2 '1's to the 2nd and 3rd Q outputs and provides another 1 to the 1st Q output so the output states show 1 1 1. [1] Because the 3rd Q output is now 1, the third Q bar output is now 0 so the next three pulses change the 1st Q output to 0 as well as shifting the bits at Q outputs along. The result is the output states count down from 111 to 000. [1]

Input pulse number	Q_1	Q_2	Q_3
0	0	0	0
1	1	0	0
2	1	1	0
3	1	1	1
4	0	1	1
5	0	0	1
6	0	0	0

5.1

1 a i (HF) radio waves [1]

ii microwaves [1]

b 100 [1]; 2.0 Mbps [2]

5.2

1 a Bandwidth is the range of frequencies of the signal [1] 97.5 MHz to 98.0 MHz. [1]

b i 85 kHz [1]

ii 102.45 MHz [1]

For reference

Data

Fundamental constants and values

quantity	symbol	value	units
speed of light in vacuo	c	3.00×10^8	$m\,s^{-1}$
permeability of free space	μ_0	$4\pi \times 10^{-7}$	$H\,m^{-1}$
permittivity of free space	ε_0	8.85×10^{-12}	$F\,m^{-1}$
magnitude of charge of electron	e	1.60×10^{-19}	C
Planck constant	h	6.63×10^{-34}	$J\,s$
gravitational constant	G	6.67×10^{-11}	$N\,m^2\,kg^{-2}$
Avogadro constant	N_A	6.02×10^{23}	mol^{-1}
molar gas constant	R	8.31	$J\,K^{-1}\,mol^{-1}$
Boltzmann constant	k	1.38×10^{-23}	$J\,K^{-1}$
Stefan constant	σ	5.67×10^{-8}	$W\,m^{-2}\,K^{-4}$
Wien constant	α	2.90×10^{-3}	$m\,K$
electron rest mass (equivalent to 5.5×10^{-4} u)	m_e	9.11×10^{-31}	kg
electron charge/mass ratio	$\dfrac{e}{m_e}$	1.76×10^{11}	$C\,kg^{-1}$
proton rest mass (equivalent to $1.007\ 28$ u)	m_p	$1.67(3) \times 10^{-27}$	kg
proton charge/mass ratio	$\dfrac{e}{m_p}$	9.58×10^7	$C\,kg^{-1}$
neutron rest mass (equivalent to $1.008\ 67$ u)	m_n	$1.67(5) \times 10^{-27}$	kg
gravitational field strength	g	9.81	$N\,kg^{-1}$
acceleration due to gravity	g	9.81	$m\,s^{-2}$
atomic mass unit (1 u is equivalent to 931.3 MeV)	u	1.661×10^{-27}	kg

Astronomical data

body	mass/kg	mean radius/m
Sun	1.99×10^{30}	6.96×10^8
Earth	5.98×10^{24}	6.37×10^6

Geometrical equations

arc length $= r\,\theta$

circumference of circle $= 2\pi r$

area of circle $= \pi r^2$

surface area of cylinder $= 2\pi rh$

volume of cylinder $= \pi r^2 h$

area of sphere $= 4\pi r^2$

volume of sphere $= \frac{4}{3}\pi r^3$

Momentum

force $\qquad F = \dfrac{\Delta(mv)}{\Delta t}$

impulse $\qquad F\Delta t = \Delta(mv)$

Circular motion

magnitude of angular speed $\omega = \dfrac{v}{r}$

$\omega = 2\pi f$

centripetal acceleration $\qquad a = \dfrac{v^2}{r} = \omega^2 r$

centripetal force $\qquad F = \dfrac{mv^2}{r} = m\omega^2 r$

Simple harmonic motion

acceleration $\qquad a = -\omega^2 x$

displacement $\qquad x = A\cos(\omega t)$

speed	$v = \pm\, \omega\sqrt{A^2 - x^2}$
maximum speed	$v_{\text{max}} = \omega A$
maximum acceleration	$a_{\text{max}} = \omega^2 A$
for a mass-spring system	$T = 2\pi\sqrt{\dfrac{m}{k}}$
for a simple pendulum	$T = 2\pi\sqrt{\dfrac{l}{g}}$

Gases and thermal physics

gas law	$pV = nRT$
	$pV = NkT$
kinetic theory model	$pV = \dfrac{1}{3} Nm(c_{\text{rms}})^2$
kinetic energy of a gas molecule	$\dfrac{1}{2}m(c_{\text{rms}})^2 = \dfrac{3}{2}kT$
	$= \dfrac{3RT}{2N_A}$
energy to change temperature	$Q = mc\Delta T$
energy to change state	$Q = ml$

Gravitational fields

force between two masses	$F = \dfrac{Gm_1 m_2}{r^2}$
magnitude of gravitational field strength	$g = \dfrac{F}{m}$
	$g = \dfrac{GM}{r^2}$
gravitational potential	$\Delta W = m\Delta V$
	$V = -\dfrac{GM}{r}$
	$g = -\dfrac{\Delta V}{\Delta r}$

Electric fields and capacitors

force between two point charges	$F = \dfrac{Q_1 Q_2}{4\pi\varepsilon_0 r^2}$
force on a charge	$F = EQ$
field strength for a uniform field	$E = \dfrac{V}{d}$
field strength for a radial field	$E = \dfrac{Q}{4\pi\varepsilon_0 r^2}$
work done	$\Delta W = Q\Delta V$

electric potential	$V = \dfrac{Q}{4\pi\varepsilon_0 r}$
capacitance	$C = \dfrac{Q}{V}$
	$C = \dfrac{A\varepsilon_0\varepsilon_r}{d}$
decay of charge	$Q = Q_0 e^{\frac{-t}{RC}}$
capacitor charge	$V = V_0(1 - e^{\frac{-t}{RC}})$
time constant	RC
time to halve	$T_{\frac{1}{2}} = 0.69RC$
capacitor energy stored	$E = \dfrac{1}{2}QV = \dfrac{1}{2}CV^2$
	$= \dfrac{1}{2}\dfrac{Q^2}{C}$

Magnetic fields

force on a current	$F = BIl$
force on a moving charge	$F = BQv$
magnetic flux	$\phi = BA$
magnetic flux linkage	$N\phi = BAN\cos\theta$
induced emf	$\varepsilon = N\dfrac{\Delta\phi}{\Delta t}$
emf induced in a rotating coil	$N\phi = BAN\cos\theta$
	$\varepsilon = BAN\,\omega\sin\omega t$
alternating current	$I_{\text{rms}} = \dfrac{I_0}{\sqrt{2}} \qquad V_{\text{rms}} = \dfrac{V_0}{\sqrt{2}}$
transformer equations	$\dfrac{N_s}{N_p} = \dfrac{V_s}{V_p}$
efficiency	$= \dfrac{I_s V_s}{I_p V_p}$

Nuclear physics

the inverse-square law for γ variation	$I = \dfrac{k}{x^2}$
radioactive decay	$\dfrac{\Delta N}{\Delta t} = -\lambda N \quad N = N_0 e^{-\lambda t}$
activity	$A = \lambda N$
half-life	$T_{\frac{1}{2}} = \dfrac{\ln 2}{\lambda}$
nuclear radius	$R = R_0 A^{\frac{1}{3}}$
energy–mass equation	$E = mc^2$

Particle physics

Rest energy values

class	name	symbol	rest energy / MeV
photon	photon	γ	0
lepton	neutrino	ν_e	0
		ν_μ	0
	electron	$e^\pm$	0.510999
	muon	$\mu^\pm$	105.659
mesons	pion	$\pi^\pm$	139.576
		π^0	134.972
	kaon	$K^\pm$	493.821
		K^0	497.762
baryons	proton	p	938.257
	neutron	n	939.551

Properties of quarks

Antiparticles have opposite signs

type	charge	baryon number	strangeness
u	$+\frac{2}{3}e$	$+\frac{1}{3}$	0
d	$-\frac{1}{3}e$	$+\frac{1}{3}$	0
s	$-\frac{1}{3}e$	$+\frac{1}{3}$	-1

Properties of leptons

lepton	lepton number
particles: e^-, ν_e; μ^-, ν_μ	$+1$
antiparticles: e^+, $\overline{\nu}_e$; μ^+, $\overline{\nu}_\mu$	-1

Photons and energy levels

photon energy	$E = hf = \dfrac{hc}{\lambda}$
photoelectric effect	$hf = \phi + E_{K(max)}$
energy levels	$hf = E_1 - E_2$
de Broglie wavelength	$\lambda = \dfrac{h}{p} = \dfrac{h}{mv}$

Waves

wave speed	$c = f\lambda$
period	$f = \dfrac{1}{T}$
first harmonic	$f = \dfrac{1}{2l}\sqrt{\dfrac{T}{\mu}}$
fringe spacing	$w = \dfrac{\lambda D}{s}$
diffraction grating	$d \sin\theta = n\lambda$
refractive index of a substance s	$n_s = \dfrac{c}{c_s}$

For two different substances of refractive indices n_1 and n_2,

law of refraction $\qquad n_1 \sin\theta_1 = n_2 \sin\theta_2$

critical angle $\qquad \sin\theta_c = \dfrac{n_2}{n_1}$ for $n_1 > n_2$

Mechanics

Moments	moment $= Fd$
velocity and acceleration	$v = \dfrac{\Delta s}{\Delta t}$; $\quad a = \dfrac{\Delta v}{\Delta t}$
equations of motion	$v = u + at$
	$s = \dfrac{(u+v)t}{2}$
	$v^2 = u^2 + 2as$
	$s = ut + \dfrac{1}{2}at^2$
force	$F = ma$
work, energy, and power	$W = Fs \cos\theta$
	$E_K = \dfrac{1}{2}mv^2$
	$\Delta E_P = mg\Delta h$
	$P = \dfrac{\Delta W}{\Delta t} \quad P = Fv$

efficiency $= \dfrac{\text{useful output power}}{\text{input power}}$

Materials

density	$\rho = \dfrac{m}{V}$
Hooke's law	$F = k\Delta L$
Young modulus	$= \dfrac{\text{tensile stress}}{\text{tensile strain}}$
tensile stress	$= \dfrac{F}{A}$
tensile strain	$= \dfrac{\Delta L}{L}$
energy stored	$E = \dfrac{1}{2}F\Delta L$

Electricity

current and pd	$I = \dfrac{\Delta Q}{\Delta t}$
	$V = \dfrac{W}{Q}$
	$R = \dfrac{V}{I}$
emf	$\varepsilon = \dfrac{E}{Q}$
	$\varepsilon = I(R + r)$
resistors in series	$R = R_1 + R_2 + R_3 + \dots$
resistors in parallel	$\dfrac{1}{R} = \dfrac{1}{R_1} + \dfrac{1}{R_2} + \dfrac{1}{R_3} + \dots$
resistivity	$\rho = \dfrac{RA}{L}$
power	$P = VI = I^2R = \dfrac{V^2}{R}$